Highly Excited Molecules

ACS SYMPOSIUM SERIES **678**

Highly Excited Molecules

Relaxation, Reaction, and Structure

Amy S. Mullin, EDITOR
Boston University

George C. Schatz, EDITOR
Northwestern University

Developed from a symposium sponsored by the
Division of Physical Chemistry at the 212th National Meeting
of the American Chemical Society,
Orlando, Florida,
August 25–29, 1996

American Chemical Society, Washington, DC

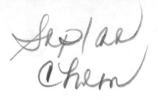

Library of Congress Cataloging-in-Publication Data

Highly excited molecules: relaxation, reaction, and structure / Amy S. Mullin, editor, George C. Schatz, editor.

p.   cm.—(ACS symposium series, ISSN 0097–6156; 678) ✔

"Developed from a symposium sponsored by the Division of Physical Chemistry at the 212th National Meeting of the American Chemical Society, Orlando, Florida, August 25–29, 1996."

Includes bibliographical references and index.

ISBN 0–8412–3534–1

1. Excited state chemistry—Congresses.

I. Mullin, Amy S., 1962–   . II. Schatz, George C., 1949–   . III. American Chemical Society. Division of Physical Chemistry. IV. American Chemical Society. Meeting (212th: 1996: Orlando, Fla.) V. Series.

QD461.5.H48   1997
541—dc21   97–34824
                                                                      CIP

This book is printed on acid-free, recycled paper.

Foreword

THE ACS SYMPOSIUM SERIES was first published in 1974 to provide a mechanism for publishing symposia quickly in book form. The purpose of the series is to publish timely, comprehensive books developed from ACS sponsored symposia based on current scientific research. Occasionally, books are developed from symposia sponsored by other organizations when the topic is of keen interest to the chemistry audience.

Before agreeing to publish a book, the proposed table of contents is reviewed for appropriate and comprehensive coverage and for interest to the audience. Some papers may be excluded in order to better focus the book; others may be added to provide comprehensiveness. When appropriate, overview or introductory chapters are added. Drafts of chapters are peer-reviewed prior to final acceptance or rejection, and manuscripts are prepared in camera-ready format.

As a rule, only original research papers and original review papers are included in the volumes. Verbatim reproductions of previously published papers are not accepted.

ACS BOOKS DEPARTMENT

Contents

INTERMOLECULAR DYNAMICS: NONREACTIVE ENERGY TRANSFER

INTRAMOLECULAR DYNAMICS: REACTIVE COLLISIONS

INDEXES

Preface

MANY IMPORTANT CHEMICAL PROCESSES in photochemistry, kinetics, and spectroscopy involve the preparation of molecules with energies comparable to the strength of chemical bonds, that is, many tens to a few hundred kilocalories per mole. Molecules in such highly excited states often have structural and spectroscopic properties that are very different from those of unexcited molecules. Because they have energies comparable to that needed to break chemical bonds, these molecules often undergo chemical transformations such as dissociation, isomerization, and reaction with other molecules to produce new chemical species.

Highly excited molecules are important in many practical chemical applications, particularly in combustion, atmospheric, laser, and interstellar chemistry, so there is much current research in this area. This research has been particularly productive in the last few years, owing in part to advances in experimental techniques such as lasers and molecular beams, which are used for preparing and detecting highly excited molecules, and in part to advances in theoretical and computational chemistry methods for modeling the properties of these molecules and describing reactive and nonreactive collisions of these molecules with other molecules.

The symposium that led to this book was the first at an ACS meeting on the topic of highly excited states. The symposium was presented at the 212th National Meeting of the American Chemical Society and was titled "Highly Excited States: Relaxation Reactions and Structure". It was sponsored by the ACS Division of Physical Chemistry in Orlando, Florida, August 25–29, 1996. Four general areas were featured, and these form the major subheadings in the book. These are highly excited bound vibrational states, unimolecular decay, collisional energy transfer from highly excited molecules, and bimolecular reactions involving highly excited molecules. The first two areas refer to the properties of isolated molecules with energies either below (bound molecules) or above (unimolecular decay) that needed for dissociation. The last two areas refer to the interaction (collision) of excited molecules with other molecules, resulting either in energy transfer (relaxation) or bimolecular reactions. Both experimental and theoretical research are important for each of the four areas; indeed, for many applications the interplay between theory and experiment plays an important role in advancing the field.

The overview chapter of this book provides a general introduction to each topic, along with enough examples of applications to provide a context for the more specialized chapters that follow. Unfortunately, space limitations have made it impossible to include some topics, particularly the description of highly excited electronic states (such as Rydberg states) and excited states in condensed-phase environments.

Acknowledgments

The original idea for this symposium was in large measure due to George Flynn (Columbia University), who was the Program Chair of the ACS Division of Physical Chemistry when this symposium was held. We thank him for getting the symposium started and for his encouragement throughout the planning stages and the subsequent meeting in Orlando. Financial support for symposium speakers from outside North America was provided by The Petroleum Research Fund of the ACS and by the ACS Division of Physical Chemistry. We also thank Marc Fitzgerald and David Orloff of the ACS Books Editorial Staff for their help in getting the book started and completed.

AMY S. MULLIN
Department of Chemistry
Boston University
590 Commonwealth Avenue
Boston, MA 02215

GEORGE C. SCHATZ
Department of Chemistry
Northwestern University
Evanston, IL 60201

July 10, 1997

OVERVIEW

Chapter 1

Dynamics of Highly Excited States in Chemistry: An Overview

Amy S. Mullin[1] and George C. Schatz[2]

[1]Department of Chemistry, Boston University, Boston, MA 02146
[2]Department of Chemistry, Northwestern University, Evanston, IL 60208–3113

This article provides an overview of chemical research that is concerned with the dynamics of highly excited states of molecules. Four broad topics are discussed, including (1) highly excited bound vibrational states of polyatomic molecules, (2) unimolecular decay of highly excited molecules, (3) collisional relaxation of highly excited polyatomic molecules, and (4) bimolecular reactions involving vibrationally excited molecules. For each topic, we present an overview of current experimental and theoretical research being done, along with a brief discussion of selected applications.

Highly excited states of molecules are important in many areas of chemistry, including photochemistry, reaction kinetics and spectroscopy. The dynamics of highly excited states in chemical processes is often quite complex, but at the same time is amenable to detailed experimental and theoretical study, and as a result the interplay between theory and experiment plays an important role. Excited state dynamics often involve intermolecular and/or intramolecular energy flow between many different kinds of motions (vibrational, rotational, translational, electronic) that span a wide range of time scales. As a result, there are many different kinds of experiments that can be brought to bear on these problems, including frequency and time-resolved spectroscopy, state-resolved kinetics, energy transfer measurements and pump-probe laser experiments. Many of these experiments require close coupling to theoretical modelling to optimize interpretation. The theoretical description of these processes can be done in many ways, including theories based on classical and quantum mechanics, statistical approximations and dynamical models. In addition, many of the theoretical applications require close interplay between electronic structure theory (for determining potential energy surfaces and surface couplings) and theories of dynamical processes (which use the surfaces to calculate measurable properties). As a result, many parts of the theoretical and experimental chemistry community are brought together in these studies.

This article considers experimental and theoretical work related to the description of highly excited states in four broad areas: highly excited bound vibrational states, unimolecular decay, collisional energy transfer and bimolecular reactions. For each topic we first give a general overview of experiment and theory used in describing the dynamics of highly excited states, and we describe some specific work currently under consideration. Our primary emphasis is on excited states of isolated molecules, particularly molecules with three or more atoms, and of the interactions and reactions of polyatomic molecules in the gas phase under "single collision" conditions. These applications to small molecule dynamical processes are the customary "laboratory" of physical chemistry where innovation and conceptualization starts. However many of the experiments and theories we discuss have also been adapted to condensed phase dynamics.

Highly Excited Bound Vibrational States

Experimental Studies. We begin by considering the vibrational states of highly excited isolated polyatomic molecules. The traditional method for experimentally determining the bound states of such molecules is based on spectral measurements of the frequencies and intensities of transitions between states.*(1)* The states are modeled to first order using the molecular normal coordinates to describe a vibrational Hamiltonian and products of harmonic (or anharmonic) oscillator states to describe the vibrational states. For small molecules at low energies, this approach works exceedingly well. However as the complexity of the molecule increases, vibrational excitation into a localized initial state does not remain stationary. The energy may flow into other regions of the molecule in the process known as IVR (intramolecular vibrational redistribution).*(2)* In the frequency domain, this is manifested in high resolution spectra as dense branching of zero-order bright states. The blossoming of spectral complexity is caused by coupling between optically "bright" states and an often very dense background of "dark" states. Thus a single transition observed with medium resolution fractionates at high resolution to a dense manifold of lines with frequency spacings and intensities which are related to the true molecular eigenstates. By analyzing the clumps of spectral features and relating the observed molecular eigenstates to the initial bright state, the coupling strengths between the states and rates of energy flow can be determined. Frequency domain experiments rely extensively on high resolution optical techniques,*(3)* since obtaining accurate intensity and transition frequency data is paramount to unraveling the vibrational relaxation dynamics.

Time Domain Experiments. It is important to recognize that while most experimental studies of IVR are done in the frequency regime, completely analogous information about the bound states of vibrationally excited molecules can be obtained from time domain experiments through the time dependent autocorrelation function for coherently excited states.*(4-6)* In a typical time-domain IVR experiment, a molecule is prepared in an initial "bright" state with localized energy using a short-pulse (or series of pulses) of light. For the simplest case of two coupled states, IVR will occur as the initial state couples to the "dark" state. The system oscillates to the mixed state and then transforms into the dark state. This is observed as a fast decay of the bright state with time dependent quantum beats. The oscillation of the quantum beats is described by the Fourier components corresponding to the energy eigenvalue differences and amplitudes corresponding to the coefficients that mix the initial wavefunction with the molecular wavefunction. In most molecules, there is more than one dark state which couples to the bright state. The coupling of more than two levels results in a time dependence which is much more complicated, since each quantum beat feature has a characteristic frequency, amplitude, phase and decay rate. The usefulness of the time dependent approach depends on the ability to detect and separate the individual oscillations. This may be difficult for the low amplitude oscillations that result from weak couplings. In addition, low frequency oscillations need to be recorded over a longer time frame.

One Photon Absorption Experiments. Experimental studies of IVR rely on the ability to access a vibrationally excited bright state. One approach to delivering vibrational energy to a molecule is one photon direct vibrational absorption. This method is particularly effective for low quanta of vibrational excitation where large transition moments and favorable harmonic oscillator selection rules ($\Delta v = +1$) can move a substantial population into vibrationally excited levels. One photon direct excitation has also been used for accessing higher overtone transitions of C-H stretches in alkanes, where high frequencies and large anharmonicities help to enhance transition probabilities.[7] Another approach for preparing vibrationally excited molecules is Franck-Condon pumping followed by either dispersed fluorescence or stimulated emission pumping (SEP).[8] These techniques can be used for molecules that have a bound excited electronic state. One photon excites the molecule to the upper electronic surface and then the excited state can either fluoresce to vibrationally excited levels in the ground electronic state or can be stimulated down with a second photon. Due to structural differences in the ground and excited electronic states, often the final vibrational states prepared in this way will be quite different from those prepared via direct one photon excitation. Thus, Franck Condon excitation followed by SEP can be complementary to direct one photon excitation. By accessing different intermediate levels in the excited electronic state, a wider range of vibrations in the ground state can be accessed. This offers greater flexibility in preparing the initial bright state. Other methods for accessing vibrational states in polyatomic molecules include a variety of two step pumping schemes based either on sequential vibrational steps or a combination of direct vibrational excitation and Franck-Condon pumping.

Most experimental IVR studies have focused on relaxation in alkanes following initial excitation of various quanta of C-H stretches. For low levels of vibrational excitation, IVR was first identified in experiments that excited $v_{CH} = 1$ and later $v_{CH} = 2$ using pulsed infrared excitation.[9,10] The subsequent fluorescence was substantially diminished due to IVR and a threshold for IVR was determined at a rovibrational state density of at least 10-100/cm⁻¹. Subsequently, high-resolution IR laser techniques coupled with supersonic pinhole and slit expansions[11,12] and optothermal detection[13] have provided sufficient spectral resolution for direct measurements of IVR-induced spectral clumps in a number of acetylene-like molecules with $v_{CH} = 1$. The uncoupled energies of the "bright" states and "dark" background states can be determined from the observed frequencies and relative intensities, and the number and density of the interacting states can be identified through deconvolution methods.[14] Some of the key findings are that the density of coupled states equals the density of molecular states (that is, almost all background states participate in the IVR process) and that IVR for these low density of states occurs on the 10^{-9} s timescale (which is surprisingly much longer lived than the CH stretching period itself). In addition, the observed IVR relaxation rates for this class of molecules are fairly insensitive to the structure and density of states of the molecule, increasing only an order of magnitude for ten orders of magnitude variation in the density of states.

Overtone Absorption Experiments. Efforts to characterize IVR in molecules with higher overtone excitation have also been based on CH stretch excitation. Broadening of spectral features is observed in benzene excited to $v_{CH} = 4$ through $v_{CH} = 7$, corresponding to enhanced IVR due to a 2:1 Fermi resonance between the CH stretch and the CCH bend.*(15-17)* In this case, the observed relaxation rate could be predicted fairly well by analysis of the local stretch-bend interactions on the same functional group. Such 2:1 Fermi resonances have been observed in a number of CH-containing molecules, such as acetylene*(18)* and the substituted methanes,*(19)* and have proven to be an important channel for IVR at short timescales. The near resonant stretch-bend states mix with the zero order bright state to form a "polyad" of spectral bands, the number of which is related to the number of ways to put x quanta of stretch into stretch-bend combination states. Polyads have proven to be a powerful pattern recognition tool in complicated overtone spectra and have helped elucidate the short time dynamics of IVR. A hierarchy of energy flow exists, first from the bright state to the Fermi resonance states and then into the higher density of background states. In acetylene, degradation of polyad quantum numbers reveal subsequent IVR from the . polyad. The resulting time domain picture is of rapid IVR among a single polyad and then slower energy flow into other polyads. Ironically, lower resolution spectra (~ 5 cm^{-1}) of acetylene using dispersed fluorescence are more useful for extracting the early time IVR dynamics than high resolution SEP, which includes both the long and short time dynamics. In addition, identification and assignment of polyads have provided useful information about the functional form of the ground electronic potential energy surface.

Theoretical Studies. The traditional theoretical approach to describing the vibrational bound states of polyatomic molecules is based on expressing the vibrational Hamiltonian in normal mode coordinates and approximating the vibrational states as products of harmonic oscillator states. This approach, and the obvious improvement to it that arises from treating the effects of anharmonic coupling using perturbation theory, works well in describing the low lying states of most polyatomic molecules. However it becomes increasingly inappropriate as energy increases, and ultimately breaks down completely when the atoms are able to exhibit large amplitude motion (isomerization, internal rotation, dissociation) The theoretical description of the highly excited states of polyatomic molecules is therefore concerned with determining the stationary states of a coupled many-body problem, which is a very formidable problem in the most general case. This problem has been of great interest during the past 10 years, as modern computational facilities provide substantially greater capabilities for treating this problem. As a result, the vibrational states of small molecules can be determined exactly (for a given potential energy surface) in some cases (triatomics and some tetraatomics). Other methods are less accurate but can treat much larger polyatomic molecules, even proteins, at a level that goes significantly beyond the normal mode/harmonic oscillator treatment. In the following paragraphs we first describe quantum mechanical approaches to the determination of vibrational energy levels of polyatomic molecules, and then semiclassical approaches. Two important general references in this field are: (a) the series of books entitled *Advances in Molecular Vibrations and Collision Dynamics* edited by J. M. Bowman *(20)* and (b) a special issue of *Computer Physics Communications* devoted to vibrational energy level calculations *(21)*.

Quantum Studies. Following the lead of electronic structure theory, the "standard" approach for describing the states of polyatomic molecules that includes for the complete intramolecular potential (beyond the harmonic approximation) to produce fully converged vibrational eigenstates involves expanding the wavefunction in products of harmonic states, and solving the nuclear motion Schrödinger equation by determining the eigenvalues of a secular equation. This secular equation has the form:

$$|H-ES| = 0$$

where H is the Hamiltonian matrix and S is the overlap matrix. This sort of calculation has been used on many occasions, but not commonly of late. Perhaps the main reason for this is that the harmonic basis is generally quite poor for describing vibrations in molecules with even modest levels of excitation.

One approach that improves on uncoupled harmonic oscillators is the vibrational *self consistent field* (SCF) method.*(22,23)* This is a mean-field approach, wherein the effective potential for each vibrational mode is obtained by averaging the intramolecular potential over all the other modes. Although normal coordinates are the obvious choice for implementing this approach, other coordinate systems are possible, and in fact it is possible to optimize results with respect to the choice of coordinates in some cases. Of course, SCF methods invariably leave out correlation between the motions of different modes, but the resulting eigenvalues are considerably improved compared to the harmonic results. Improvements to this method based on configuration interaction (CI) methods have also been considered, *(24,25)* and so have multireference SCF methods.

An approach mentioned earlier that improves on the harmonic description is *perturbation theory*. This approach traditionally starts with the uncoupled harmonic oscillator description, and includes for the influence of anharmonicity using at least second order perturbation theory, and often fourth or sixth order. A detailed description of this approach has been given by McCoy and Sibert,*(26)* and the modern implementation of this includes improvements such as optimized coordinates, treatment of Fermi resonant states, and the treatment of Coriolis effects.

Wavepackets have become increasingly popular in recent years as an approach to determine quantum energy levels.*(27)* The basic idea is to determine the time dependence of a wavepacket by solving the time dependent Schrödinger equation using the full molecular Hamiltonian, and then calculate the Fourier transform of its autocorrelation function:

$$C(E) = \int e^{iEt/\hbar} <\psi(0)|\psi(t)>dt$$

In the limit that the time dependence of $\psi(t)$ is determined for a long enough period, the function C(E) consists of a series of peaks corresponding to the vibrational energy levels. Ideally the time duration needed for this calculation should be much larger than $\hbar/\Delta E$, where ΔE is the smallest energy level spacing, however algorithms which make optimal use of shorter time integrations (the so-called Prony method) have been developed.*(28)*

It is also possible to use a short time integration but still retain high accuracy using a method known as *filter diagonalization* that has been developed by Neuhauser*(29)*. In this approach one transforms the time dependent wavepacket $\psi(t)$

into an approximate energy eigenfunction using a filtering function as follows:

$$\psi(E_k) = \int e^{-t^2/2T^2} e^{iE_k t/\hbar} \psi(t)dt$$

Here, the filtering function $\exp\{-t^2/2T^2 + iE_k t/\hbar\}$ serves to select only eigenfunctions having energies near E_k, with the parameter T (the integration period) determining the width of this distribution. If T isn't too large, then $\psi(E_k)$ is not an eigenfunction, but it can be used as a basis function to determine accurate energy levels near E_k by diagonalizing a secular equation. To do this one generates several $\psi(E_k)$'s for energies that cover an energy window in the vicinity of the desired energy. The resulting functions are not generally orthogonal, but can nevertheless be used as basis functions in a secular equation, and the resulting generalized eigenvalue problem can be solved to determine eigenvalues within that range.

Semiclassical Studies. Traditional WKB (Wentzel, Kramers, Brillouin) theory (otherwise known as primitive semiclassical theory) for one-dimensional problems determines quantum energy levels by requiring that the classical actions which govern molecular vibrational motions have integer values (in units of $\hbar$) after a Maslov index (related to the number of turning points per period of oscillation) is subtracted. For multidimensional problems the generalization of this theory is known as EBK (Einstein, Brillouin, Kramers) theory, and it requires that the actions obey the formula:

$$\frac{1}{2\pi}\oint_{C_i}\mathbf{p}\cdot d\mathbf{r} = (n_i + \frac{1}{2})\hbar \quad i = 1,...,3N-6$$

Here C_i represents a closed contour in phase space that encircles the torus that is swept out by the trajectory as it undergoes vibrational motion. There will be one such contour for each of the degrees of freedom in the molecule, leading to the definition of 3N-6 actions n_i for a molecule with N atoms. In order to apply EBK theory, it is necessary that trajectory motion for the triatomic molecule be *quasiperiodic*, which means that trajectory motion repeats itself in a regular, though not periodic, way, as this leads to the formation of a torus in phase space. This type of motion is commonly found at low vibrational energies, but at high energies, substantial portions of the available phase space are chaotic, and in this case, rigorous EBK eigenvalues cannot be defined.

Several nonperturbative methods for determining EBK eigenvalues have been developed, and the subject has been the subject of several reviews.*(30-33)* Semiclassical eigenvalues as defined by the EBK formula are not accurate enough to be of use in interpreting spectroscopic measurements, but are good for qualitative interpretation of intramolecular dynamics, and they are also used in trajectory studies of bimolecular collisions*(34)* (where they define the vibrational states of the reactant and product molecules).

Unimolecular Decay.

Experimental Studies. At high enough levels of overtone excitation, IVR into dissociative modes can result in molecular decomposition. A primary goal in experimental studies of unimolecular decay is to identify dissociation channels and determine lifetimes as a function of energy above the dissociation limit. Competition between IVR and decomposition is a key factor in determining whether energy is randomized prior to dissociation. A wide range of experimental techniques have been developed and applied to unimolecular decomposition of isolated molecules. Classical kinetics experiments determine unimolecular decomposition rates indirectly through pressure dependent measurements and rely on competition from collisional deactivation. Photo-initiated dissociation experiments use a number of optical techniques for vibrational activation and are capable of determining state-to-state pathways and lifetimes. Small molecules with low energy barriers to dissociation have low state densities near the barrier, and as such the isolated resonances are amenable to quantum state-resolved experimental methods. Data from these experiments provide excellent comparison with quantum and classical dynamics calculations. Larger molecules with high dissociation energies possess initial states that can result from many overlapping dissociative resonances embedded in a high density of states. IVR occurs readily in these cases so that there are no readily identifiable quantum numbers to describe the initial state. In general, the measurable properties from such a state represent an average of the possible dissociation channels and lifetimes and hence statistical theories best describe the dissociation.

Dissociation in Gas Phase Kinetics. The classical approach to unimolecular decomposition dates back to the early 1920's when Lindemann*(35)* proposed his mechanism for unimolecular decomposition. This mechanism involves collisional activation to prepare a highly excited reactant which then undergoes unimolecular decomposition. Of course, collisions can also deactivate the excited reactant so that a pre-equilibrium exists before the decomposition step. At low pressures, collisional activation is rate limiting and the observed decomposition rates are governed by the energy transfer rate for activating the reactant molecule. At high pressures, the collisional activation/deactivation equilibrium occurs rapidly and the observed rate reduces to the unimolecular decomposition rate. The classical collisional approach is extremely useful for establishing overall lifetimes and can be used to complement laser-based techniques.*(36)*

Multiphoton Photodissociation. Optical methods of chemical activation have been employed in low pressure environments to prepare vibrationally excited reactants without relying on collisional activation. In infrared multiphoton excitation, sequential infrared photons from an intense laser pulse are absorbed by a molecule.*(37)* This process results in high levels of vibrational excitation on the ground electronic surface and is an effective means of preparing a large number of internally excited molecules. One disadvantage of this approach is its lack of specificity in the number of photons absorbed by each molecule. However, it has been used extensively to measure dissociation lifetimes from polyatomic molecules with reasonably well defined initial energies and as such has proven to be valuable in determining the appropriateness of statistical rate theories like RRKM theory where internal energy is assumed to randomize prior to molecular decomposition. It is also noteworthy that IR multiphoton excitation was used for the first experiments to demonstrate that a laser could be used to dissociate a molecule through vibrational excitation. This early work laid the foundations for a very active area of current research in controlling chemical reactions with light both in the time and frequency regimes.

State-Resolved Measurements. State-to-state studies can provide detailed information about the unimolecular decomposition of small molecules with low barriers to dissociation which possess isolated resonances near dissociation. For these experiments, a specific initial state is prepared using either SEP or IR-visible double resonance excitation and individual decay channels are probed by monitoring the appearance of reaction products typically using LIF or REMPI (resonance-enhanced multiphoton ionization) detection.*(38)* By collecting photofragments as function of the excitation wavelength, an action spectrum is obtained which is representative of the dissociating state. This technique is called photofragment yield (PHOFRY) or photofragment excitation (PHOFEX) spectroscopy. Initial states of specific decomposition channels are identified by assigning resonances in the action spectrum. In the lack of other broadening mechanisms, lifetimes τ for individual dissociation channels can be determined from the lifetime broadened Lorentzian line profiles where the line width Γ is given by $\Gamma = (2\pi c\tau)^{-1}$.

IVR. The observed unimolecular dissociation dynamics often depends on whether vibrational energy remains localized in specific bonds or randomizes through the molecule. In the dissociation of HCO → H + CO, IVR is slower than dissociation and a remarkable degree of mode specificity is observed in the decomposition rates.*(39-41)* Initial states prepared with excitation in the CO stretch are long-lived while the shortest lifetimes result from excitation of the CH stretch. Accurate potential energy surfaces exist for HCO and the experimental findings are in very good agreement with theoretical predictions (as further discussed below). In the reaction of HFCO → HF + CO, there is evidence for restricted IVR.*(42)* The spectrum of the initial dissociating state exhibits some fractionation due to IVR in the stretching modes, while the out-of-plane bend is reasonably stable against IVR. The out-of-plane bend does not couple strongly to the motion of the reaction coordinate and excitation in the bend results in the longest lifetimes. In addition, Coriolis-induced mixing results in enhanced IVR for high J states. It is noteworthy that dissociation from states of HFCO which do demonstrate rapid IVR is well described by RRKM theory, in spite of the small number of modes and low reaction barrier.

In the limit where IVR is rapid and unrestricted compared to decomposition, the dissociation spectrum consists of contributions from (often numerous) overlapping resonances, and statistical lifetimes are observed. One experimental method for studying dissociation from a quasi-continuum of states near the dissociation threshold involves vibrational excitation resulting from electronic excitation followed by rapid radiationless transitions to the ground electronic state.*(43)* If the photon energy is above the threshold for dissociation, unimolecular reaction products can be detected in real time using direct absorption, LIF or REMPI. Rates and product distributions can be obtained, from which the dissociation dynamics can be inferred. Photofragment translation energy spectroscopy is another approach for identifying unimolecular reaction channels and the kinetic energy release of the photofragments.*(44)* Efforts are underway in the quasi-continuum regime to identify how the initial energy and density of states influence decomposition rates and channels. One difficulty for interpreting dissociation results in this high energy and state density regime is in our inability to determine accurate densities of states and potential energy surfaces for polyatomic molecules.

Vibrationally Mediated Photodissociation. A related example of unimolecular decomposition involving vibrational excitation is demonstrated by two photon vibrationally mediated photodissociation experiments.*(45-47)* In these experiments, one photon is used for direct overtone pumping to a high vibrational level which, due to anharmonicity, has significant geometric deviations from the vibrationless state. From this intermediate vibrational level, a second photon promotes the system to an excited electronic state from which dissociation can occur. The key to these experiments is that the second photon would not have sufficient energy for dissociation if the molecule remained in its original geometry. Vibrational excitation extends the Franck-Condon overlap between the initial and the dissociative wavefunction so that a lower energy threshold dissociation can be accessed.

Theoretical Studies. The primary quantity of interest in unimolecular decay is the lifetime of the excited molecule as a function of the energy of excitation above the dissociation limit. Unimolecular decay in small molecules at low energies often involves *isolated* initial states, where the decay width is small compared to the average energy level spacing. These correspond to nonoverlapping scattering resonances in collisions between the dissociating fragment molecules. In this limit, one generally needs to use a quantum or classical dynamical theory to determine accurate state-resolved lifetimes. For larger molecules or even small molecules at high energies, the initial states are strongly overlapped, and statistical theory becomes more useful.

Quantum Dynamics. Lifetimes for unimolecular decay may be calculated quantum mechanically using either time-independent or time-dependent methods. There are many different approaches in both cases, so the discussion here will be representative rather than comprehensive. A limitation of all of these methods is that they are restricted to problems with few enough states that it is possible to calculate and/or store matrix elements of the Hamiltonian matrix, or of related operators.

One example of a *time independent* method has been described by Bowman*(48)*. In this method one solves the time independent Schrödinger equation as if the states were bound, using an expansion in square integrable basis functions, but with the Hamiltonian modified by adding in a negative imaginary potential in the regions corresponding to dissociation. Such imaginary potentials produce eigenstates with outgoing flux in the asymptotic region, which is the proper boundary condition for dissociation. As a result, the eigenvalues obtained by this procedure have an imaginary part Γ that can be used to determine the lifetime τ. Some of the eigenfunctions will not correspond to resonances; these are approximations to direct scattering between the dissociating fragments, and the lifetimes are very short. It is usually straightforward to distinguish between resonant and direct states by examining the degree of localization of the eigenfunctions to the well region of the potential. Bowman and others *(48)* have used this method to evaluate resonance energies and widths for HCO unimolecular decay.

Another time-independent method has been described by Dobbyn and Schinke *(49)*. In this approach one solves the Schrödinger equation as a scattering problem using methods which we discuss below in the section on *Collisional Energy Transfer*. The overlap of the resulting scattering wavefunction with a square integrable wavefunction that is localized in the well region is then evaluated. This can be thought of as an approximation to the Franck-Condon factor for photodissociation of the molecule, but actually the dynamical process can be totally fictitious in this case. This procedure is then repeated for many different energies, yielding a photodissociation spectrum as a function of E. Any resonance states that are localized in the well region will produce peaks in this spectrum, and by fitting these to a Lorentzian lineshape, it is possible to extract resonance energies and widths. This approach has been applied to calculate the resonance energies and widths in the dissociation of HCO, giving results that are in good agreement with results from the imaginary potential method.

A commonly used *time-dependent* method for calculating unimolecular lifetimes has been described by Gray and coworkers. *(50)* In this method one first determines an approximation to the initial dissociative state, usually by diagonalizing the Hamiltonian in a square integrable basis set (and without any imaginary potential in the Hamiltonian). Starting with one of the resulting eigenfunctions as initial conditions, the wavepacket is then propagated for a period of time which needs to be long enough for the wavepacket to undergo some decay. The time dependent overlap C(t) between the evolving wavepacket and its initial value is accumulated during this propagation, and the resulting function is analyzed for exponential decay to determine the lifetime. This method has been applied to the decay of HCO, yielding results in agreement with the time-independent methods.

Classical Dynamics. Trajectories have long been used to evaluate unimolecular lifetimes. *(51-52)* The usual procedure is to run a batch of trajectories, with initial conditions selected from a microcanonical ensemble at the energy of interest. By accumulating information about the time duration of each trajectory prior to dissociation, one can construct the survival probability function (the probability that the molecule has not decayed) as a function of time. Often this function has a simple exponential dependence on time, so the lifetime is then determined from the slope of a semilog plot. Sometimes a short time transient is observed in the semilog plot prior to exponential decay, and sometimes nonexponential decay can occur. This more complicated dependence is indicative of nonstatistical effects, as has been discussed in the literature, *(53)* but in most cases it is still possible to determine unimolecular decay lifetimes from those trajectories which are decaying exponentially.

The lifetimes that are obtained from survival probabilities are generally in very good agreement with the *average* lifetimes obtained from quantum dynamics calculations, and often they are also in good agreement with statistical estimates. Quantum effects can show up at energies just above the dissociation threshold where tunnelling or zero point energy violation can be important. Also, quantum microcanonical dissociation rates can show steps as a function of energy due to quantum energy level structure that is not seen in the classical calculations.

The primary virtue of the use of classical trajectories to calculate lifetimes for unimolecular decay is that provided one has a potential energy surface, the calculation is quite straightforward, and can be applied to polyatomic molecules with many atoms (dozens). Disadvantages are that a potential surface is needed which describes all the degrees of freedom, and the computation time is excessive if the lifetime is longer than approximately 100ps. Both of these problems can be reduced by using statistical theory.

Statistical Theories. The most commonly used statistical theory of unimolecular decay is *Rice-Ramsberger-Kassal-Marcus* (RRKM) theory.*(54)* This transition state theory method gives the following expression for the microcanonical decay rate (the inverse of the lifetime):

$$k(E) = \frac{N(E)}{h\rho(E)}$$

where $N(E)$ is the number of states accessible at the transition state for unimolecular decay, h is Planck's constant and $\rho(E)$ is the density of states of the molecule which is undergoing dissociation. This expression may be evaluated using either classical or quantum mechanics, and in the classical application it provides an upper bound to the exact classical rate (due to the neglect of recrossing of the transition state).

The location of the transition state where $N(E)$ is calculated is often taken to be the top of the barrier to dissociation, assuming that such a barrier exists. However a better procedure which also takes care of the case where there is no barrier is to locate the transition state at the point where $N(E)$ is minimized as a function of the dissociation coordinate. This leads to a theory known as *variational RRKM* which has proven successful in many applications.*(55)* Additional improvement may be obtained by carefully defining the "dissociation coordinate". In particular, in many applications $N(E)$ is smaller if bond distances are used rather than interfragment separations.*(56)*

The accurate evaluation of $N(E)$ and ρ requires great care. At the transition state, there are two classes of vibrational modes, those which correlate to vibrational modes of the separated fragments (the "conserved modes") and those which correlate to translations and rotations (the "transitional modes"). The evaluation of the contribution to $N(E)$ from the conserved modes is relatively straightforward since these modes tend to have small amplitude motions that are well described by the (quantum) harmonic approximation. However the contribution to $N(E)$ from the conserved modes is tricky since motion in these modes is often highly anharmonic, with large amplitude displacements and isomerization possible. Fortunately these modes generally have low vibrational frequencies, so the contribution to $N(E)$ from these modes may be evaluated using classical mechanics with reasonable accuracy.*(57)* Another difficulty in the RRKM calculation arises in the evaluation of $\rho(E)$, as the effects of anharmonicity are often quite important (due to the fact that the molecule is highly enough excited that it can dissociate). Recently Peslherbe and Hase *(58)* have demonstrated that accurate evaluation of $\rho(E)$ may be implemented using classical phase space quadratures.

Other commonly used statistical methods are *phase space theory* (PST) *(59)*, and the *statistical adiabatic channel model* (SACM) *(60)*. PST is a simpler type of statistical theory than RRKM that is more easily applied but often makes serious errors. It is most accurate when applied to reactions with extremely loose transition states. SACM is an alternative to RRKM that uses separate bottlenecks for each asymptotic state. This gives very similar results to RRKM for many kinds of problems, such as for tight transition states, but is different for others. Assessing the relative accuracy of RRKM and SACM is an active research topic.

Collisional Energy Transfer in Highly Excited Molecules

Experimental Studies. The importance of collisional energy transfer in chemical environments was demonstrated early on when it was shown that collisional activation and deactivation form the rate limiting step in unimolecular decomposition reactions at low pressures. Experimental efforts to understand collisional energy transfer of highly excited molecules*(61)* naturally fall into two distinct categories depending on the complexity of the excited molecule: 1) vibrationally excited small molecules with low complexity of the excited molecule: 1) vibrationally excited small molecules with low state densities which are amenable to state-resolved studies and 2) larger polyatomic molecules at energies where the vibrational motion is non-separable and individual quantum-resolved states do not exist. We begin our discussion of collisional energy transfer with experiments at the small molecule, state-resolved limit, then consider the intermediate region of larger polyatomic molecules where individual quantum states are still meaningful and finally conclude with large molecules in the dense state limit. An overall goal in each of these regimes is to develop physically meaningful mechanisms for energy transfer in highly excited molecules.

Small Molecule Energy Transfer Experiments. Experiments on collisional energy transfer in highly excited small molecules can utilize many of the same techniques for preparing vibrationally excited states as in unimolecular decay and IVR studies. In particular, SEP provides effective access to very high vibrational levels for a wide range diatomic molecules, and collision-induced depletion of the excited initial state can be monitored using laser-induced fluorescence (LIF). Remarkably, experiments on I_2,(62) NO,(63) and O_2(64) in very high vibrational levels (as high as $v = 42$ in I_2, $v = 25$ in NO and $v = 26$ in O_2) reveal that the dynamics of energy transfer from highly excited diatomic molecules has many features in common with energy transfer from only slightly excited diatomic molecules ($v = 1$ or 2). For example, single quantum changes of $\Delta v = -1$ in the energy donor are favored in both the high and low excitation regimes for collisions that excite rotations and translation in the collision partner ($V \rightarrow R/T$). Cross sections for these single quantum transitions in both energy regimes increase as the mass of the collision partner decreases. The propensity for resonant and near-resonant vibration to vibration ($V \rightarrow V$) energy transfer is apparent for molecules with high levels of vibrational excitation, just as it is for molecules with few vibrational quanta excited. The mechanisms for energy transfer do not appear to depend crucially on energy content, either. Collisions between O_2 ($v = 26$) and unexcited O_2 result in $V \rightarrow R/T$ energy transfer with a strong positive temperature dependence, reminiscent of impulsive energy transfer in low energy molecules. Near-resonant $V \rightarrow V$ energy exchange from O_2 ($v = 26$) to both CO_2 and N_2O shows the inverse temperature dependence which is similar to that observed for long-range $V \rightarrow V$ processes at low levels of vibrational energy. However, a notable difference is that highly excited molecules have greater possibilities for satisfying the resonance condition between energy donor and acceptor because of vibrational anharmonicity. Both 1:1 and 2:1 vibrational resonances have been observed for a number of collision partners. Another distinguishing and intriguing aspect of collisional energy transfer from highly excited small molecules is that reactive regions of the potential energy surface may be accessed at high enough levels of vibrational excitation. For example,(65) collisions of NO ($v > 15$) + NO have enough energy to reach the lowest energy transition state of NO + NO $\rightarrow$ N_2O + O. This remains an active area of research, for both fundamental reasons and applications to atmospheric chemistry.

Large Molecule Energy Transfer Experiments. Developing a detailed and concise picture of collisional energy relaxation of excited polyatomic molecules poses a formidable challenge for the experimentalist. Preparation of specific vibrationally excited states is hampered by spectral congestion, thermal populations of low frequency modes and rapid IVR. State specific excitation for low vibrational energies can be accomplished using a combination of jet expansions and direct IR excitation.(68)

Windows in intersystem crossing lifetimes can allow collision dynamics of vibrationally excited triplet states to be studied.*(67)* Another approach which has proven highly effective for large aromatic molecules is to excite specific quantum levels in an excited electronic state with direct visible or UV absorption. The collisional dynamics are then probed using dispersed fluorescence. Vibrational energy transfer from low lying vibrational levels of the S_1 state in para-difluorobenzene has been studied with this technique.*(68)* In this case, the energy transfer is highly selective, despite relatively high densities of neighboring states, and is dominated by small quantum changes in the low frequency modes. Extensive studies*(69)* using SEP to prepare the same initial states on the S_o surface yield remarkably similar results, indicating that at least for low levels of excitation, the collisional energy transfer is electronically adiabatic. It is noteworthy that temperature dependent cross sections (1-12K) are best explained by scaling to Lennard-Jones cross sections rather than hard sphere cross sections and that this correlation persists up to temperatures of 300K. Thus, the attractive interaction between collision partners is quite important in the vibrational energy transfer of large molecules.

Efforts to understand the collisional dynamics of highly excited polyatomic molecules are linked to their prevalence in unimolecular reactions and chemistry near reaction barriers. Preparing polyatomic molecules with very large amounts of internal energy is most effectively accomplished by utilizing the photophysical properties of a large class of aromatic molecules. Following optical excitation to a bound electronic state, the molecule relaxes to the ground electronic state via rapid radiationless transitions, and is left with an internal energy comparable to the photon energy, often enough to break chemical bonds. IVR occurs readily and this "chemically significant" amount of internal energy randomizes throughout the excited molecule. A very high density of vibrational states results, although translations and rotations remain thermal through momentum conservation. Two techniques, UV absorption*(70)* and IR fluorescence*(71)* have been used extensively in collisional environments to monitor the overall time evolution of such highly excited molecules following photo-activation. Experiments of this type have been used to study the collisional relaxation of excited molecules such as toluene, azulene, and benzene with a number of collision partners. The experimental probes for these techniques relies solely on the spectral characteristics of highly excited molecules so a wide range of collision partners can be studied. Results of these studies show that the average energy lost per collision <ΔE> depends approximately linearly on the average energy content <E> of the donor and that <ΔE> increases with the size and complexity of the collision partner. An additional method is to monitor the collisional energy transfer using time- and frequency-resolved FTIR emission spectroscopy.*(72)* This method is sensitive to any species which has an IR emission; as such it has the possibility of identifying active modes in the energy depletion pathways. Another approach for studying the collisional quenching of highly excited polyatomic molecules is "kinetically controlled selective ionization" (KCSI).*(73)* In this method, the collisional energy loss from an excited donor is monitored using a two photon ionization scheme that defines the energy content of the donor. The ion current is collected as a function of time for a given energy window and the energy transfer distribution is obtained. Notably, this technique is sensitive to the magnitude of the energy loss in time and can distinguish collisions that involve the large energy exchanges, by the early time arrival of molecules in a low energy window. Large energy transfer collisions (with ΔE as large as ~10,000 cm^{-1}) such as these have been referred to as "supercollisions" and there is support for their existence in the KCSI results. These techniques have provided extensive data for a large number of donor-acceptor collision partners.

Probes of Energy Transfer to the Bath. An alternate approach is to focus attention on the energy gain of the collision partner, which can be chosen to have a well resolved IR spectrum. By measuring high resolution IR transient absorption,*(74)* the vibrational and rotational states of the scattered molecules and their distributions are well determined. In addition, through measurement of Doppler broadened transient line shapes, the average recoil velocities associated with individual rotational states can be obtained. Thus a complete energy profile of the scattered molecule can provide substantial insight into the mechanisms of energy transfer at a quantum-state-resolved level even though the highly excited energy donor does not have well defined vibrational states. Notably, long range near-resonant $V \rightarrow V$ energy transfer results in vibrationally excited collision partners with very little excitation in rotation or translation.*(75)* Energy exchanging encounters such as this exhibit the inverse temperature dependence which is the signature of long range processes. In contrast, there can be substantial energy transferred from the donor vibrations solely into rotations and translation, but not vibrations, of the collision partner. This $V \rightarrow R/T$ energy transfer displays a positive temperature dependence which is associated with short range interactions.*(76)* The temperature dependent results and the large recoil velocities offer strong evidence that the $V \rightarrow R/T$ energy transfer occurs through impulsive collisions, similar to results for low energy molecular scattering. Because of the large amounts of energy going into rotation and translation in these short-range collisions, it is likely that $V \rightarrow R/T$ energy transfer contributes significantly to "supercollisions." Efforts are currently underway to determine the energy dependence of both long- and short-range energy changing mechanisms using quantum resolved scattering techniques.*(77)*

Theoretical Studies. Collisional relaxation has been studied theoretically since the 1930's, but its application to highly excited states of polyatomic molecules has mostly arisen in the last 10 years. One reason for this is that it has only been during this period that experimental methods for studying collisional relaxation have been able to provide *direct* information about the flow of energy from the highly excited molecule to the bath. Prior to that, most information about collisional relaxation was either obtained indirectly from an analysis of kinetics experiments, or directly from laser excitation of low lying states of molecules. The kinetics experiments are generally sensitive to only limited details of the relaxation process, so the traditional theoretical analysis has been based on simple models of the energy transfer dynamics (i.e., step ladder models, exponential models). The relaxation of low lying states of molecules is, by contrast, very sensitive to the quantum energy level structure of the excited molecules and the colliding partner, and to the collision dynamics, so often the only successful theories have been based on quantum scattering theory with accurate intermolecular potentials. The recent experimental work on highly excited molecules has provided important new challenges to theory, as the simple models used in classical kinetics are not sophisticated enough to describe the details of the experiments, but the quantum scattering calculations are too difficult to apply to highly excited polyatomics without severe approximation. As a result, much of the initial interest in this field has been in the use of an intermediate level of theory, namely classical molecular dynamics methods, so we consider this first.

Classical Molecular Dynamics. In classical molecular dynamics methods, one simulates collisional energy transfer directly by integrating trajectories. To do this one needs both intermolecular and intramolecular potentials, and for most problems of interest the potentials used are approximate (usually empirical) functions, and therefore are a major source of uncertainty in the comparison with experiment. The basic procedure used in classical simulation of collisional energy transfer is described in several places *(78-80)* so here we give a brief outline. In the usual application, one assumes that the excited molecule is in a microcanonical ensemble, and the collision partner properties (the bath gas) are sampled from a classical Boltzmann distribution. No attempt is made to mimic quantum effects as it is important to do the simulations under circumstances where the long time evolution of the excited molecule would produce a classical Boltzmann distribution if relaxation were allowed to proceed to completion. In the simplest applications, each trajectory is analyzed to determine the vibrational energy transfer resulting from the collisional interaction, and then an average of this over impact parameters and other initial conditions is performed to determine the average energy transfer per collision $<\Delta E>$. Higher energy transfer moments can also be determined, as can the energy transfer distribution function $P(E,E')$. There are some ambiguities in performing the average over impact parameters and in separating vibration from rotation that have been discussed in the literature.*(78)* In addition, the use of a microcanonical ensemble to define initial conditions is uncertain for simulating experiments where intramolecular vibrational redistribution may be incomplete between collisions. Alternative procedures have been considered *(81)*, but in spite of these problems the agreement between theoretical results based on microcanonical initial conditions and experiment is generally within a factor of two if realistic potentials are available.

Quantum Studies. A legitimate concern about the classical molecular dynamics methods is the importance of quantum effects on the results. This has provided motivation to extend the many past quantum studies of collisional energy transfer from low lying states of molecules *(82)* to studies of highly excited polyatomics. However such quantum studies are extremely difficult, and as a result, there have only been a few quantum studies of collisional energy transfer in highly excited polyatomic molecules *(83-85)*.

The traditional approach to quantum scattering theory for collisional energy transfer is the *coupled channel method*. In this approach, the scattering wavefunction is expanded in a basis of molecular vibrational states, with the expansion coefficients regarded as functions of the intermolecular separation coordinate. Substitution of this expansion into the Schrödinger equation yields the coupled equations for the coefficients (the coupled channel equations) which can be solved by a variety of methods to give the scattering matrix. From this one can determine energy transfer moments and probability distributions, although it has only been in the recent calculations of Schatz and Lendvay *(84)* that microcanonical moments and probability distributions have been calculated, and compared to results from trajectory calculations. An article describing these results is given elsewhere in this volume. These results demonstrate good classical-quantum correspondence for a model of He + CS_2 collisions under circumstances where many hundreds of states are accessible, and the zero point energy is a small fraction of the available energy. It is unclear from these results whether this agreement generalizes to other molecule/collision systems, so this is an important topic for further study.

Models. The use of dynamical or statistical models to describe collisional energy transfer in highly excited polyatomic molecules is just starting to get significant attention. Barker and coworkers *(86)* have begun studying the use of impulsive collision models such as SSH theory and ITFITS to describe energy transfer. These models have been successfully used in studies of the relaxation of low-lying states of molecules, and Barker's initial applications (see elsewhere in this volume) have been based on their direct application to higher levels of excitation. Nordholm and coworkers *(87)* have developed statistical models, which are based on the idea that the collision leads to a complete redistribution of the initial excitation into all of the degrees of freedom available to the collision complex. These models always overestimate the amount of energy transfer, but improved results can be obtained if one assumes that only a limited number of degrees of freedom are able to participate in the energy transfer process. Still another type of model has been developed by Gilbert and coworkers *(88)*, who have studied the use of stochastic models (the biased random walk model) to describe energy transfer from large molecules.

Bimolecular Reactions Involving Highly Excited Molecules

Experimental Studies. Bimolecular reactions often require that one or both reactants supply energy at some point along the reaction coordinate. Experiments to observe, understand, and potentially influence chemical reactivity have several essential prerequisites: 1) the ability to excite the reactant(s) selectively and with well defined energy, 2) the need for energy to remain well defined until a (possibly reactive) collision occurs and 3) the ability to detect a specific outcome of the chemical reaction. Because of these experimental requirements, much less is known about the bimolecular reactions of highly excited polyatomic molecules*(89)* as compared with our current understanding of IVR, unimolecular decay and collisional energy transfer. In this section, we discuss experimental methods for investigating the bimolecular reactions of highly excited molecules.

Excitation of Reagents. The earliest experimental indications of the energy requirements of bimolecular reactions come from the temperature dependence of rate constants.*(90)* Arrhenius plots of reaction rates are commonly used to establish average barrier heights for bimolecular reactions. On a more detailed level, experimentalists aim to understand competition between reactive channels and identify the features of the potential energy surface that control chemical reactivity. This can be accomplished by studying reactivity as a function of specific initial excitation. The initial energy can be in the form of translation, vibration or a combination of both. Hot atom experiments that use photolysis to create translationally hot hydrogen atoms have demonstrated the importance of translational energy in assisting endothermic reactions.*(91)* This approach, in combination with LIF detection of the OH reaction product, has been used to study the dynamics of the $H + H_2O \rightarrow H_2 + OH$ reaction*(92)* and of its isotopes.*(93)* High resolution spectroscopic detection of the spatially-oriented products can be combined with initial translational excitation to measure the translational energy dependence of the state-to-state cross sections, which are determined from the measured spectral widths.*(94)* The role of reactant vibrational energy can be explored in a similar manner by preparing the reactant using direct IR excitation and then extracting the vibrational dependent state to state cross sections.

Bond Selective Reactions. For small polyatomic molecules, a remarkable degree of bond selectivity is observed when the initial excitation of a well-defined state remains confined to a specific region of the molecule. A striking example of mode-selective vibrationally-driven chemistry has been demonstrated using vibrational overtone excitation for reactant preparation.*(95)* The reaction of H + HOD has two chemically distinct channels, H_2 + OD and HD + OH. The reaction barrier can be overcome with either 4 quanta of the OH stretch or 5 quanta of the OD stretch, each which can be populated selectively. Because the initial vibrational excitation remains localized in HOD until the reaction begins, the optically excited mode preferentially breaks; it is favored over the other channel by a factor of over 200! Analogous experiments*(96)* using Cl atoms in place of H atoms show similar results. These experimental results provide an excellent test for classical and quantum scattering theories.

For larger polyatomic molecules where IVR occurs rapidly, it is extremely unlikely that a similar degree of specificity will be observed in bimolecular reactions. It is possible however that high levels of vibrational energy will permit reactants to sample new types of reactivity in previously-unexplored regions on a reactive potential energy surface. This possibility is particularly interesting at energies near dissociation where large amplitude vibrational motion occurs. Evidence for large energy transfer ("supercollisions") from highly excited polyatomic molecules indicate that substantial amounts of energy are available for exchange in a bimolecular encounter. How this influences bimolecular reactions is unknown at the present time, but research is currently underway to address this issue.*(97)*

Theoretical Studies. The theory of bimolecular reactions has been an extremely active area of research for many years, but research on reactions of excited polyatomic molecules has only been of significant interest in the last 10 years. Both classical and quantum theories have been used, and each has limitations as will be described below. The quality of the potential surfaces has also been a problem in making comparisons with experiment, but in spite of these limitations, theory has played an important role in the interpretation and prediction of experiments, as recently reviewed by Bowman and Schatz *(98)* and by Schatz *(99)*.

Quasiclassical Studies. The *quasiclassical trajectory* (QCT) method was the first method to be applied to the description of bimolecular reactions involving highly excited polyatomic molecules *(100)* and is still the mostly commonly used method. In this method, one uses a semiclassical quantization scheme to determine the initial vibrational state of the molecule being considered. The quasiperiodic orbit associated with this state can be used to define initial conditions in a trajectory simulation of bimolecular collisions, from which the reactive cross section and rate constant can be calculated. Semiclassical eigenvalues are determined using methods that were described earlier in this article, and thus require that molecular vibrational motion be quasiperiodic in order to produce rigorous results. However, the most commonly used methods for doing these calculations are able to "tolerate" a certain amount of chaotic vibrational motion. Most of the applications of these methods have been done by Schatz and his group *(99)*, however studies have also been reported by Hase and coworkers *(101)* and by Muckerman, Valentini and coworkers *(102)*.

The ability of quasiclassical methods to describe bimolecular reactions involving excited polyatomic molecules is not yet known. Schatz *et al (30)* and

Takayanagi and Schatz *(103)* have recently compared QCT results and reduced dimensionality quantum scattering results for the reaction $CN + H_2 \rightarrow HCN + H$, including an analysis of the product vibrational distributions in the HCN product. This is the reverse of the type of reaction that we are considering here, but the comparison is still useful. These studies find good correspondence between QCT and quantum results for the overall thermal rate constant, and for the average quantum numbers for each vibrational mode of HCN, but the distribution of individual quantum states is less accurate, with some state-resolved cross sections differing by factors of 2-3. This suggests that QCT methods may only be capable of semiquantitative predictions of state specific reactivity, but for many applications this may be sufficient.

One application where QCT calculations have already proven to be useful in studies of highly excited polyatomics is the reaction H + HOD *(100,104)*. QCT studies of this reaction were originally done several years before the experiments, and they demonstrated that there was a considerably larger reactivity to produce the OH product when the OH stretch mode of HOD was initially excited with several quanta of excitation than when a comparable amount of energy was put into the OD stretch mode. Once the experiments were done *(95)*, more refined calculations were used to model the experiments *(104)*, and the predicted enhancements in cross sections were in generally good agreement.

Quantum Studies. Although there are numerous methods for using quantum scattering theory to describe bimolecular reactions, only a small group of these methods have been used to describe reactions involving excited triatomic and larger molecules at the state-resolved level. An extensive review of these methods has recently been given *(30)*, so we will only present a brief discussion here.

Until very recently, all quantum studies of reactions involving four or more atoms have been based on dynamical approximations where the motions of certain degrees of freedoms are approximated either using adiabatic or sudden approximations. There are many different types of these so-called *reduced dimensionality approximations*, but the general goal of all of them is to reduce the number of actively treated coordinates to three or four, as this tends to be the maximum number of variables that can be treated accurately using quantum scattering methods. One constraint on the choice of active variables is that it takes at least two variables to describe the basic bond breaking and bond forming step in an atom transfer reaction, and these must be treated actively to describe energy disposal and tunnelling. This means that only one or two of the "spectator" coordinates (i.e., coordinates which are preserved in going from reagents to products) can be included in the active variable list.

The two most commonly used reduced dimensionality methods are the *adiabatic bend* method of Bowman and coworkers *(105)* and the *rotating bond approximation* method of Clary and coworkers *(106)* In the adiabatic bend method the three stretch variables in a four-atom reaction of the type $AB+CD \rightarrow ABC + D$ are treated actively while the bends are treated adiabatically. In the rotating bond approximation, two stretch variables and spectator rotation are treated actively, and the remaining variables are treated in the sudden approximation. It is interesting to note that results derived from adiabatic and sudden approximations are generally similar. One exception is that to calculate accurate rate constants it is essential to include for zero point energy in the degrees of freedom not actively treated. This is automatically included in the adiabatic approximation, but needs to be added as an *ad hoc* procedure in the sudden approximation.

The solution of the Schrodinger equation for the active variables in the reduced dimensionality methods has almost always been accomplished using coupled channel methods based on *hyperspherical coordinates.* Hyperspherical coordinates are a generalization of polar coordinates to many degrees of freedom, and there is only one continuous variable, the hyperradius. All other coordinates are angles, so the eigenfunctions associated with motion in the hyperangles are bound states, and these states provide the discrete basis expansion needed in a coupled channel calculation, with the hyperradius taken to be the propagation variable.

Very recently, several quantum scattering methods have been presented which include either 5 or 6 degrees of freedom in the description of a four atom reaction. *(107)* These methods are based on either coupled channel hyperspherical calculations or wavepacket calculations, and only limited applications presented so far have only considered the $OH + H_2$ reaction.

Acknowledgments

A. S. M. is supported by a Clare Boothe Luce Professorship from the Henry Luce Foundation. Research support for A. S. M. comes from NSF grant CHE-9624533 with equipment funding from ONR grant N00014-96-1-0788. G. C. S. was supported by NSF Grant CHE-9527677.

Literature Cited

1.   Herzberg, G., *Molecular Spectra and Molecular Structure*, Vols. 1-3, D. Van Nostrand Co., **1945**.
2.   Nesbitt, J. D.; Field, R. W. *J. Phys. Chem.* **1996**, *100*, 12735.
3.   Lehman, K. K.; Scoles, G.; Pate, B.H. *Annu. Rev. Phys. Chem.* **1994**, *45*, 241.
4.   Felker, P. M.; Zewail, A. H. in *Jet Spectroscopy and Molecular Dynamics*; Hollas, J. M.; Phillips, D., Eds.; Blackie A& P: London, **1995**.
5.   Chaiken, J.; Gurnick, M.; McDonald, J. D. J. Chem. Phys. **1981**, *74*, 106; *J. Chem. Phys.* **1981**, *74*, 117.
6.   Brucat, P.J.; Zare, R. N. *J. Phys. Chem.* **1983**, *78*, 100; J. Chem. Phys. **1984**, *81*, 2562.
7.   Special Issue on "Overtone Spectroscopy and Dynamics" Lehmann, K. K.; Herman, M. Eds., *Chem. Phys.* **1995**, *190*.
8.   Hamilton, C. E.; Kinsey, J. L.; Field, R. W. *Annu. Rev. Phys. Chem.* **1986**, *37*, 493.
9.   McDonald, J. D. *Annu. Rev. Phys. Chem.* **1979**, *30*, 29; Stewart, G. M.; McDonald, J. D. *J. Chem. Phys.* **1983**, *78*, 3907; Kulp, T. J.; Ruoff, R. S.; McDonald, J. D. *J. Chem. Phys.* **1985**, *82*, 2175.
10.   Nesbitt, D. J.; Leone, S. R. *Chem Phys. Lett.* **1982**, *87*, 123.
11.   McIlroy, A.; Nesbitt, D. J. *J. Chem. Phys.* **1989**, *91*, 104; McIlroy, A.; Nesbitt, D. J. *J. Chem. Phys.* **1990**, *92*, 2229
12.   de Souza, A. M.; Kaur, D.; Perry, D. S. *J. Chem. Phys.* **1988**, *88*, 4569; Perry, D. S. *J. Chem. Phys.* **1993**, *98*, 6665
13.   Lehmann, K. K.; Scoles, G.; Pate, B. H. *Annu. Rev. Phys. Chem.* **1994**, *45*, 241.
14.   Lawrence, W.; Knight, A. E. W. *J. Phys. Chem.* **1985**, *89*, 917.
15.   Reddy, K. V.; Berry, M. J. *Chem. Phys. Lett.* **1977**, *52*, 111; Bray, R. G.; Berry, M. J. *J. Chem. Phys.* **1979**, *71*, 4909.

16. Sibert, E. L.,III; Reinhardt, W. P.; Hynes, J. T. *Chem. Phys. Lett.* **1982**, *92*, 455; Sibert, E. L.,III; Reinhardt, W. P.; Hynes, J. T. *J. Chem. Phys.* **1984**, *81*, 1115; Sibert, E. L.,III; Reinhardt, W. P.; Hynes, J. T. *J. Chem. Phys.* **1984**, *81*, 135.
17. Iung, C.; Leforestier, C.; Wyatt, R. E. *J. Chem. Phys.* **1993**, *98*, 6722.
18. Jonas, D. M.; Solina, S. A. B.; Rajaram, B.; Silbey, R. J.; Field, R. W.; Yamanouchi, K,; Tsuchiya, S. *J. Chem. Phys.* **1992**, *97*, 2813.
19. Amrein, A.; Duebal, H.-R.; Duebal, H.-R.; Quack, M. *J. Chem. Phys.* **1984**, *81*, 3779; Segall, J.; Zare, R. N.; Duebal, H.-R.; Lewerenz, M.; Quack, M. *J. Chem. Phys.* **1987**, *86*, 634; Duebal, H.-R.; Quack, M. *J. Chem. Phys.* **1988**, *88*, 5408.
20. *Advances in Molecular Vibrations and Collision Dynamics*, ed. Bowman, J. M. JAI Press, Greenwich, CT. Vol. 1A and 1B, 1991, and subsequent volumes.
21. *Comp. Phys. Comm.* **1988**, *51*.
22. Carney, G. D.; Sprandel, L. I.; Kern, C. W. *Adv. Chem. Phys.* **1978**, *37*, 305.
23. Gerber, R. B. ; Ratner, M. A. *Adv. Chem. Phys.* **1988**, *20*, 97; Ratner, M. A.; Gerber, R. B.; Horn, T. R.; Williams, C. J.; See. Ref. 1, Vol. 1A, p. 215.
24. Romanowski, H.; Bowman, J. M. *Chem. Phys. Lett.* **1984**, *110*, 235.
25. Ratner, M. A.; Buch, V.; Gerber, R. B. *Chem. Phys.* **1980**, *53*, 345.
26. McCoy, A. B.; Sibert III; E. L; See Ref. 1, Vol. 1A, p. 255.
27. Heller, E.J.; Davis, M.J. *J. Phys. Chem.* **1980**, *84*, 1999
28. Noid, D.W.; Brocks, B.T.; Gray, S. K.; Marple, S. L. *J. Phys. Chem.* **1988**, *92*, 3386.
29. Neuhauser, D. *J. Chem. Phys.* **1990**, *93*, 2611; **1994**, *100*, 5076.
30. Schatz, G. C.; Takayanagi, T.; ter Host, M, in *Methods for Computational Chemistry*, ed. D. Thompson, JAI Press, **1997**, in press.
31. Martens, C.C.; Ezra, G. S. *J. Chem. Phys.* **1985**, *83*, 2990.
32. Skodje, R. T.; Borondo, F.; Reinhardt, W. P. *J. Chem. Phys.* **1985**, *82*, 4611.
33. Percival, I. C. *Adv. Chem. Phys.* **1977**, *36*,1; Noid, D. W.; Koszykowski; M. L. Marcus, R. A. *Annu. Rev. Phys. Chem.* **1981**,*32*,267; Reinhardt, in *Mathematical Analysis of Physical Systems*, ed. Mickens, R. Van Nostrand Reinhold, New York, 1984; Tabor, M. *Adv. Chem. Phys.* **1981**, *46*, 73.
34. Schatz, G. C. *J. Phys. Chem.* **1996**, *100*, 12839
35. Lindemann, F. A. *Trans. Faraday Soc.* **1922**, *17*, 598.
36. Michaels, C,; Tapalian, C.; Lin, Z.; Sevy, E.; Flynn, G. W. *Faraday Discuss. Chem. Soc.* **1995**, *102*, 405.
37. Schulz, P. A.; Sudbo, Aa. S.; Krajnovich, D. J.; Kwok, H. S,; Shen, Y. R.; Lee, Y. T. *Annu. Rev. Phys. Chem.* **1979**, *30*, 379.
38. S. A.; Reisler, H. *Annu. Rev. Phys. Chem.* **1996**, *47*,495.
39. Williams, S.; Tobiason, J. D.; Dunlop, J. R.; Rohlfing, E. A. *J. Chem. Phys.* **1995**, *102*, 8342; Tobiason, J. D.; Dunlop, J. R.; Rohlfing, E. A. *J. Chem. Phys.* **1995**, *103*, 1448.
40. Adamson, G. W.; Zhao, X.; Field, R. W. *J. Mol. Spectros.* **1993**, *160*, 11.
41. Neyer, D. W.; Luo, X.; Houston, P. L.; Burak, I. *J. Chem. Phys.* **1993**, *98*, 5095; Neyer, D. W.; Luo, X.; Houston, P. L.; Burak, I. *J. Chem. Phys.* **1995**, *102*, 1645.
42. Choi, Y. S.; Moore, C. B. *J. Chem. Phys.* **1991**, *94*, 5414; Choi, Y. S.; Moore, C. B. *J. Chem. Phys.* **1992**, *97*, 1010; Choi, Y. S.; Moore, C. B. *J. Chem. Phys.* **1995**, *103*, 9981.

43.   Hippler, H.; Luther, K.; Troe, J.; Walsh, R. *J. Chem. Phys.* **1978**, *68*, 323.
44.   See, for example, Zhao, X.; Miller, W. B.; Hintsa, E. J.; Lee., Y. T. *J. Chem. Phys.*, **1989**, *90*, 5527; Chesko, J. D.; Stranges, D.; Suits, A. G.; Lee, Y. T. *J. Chem. Phys.* **1995**, *103*, 6290.
45.   Rizzo, T. R.; Hayden, C. C.; Crim, F. F. *Faraday. Discuss. Chem. Soc.* **1975**, *75*, 223.
46.   Fleming, P. R.; Li, M.; Rizzo, T. R. *J. Chem. Phys.* **1991**, *94*, 2425; Luo, X.; Rizzo, T. R. *J. Chem. Phys.* **1992**, *96*, 5659.
47.   Utz, A.; Carrisquillo, E.; Tobiason, J. D.; Crim, F. F. *Chem. Phys.* **1995**, *190*, 311; Crim, F. F. *Annu. Rev. Phys. Chem.* **1993**, *44*, 397.
48.   Wang, D.; Bowman, J. M., *J. Chem. Phys.* **1994**, *100*, 1021.
49.   Stumpf, M.; Dobbyn, A.J.; Mordaunt, D. H.; Keller, H.-M.; Fluethmann, .; Schinke, R.; Werner, H.-J.; Yamashita, K. *Far. Disc. Chem. Soc.* **1995**, *102*, 193.
50.   Gray, S.K. *J. Chem. Phys.* **1992**, *96*, 6543.
51.   Sewell, T. D. Schranz, H. W.; Thompson, D. L.; Raff, L. M. *J. Chem. Phys.* **1991**, *95*, 8089; Woodruff, S. B.; Thompson, D. L. *J. Chem. Phys.* **1979**, *71*, 376.
52.   Hase, W. L. In *Dynamics of Molecular Collisions*, Part B, ed. Miller, W. H. (Plenum, New York, 1976), p. 121.
53.   Davis, M. J.; Gray, S.K. *J. Chem. Phys.* **1986**, *5389* (1986).
54.   Marcus, R. A. *J. Chem. Phys.* **1952**, *20*,359; Wardlaw, D. M.; Marcus, R. A. *Adv. Chem. Phys.* Part I **1988**, *70*, 231.
55.   Wardlaw, D. M.; Marcus, R. A. *J. Phys. Chem.* **1986**, *90*, 5383.
56.   Klippenstein, S. J. *J. Chem. Phys.* **1991**, *94*, 6469; **1992**, *96*, 367.
57.   Wardlaw, D. M.; Marcus, R. A. *J. Chem. Phys.* **1985**, *83*, 3462;
58.   Peslherbe, G. H.; Hase,W. L. *J. Chem. Phys.* **1994**, *101*, 8535.
59.   Pechukas, P.; Light, J. C. *J. Chem. Phys.* **1965**, *42*, 3281; Pechukas, P.; Rankin, R.; Light, J. C. *J. Chem. Phys.* **1966**, *44*, 794; Klots, C. E. *J. Phys. Chem.* **1971**, *75*, 1526.
60.   Quack, M.; Troe, J. *Ber. Bunsenges. Phys. Chem.* **1974**, *78*, 240; **1975**, *79*, 170, 469.
61.   Flynn, G. W.; Parmenter, C. S.; Wodtke, A. M. *J. Phys. Chem.* **1996**, *100*, 12817.
62.   Nowlin, M. L.; Heaven, M. C. *J. Chem. Phys.* **1993**, *99*, 5654; Nowlin, M. L.; Heaven, M. C. *Chem. Phys. Lett.* **1995**, *239*, 1.
63.   Yang, X.; Kim, E. H.; Wodtke, A. M. *J. Chem. Phys.* **1992**, *96*, 5111.
64.   Mack, J. A.; Mikulecky, K.; Wodtke, A. M. *J. Chem. Phys.* **1996**, *105*, 4105.
65.   Cousins, L. M.; Leone, S. R. *J. Chem. Phys.* **1987**, *86*, 6731.
66.   Orr, B. *J. Chem. Phys.* **1995**, *190*, 261.
67.   Bevilacqua, T. J.; Weisman, R. B. *J. Chem. Phys.* **1993**, *98*, 6316.
68.   Catlett, D. L., Jr.; Parmenter, C. S.; Pursell, C. J. *J. Phys. Chem.* **1994**, *98*, 7725.
69.   Lawrence, W. D.; Knight, A. E. W. *J. Chem. Phys.* **1982**, *77*, 570; Muller, D. J.; Lawrence, W. D.; Knight, A. E. W. *J. Phys. Chem.* **1983**, *87*, 4952; Kable, S. H.; Knight; A. E. W. *J. Chem. Phys.* **1987**, *86*, 4709; Kable, S. H.; Knight, A. E. W. *J. Chem. Phys.* **1990**, *93*, 3151.
70.   Hippler, H.; Troe, J.; Wendelken, H. J. *Chem. Phys. Lett.* **1981**, *84*, 257; Hippler, H.; Troe, J.; Wendelken, H. J. *J. Chem. Phys.* **1983**, *78*, 6709.

71.   Smith, G. P.; Barker, J. R. *Chem. Phys. Lett.* **1981**, *78*, 253.
72.   Hartland, G. V.; Xie, W.; Dai, H. L.; Simon, A; Anderson, M. *J. Rev. Sci. Instrum.* **1992**, *63*, 3261; G. V. Hartland; Qin, D.; Dai, H. L. *J. Chem. Phys.* **1994**, *100*, 7832.
73.   Löhmannsröben, H. Luther, K. *Chem. Phys.* **1993**, *175*, 99; Luther, K.; Reihs, K. *Ber. Bunsen-Ges. Phys. Chem.* **1988**, *92*, 442.
74.   Chou, J. Z.; Flynn, G. W. *J. Chem. Phys.* **1990**, *93*, 6099; Mullin, A. S.; Park, J.; Chou, J. Z.; Flynn, G. W.; Weston, R. E., Jr. *Chem. Phys.* **1993**, *175*, 53.
75.   Michaels, C. A.; Mullin, A. S.; Flynn, G. W. *J. Chem. Phys.* **1995**, *102*, 6682.
76.   Mullin, A. S.; Michaels, C. A.; Flynn, G. W. *J. Chem. Phys.* **1996**, *102*, 6032
77.   Wall, M. C.; Lemoff, A. and Mullin, A. S. results to be published.
78.   Lendvay, G.; Schatz, G. C. in **Vibrational Energy Transfer Involving Large and Small Molecules**, ed. Barker, J. A., *Advances in Chemical Kinetics and Dynamics*, Vol. 2B (JAI Press, 1995), pp. 481.
79.   Gilbert, R. G. *Int. Rev. Phys. Chem.* **1991**, *10*, 319.
80.   Lenzer, T.; Luther, K.; Troe, J.; Gilbert, R. G.; Lim, K. F. *J. Chem. Phys.* **1995**, *103*, 626.
81.   Bruehl, M.; Schatz, G. C. *J. Chem. Phys.* **1988**, *89*, 770; Lendvay, G.; Schatz, G. C. *J. Phys. Chem.* 1991, **95**, 8748.
82.   Clary, D. C. *J. Phys. Chem.* **1987**, *91*, 1718; Clary, D. C. *J. Chem. Phys.* **1981**, *75*, 2023.
83.   Clary, D. C.; Gilbert, R. G.; Bernshtein, V.; Oref, I. *Far. Disc. Chem. Soc.* **1995**, *102* ,423; Nalewajski, R. F.; Wyatt, R. E. *Chem. Phys.* **1983**, *81*, 357; **1984**, *85*, 117; **1984**, *89*, 385.
84.   Schatz, G.C.; Lendvay, G. *J. Chem. Phys.*, **1997**, *106*, 3548.
85.   Pan, B.; Bowman, J. M. *J. Chem. Phys.* **1995**, *103*, 9668; Bowman, J. M.; Padmavathi, D. A. *Mol. Phys.* **1996**, *88*, 21.
86.   Barker, J. R.; To be published.
87.   Nordholm, S.; Freasier, B. C.; Jolly, D. L. *Chem. Phys.* **1977**, *25*, 433; Schranz, H. W.; Nordholm, S.; Andersson, L. *Chem. Phys. Lett.* **1991**, *186*, 65; Borjesson, L.E.B.; Nordholm, S. *J. Phys. Chem.* **1995**, *99*, 938.
88.   Lim, K.F.; Gilbert, R. G., *J. Chem. Phys.* **1986**, *84*, 6129.
89.   Crim, F. F. *J. Phys. Chem.* **1996**, *100*, 12725.
90.   van't Hoff, J. H. *Etudes de Dynamique Chimique*, Muller, Amsterdam, **1884**; Arrhenius, S, Z. *Phys. Chem.* **1889**, *4*, 226.
91.   Flynn, G. W.; Weston, R. E., Jr. *Annu. Rev. Phys. Chem.* **1986**, *37*, 551.
92.   Jacobs, A.; Volpp, H. R.; Wolfrum, J. *J. Chem. Phys.* **1994**, *100*, 1936.
93.   Bronikowski, M. J.; Simpson, W. R.; Zare, R. N. *J. Phys. Chem.* **1993**, *97*, 2204.
94.   Shafer, N. E.; Xu, H.; Tuckett, R. P.; Springer, M.; Zare, R. N. *J. Phys. Chem.* **1994**, *98*, 3369.
95.   Sinha, A.; Hsiao, M. C.; Crim. F. F. *J. Chem. Phys.* **1990**, *92*, 6333; Sinha,A. *J. Phys. Chem.* **1990**, *94*, 4391; Sinha, A.; Hsiao, M. C.; Crim. F. F. *J. Chem. Phys.* **1991**, *94*, 4928; Sinha, A.; Hsiao, M. C.; Crim, F. F. *J. Phys. Chem.* **1991**, *94*, 4928; Hsiao, M. C.; Sinha, A.; Crim. F. F. *J. Chem. Phys.* **1991**, *95*, 8263.

96.   Sinha, A.; Thoemke, J. D.; Crimm, F. F. *J. Chem. Phys.* **1992**, *96*, 372; Metz, R.B.; Thoemke, J.D.; Pfeiffer, J.M.; Crim, F.F. *J. Chem. Phys.* **1993**, *99*, 1744; Thoemke, J. D.; Pfieffer, J. M.; Metz, R. B.; Crim, F. F. *J. Phys. Chem.* **1995**, *99*, 13748.

97.   Fraelich, M.; Elioff, M. E.; Mullin, A. S. results to be published.

98.   Bowman, J. M; Schatz, G.C. *Annu. Rev. Phys. Chem.* **1995**, *46*, 169.

99.   Schatz, G. C. *J. Phys. Chem.* **1995**, *99*, 516.

100.   H. Elgersma and G.C. Schatz, *Int. J. Quant. Chem. Symp.* **1981**, *15*, 611; G.C. Schatz, M.C. Colton and J.L. Grant, *J. Phys. Chem.* **1984**, *88*, 2971.

101.   Nizamov, B.; Setser, D. W.; Wang, H.; Peslherbe, G. H.; Hase, W. L. *J. Chem. Phys.* **1996**, *105*, 9897.

102.   Huang, J.; Valentini, J. J.; Muckerman, J. T. *J. Chem. Phys.* **1995**, *102*, 5695.

103.   Takayanagi, T; Schatz, G. C., J. Chem. Phys. **1997**, *106*, 3227.

104.   Kudla, K.; Schatz, G. C. *Chem. Phys. Lett.***1992**, *193*, 507; Kudla, K.; Schatz, G. C. *Chem. Phys.***1993**, *175*, 71.

105.   Bowman, J. M. *J. Phys. Chem.***1991**, *95*, 4960; Sun, Q.; Bowman, J. M. *J. Chem. Phys.* **1990**, *92*, 1021; **1990**, *92*, 5201; Sun, Q.; Yang, D. L.; Wang, N. S. Wang, Bowman, J. M.; Lin, M. C. *J. Chem. Phys.***1990**, *93*, 4730.

106.   Clary, D. C. *J. Phys. Chem.* **1995**, *99*, 13664; Clary, D. C. *J. Chem. Phys.* **1991**, *95*, 7298; Clary, D. C. *J. Chem. Phys.* **1992**, *96*, 3656; Nyman; G.; Clary, D. C. *J. Chem. Phys.* **1993**, *99*, 7774; Clary, D.C. *J. Phys. Chem.* **1995**, *99*, 13664.

107.   Clary, D. C. private communication; Zhang, D. H.; Light, J. C. *J. Chem. Phys.* **1996**, *105*, 1291; Zhang, D. H.; Light J. C. *J. Chem. Phys.* **1997**, *106*, 551; Zhu, W.; Dai, J.; Zhang, J. Z. H.; Zhang, D. H. *J. Chem. Phys.* **1996**, *105*, 4881.

INTRAMOLECULAR DYNAMICS: BOUND MOTIONS

Chapter 2

Filter Diagonalization

A General Approach for Calculating High-Energy Eigenstates and Eigenfunctions and for Extracting Frequencies from a General Signal

Daniel Neuhauser

Department of Chemistry, University of California, Los Angeles, CA 90095–1569

Filter-Diagonalization was introduced in 1990 (D. Neuhauser, J. Chem. Phys. **93**, 2611) as a general approach for extracting high-energy eigenstates. The method combines simultaneous calculation of a rough Filter at many energies with a subsequent small-matrix diagonalization stage. The hybrid method combines the advantages of the pure-matrix and the diagonalization approaches, without their deficiencies.

Recently, we have been able to reformulate the method so that it can extract frequencies and eigenfunctions at any energy range from a fixed set of residues, without large storage requirements. Given the residues, repetitive calculations of spectra and eigenfunctions at arbitrary multiple energy ranges are trivial.

We review the achievements of the approach, as well as a a new feature of it, the ability to extract frequencies from a short-time segment of any signal, even if the signal is not due to a quantum correlation function.

Introduction

A general problem in chemistry and other fields is the extraction of highly excited eigenstates and eigenvalues of large sparse matrices. Recently, the author developed the Filter-Diagonalization approach [1]-[6], which furnishes a new way to extract eigenstates for the sparse molecular Hamiltonians encountered in quantum molecular spectra and dynamics simulations. The method has been adapted by several authors [8]-[17]. It combines the good features of the filter-technique for extracting highly excited eigenstates, while avoiding long propagation times by adding a diagonalization step. Further, Filter-Diagonalization solves another problem: efficient extraction of frequencies and damping factors out of a general time-dependent signal, without long time-propagations or large-matrix diagonalizations.

In this short review we outline the Filter-Diagonalization method and discuss recent developments and applications. For concreteness, we start with the problem of eigenvalue and eigenfunction extraction for an Hamiltonian H (or simply a

26

general matrix) acting on functions with N points. For concreteness, we assume that H is either real-symmetric, or, more generally, complex symmetric:

$$H^T = H. \tag{1}$$

(The final results are also valid for Hermitian Hamiltonians.) The one feature we need to recall on complex symmetric Hamiltonians is that they have orthogonal eigenstates, ϕ_n, fulfilling

$$(\phi_n|\phi_m) = \delta_{nm}, \tag{2}$$

where we introduced the c-product [18], which is equivalent to the Hermitian product but has no complex conjugation:

$$(\phi|\chi) \equiv \langle \phi^*|\chi \rangle = \sum_j \phi(x_j)\chi(x_j), \tag{3}$$

where the sum extends over all grid points (or simply over all N points). The problem of diagonalizing this Hamiltonian, i.e., the general problem of diagonalizing an $N \times N$ sparse matrix, amounts to finding the complex eigenvalues (ϵ_n) and eigenfunctions in the following representation:

$$H = \sum_n |\phi_n)\epsilon_n(\phi_n|. \tag{4}$$

The quest is either for all eigenfunctions and eigenvalues, or for those within a given interval of energies, $[E_{\min}, E_{\max}]$. The desired range of energies need not necessarily be at the edge of the spectra of H.

The first step in Filter-Diagonalization is the filter. On an arbitrary initial function, ψ_0, we define a filter $F(E)$ at an arbitrary energy E. The filter is a linear operator, such that the result of it is another vector, $|\zeta)$:

$$|\zeta) = F(E)|\psi_0). \tag{5}$$

There are different filters one can choose and different approaches for calculating the action of each on ψ_0, as elucidated later. More important however is the general property of a filter: a high quality filter, $F_{accurate}(E)$, chooses out of ψ_0 the one eigenfunction ϕ_n which has an associated ϵ_n that is the closest in energy to E:

$$\phi_n \approx F_{accurate}(E)\psi_0, \tag{6}$$

and the eigenvalue ϵ_n can then be determined from the expectation value of H,

$$\epsilon_n \equiv (\phi_n|H|\phi_n)/(\phi_n|\phi_n). \tag{7}$$

It is clear however that for $F_{accurate}(E)$ to choose accurately an eigenfunction of H which happens to be close in energy to other eigenfunctions, it needs to be a high quality filter; this generally would translate to a requirement many terms are necessary in the evaluation of $F_{accurate}$. For example, for typical iterative filters (as discussed below) the numerical effort is inversely proportional to the the spacing between the levels that are the closest to the sampling energy. This is simply a manifestation of the uncertainty principle: high accuracy requires long "times", i.e., much numerical effort.

Filter-Diagonalization avoids this caveat by first noting that typically, the effort in producing a filter at several energies is only negligibly larger than the effort of producing the filter at one energy, and therefore it pays to use a filter at multiple energies. We pick a filter, $F(E)$, which is now "rough" (as explained below) and apply it on the same initial wavefunction $|\psi_0\rangle$ but at several ("L") energies; this produces a set of vectors, each of size N **(see Figure 1)**:

$$|\zeta_j\rangle \equiv F(E_j)|\psi_0\rangle. \tag{8}$$

Here, E_j would be an arbitrary set of nearby energies; typically, an equi-spaced set within a desired energy range $[E_{\min}, E_{\max}]$ is used. The roughness of the filter means that each vector, ζ_j, encompasses many near-by eigenfunctions ϕ_n rather than one isolated eigenfunction, i.e.,

$$\zeta_j = \sum_n D_{jn}\phi_n. \tag{9}$$

where the D_{jn} are numerical coefficients. In most of the discussion the filter $F(E_j)$ is a function of the difference between the energy and the Hamiltonian,

$$F(E_j) \equiv f(E_j - H). \tag{10}$$

For such a filter, the numerical coefficients in Eq.(9) are simply equal to the strength of the filter at the difference between the eigenfunctions and the sampled energies:

$$D_{jn} = f(E_j - \epsilon_n). \tag{11}$$

The crucial point is then that *even a rough filter restricts the sum in Eq. (9) to extend only over a few eigenfunctions*, i.e., the D_{jn} coefficients are appreciable only when ϵ_n are relatively close in energy to E_j even though they may not be peaked at one frequency. Filtered functions at different nearby energies (E_j) would be covered by approximately the same subset of eigenfunctions. When the number of vectors ζ_j is sufficiently large, they therefore *span the desired energy range* and we can then use them as a *basis* for those eigenfunctions which have eigenvalues in the energy range $[E_{\min}, E_{\max}]$:

$$\phi_n = \sum_j B_{jn}\zeta_j, \tag{12}$$

and the expansion coefficients can be determined from the requirement that ϕ_n is an eigenvalue of H.

The basic algorithm is then simple: pick an energy range and an initial wavefunction and construct with the filter a set of vectors ζ_j. When the set of vectors is sufficiently large, orthogonalize it explicitly, and construct an $L \times L$ Hamiltonian matrix,

$$\mathbf{H}_{j,j'} = \langle \zeta_j | H | \zeta_{j'} \rangle.$$

The eigenvalues of the small matrix $\mathbf{H}$ are exactly those eigenvalues of H in the desired energy range, and the eigenfunctions of it are the B coefficients in Eq.(12).

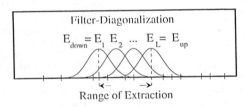

Figure 1. A schematic outline of Filter-Diagonalization. Each Filtered function encompasses a large range of eigenstates.

The advantage of the combined Filter-Diagonalization approach lies in the "divide and conquer" strategy: it is relatively easy to act with a rough filter which has a broad energy resolution, and the next stage, the entangling of eigenstates from the broad filter, is handled very efficiently by a small-matrix diagonalization.

The basic algorithm suffers however from one seeming deficiency, i.e., the calculation has to be repeated for each new desired energy range. In order to avoid this problem, we have recently recasted the algorithm [4] as follows. We leave the filtered vectors unorthogonalized. The statement that ϕ_n is an eigenstate of H, coupled with the expansion (12), leads to the generalized eigenvalue problem:

$$\mathbf{HB} = \mathbf{SB}\epsilon, \tag{13}$$

where ϵ is the diagonal $L \times L$ eigenvalue matrixes, and

$$\mathbf{S}_{ij} = (\psi_0|F(E_i)F(E_j)|\psi_0) \tag{14}$$

$$\mathbf{H}_{ij} = (\psi_0|F(E_i)HF(E_j)|\psi_0). \tag{15}$$

There are several filters for which the product $F(E_i)F(E_j)$ can be simplified, leading to eventual expressions involving terms of the form $(\psi_0|F(E_i)|\psi_0)$. These forms are easy to evaluate even when multiple energy windows are used, and they generally require very little storage (explicit vectors are not required; only correlation functions). The more technical details of the formalism are discussed in the next sections.

The ability to extract explicit eigenstates and eigenvalues at any energy range makes Filter-Diagonalization favorable also for other situations. One case is reactive scattering in the presence of resonances. Straightforward iterative calculations can run in such cases into difficulties due to the presence of the narrow resonances. The long calculation are avoided [2, 6] by noting that at late times, a scattering wavefunction transforms to a sum of a few resonance wavefunctions which are efficiently extracted by Filter Diagonalization.

Finally, the mathematical derivations of Filter-Diagonalization lead,as shown below, to an unexpected feature. The formalism allows the determinations of frequencies and damping factors for a time-dependent signal, *regardless of whether the signal is inherently due to a time-dependent Hamiltonian* or whether it is due to, e.g., a classical simulation. The normal mode frequencies can be extracted even when a very long signal containing possibly millions of frequencies is used, since only a selective set of sampling energies (at the desired energy range) is used. This is done without the long times required for extracting signals by pure Fourier transforms, and the method is equally efficient for signals containing overlapping resonances.

Filters and Propagators

Different filters were employed by us and other group. Most filters are a function of the Hamiltonian (Eq. 10). We note however that other choices are possible, e.g., the pre-Lancosz filter by Wyatt and his colleagues [17]:

$$F(E_j) = \frac{1}{E_j - H_0}, \tag{16}$$

where H_0 is a solvable zero-order approximation to the Hamiltonian H.

Of the filters of form $f(E - H)$, we have employed a Gaussian filter:

$$f(E_j - H) = e^{-(H-E_j)^2 T^2/2}, \tag{17}$$

where the parameter T is chosen relatively short (much shorter than the minimum of the inverse level spacing, $1/\Delta E_{\min}$). An additional filter is the inverse Fourier transform: [8, 14, 19]

$$f(E_j - H)\text{"}=\text{"}\pi\delta(E_j - H) = ImG^+(E_j)$$

$$\sim \frac{i}{E_j - H} - \frac{i}{E_j - H^*},$$

where G^+ is the Green's function. The delta-function equality sign is symbolic here; in practice one does not evaluate the full delta function, but rather a limited series approximation to it. A third filter would be simply

$$f(E_j - H)\text{"}=\text{"}\frac{1}{E_j - H},$$

where again one stops at a finite series approximation to f.

The $f(E_j - H)$ filters are easily evaluated by iterative approaches: Lancosz [20], Chebyshev [19], Newton [21], Modified Chebyshev as pursued by Kouri and Hoffmann [14] or the Damped Chebyshev approach of Mandelshtam and Taylor [8]. In these approaches, the action of the filter on the Hamiltonian is described by a polynomial expansion:

$$f(E_j - H)\psi_0 = \sum_n a_n(E) P_n(H)\psi_0, \tag{18}$$

where $a_n(E)$ is a set of numerical coefficients. $P_n(H)$ is a polynomial in H and its action on ψ_0 is evaluated by recursion, e.g.,

$$P_n(H)\psi_0 = \alpha_{n-1} H P_{n-1}(H)\psi_0 + \beta_{n-1} P_{n-2}\psi_0, \tag{19}$$

where α_n are numerical coefficients (or damping functions [8]). This form is feasible for describing Chebyshev-type and Lancosz series; other, more general expressions are possible. The terms $a_n(E)$ can be easily obtained in different formalisms.

For the explicit orthogonalization schemes, the working formulae are simple: the vectors $f(E_j - H)\psi_0$ are constructed for several energies E_j in a desired range, the vectors are then orthogonalized, and the Hamiltonian matrix is then obtained from orthogonalizing.

A more efficient use of the formalism is, as mentioned, the construction of the $\mathbf{H}$ and $\mathbf{S}$ matrices, involving products of the form $f(E_i - H)f(E_j - H)$. There are several filters for which the calculation of this product can be simplified. First, a Gaussian filter (Eq. 17) for which the product of two filters yields another Gaussian filter, leading to:

$$\mathbf{S}_{ij} = \exp(-(E_i - E_j)^2 T^2/4)(\psi_0|e^{-(H-\frac{E_i+E_j}{2})^2/T^2}|\psi_0)$$

$$= \exp(-(E_i - E_j)^2 T^2/4)\sum_n \bar{a}_n((E_i + E_j)/2)R_n, \tag{20}$$

where R_n are the residues defined as

$$R_n \equiv (\psi_0|P_n(H)|\psi_0),$$

and $\bar{a}_n$ are the coefficients associated with a Gaussian filter with a parameter $\sqrt{2}T$.
   Other filters can also lead to simple expressions; for example, for a filter of the form $f(E_j - H) = 1/(E_j - H)$,

$$(\psi_0|f(H - E_i)f(H - E_j)|\psi_0) = (\psi_0|\frac{1}{E_i - H} - \frac{1}{E_j - H}|\psi_0)\left(\frac{1}{E_i - H} - \frac{1}{E_j - H}\right),$$

and similar expressions in terms of the residues can be obtained if an iterative approach is used. (Note that this particular equation is not limited to iterative evaluations of the Hamiltonian, and can be used, e.g., also with $1/(H_0 - E)$ filters.)
   Even for filters in which the product $f(E - H_j)f(E_i - H)$ cannot be simplified, the method can still be used when the Chebyshev polynomials [19] $T_n(H)$ are employed in Eq. (18) [4]. The property of the Chebyshev polynomials:

$$2T_n(H)T_m(H) = T_{n+m}(H) + T_{|n-m|}(H),$$

implies readily that for a general filter:

$$\mathbf{S}_{ij} = \frac{1}{2}\sum_{nm} a_n(E_i)a_m(E_j)(R_{n+m} + R_{|n-m|}), \tag{21}$$

a sum which can be evaluated efficiently by convolutions (or in selected cases can be evaluated analytically, as recently shown by Mandelshtam and Taylor [8]).
   The analytical equations for $\mathbf{S}$, and similar equations for $\mathbf{H}$, lead thus to a very simple prescription for Filter Diagonalization when used with iterative procedures. First, a set of residues, R_n, is prepared. Next, for any desired energy window one selects a small (typically L=100) set of energies, evaluates the $a_n(E_j)$ (or $a_n((E_j + E_l)/2)$) coefficients, and uses these to calculate the small $\mathbf{S}$ and $\mathbf{H}$ matrices, which are solved by canonical diagonalization, and used to extract the eigenvalues and eigenfunctions.
   Applications of this formalism included studies by our group of a two-dimensional LiCN model [3] and an HCN triatomic molecule [7]. Mandelshtam and Taylor have used their modified Chebyshev propagator with this approach, and produced large scale simulations of HO_2 [9] as well as H_3^+ [10]. The approach emerges as a successful alternative to Lancosz approaches, and is especially useful for resonance studies (where the Hamiltonian contains non-hermitian absorbing-potential parts simulating the outgoing nature of the wavefunction) since it allows the extraction of overlapping resonances (where the difference betweem the real part of the eigenvalue is smaller than the imaginary part).

Time-Dependent Series

   An unexpected feature of Filter-Diagonalization is that it can be used to extract frequencies from any signal. Thus, assume we are given a signal $C(t)$ which can be written as a sum of fluctuating terms, possibly with a decaying magnitude:

$$C(t) = \sum_n d_n e^{-i\omega_n t - \Gamma_n t/2}, \tag{22}$$

where d_n are (possibly complex) constants. Our goal would be to extract the ω_n and the Γ_n for a given frequency range.

We first note that the following Hamiltonian

$$H = \sum_n |\phi_n)\epsilon_n(\phi_n|, \tag{23}$$

(where ϕ_n is an arbitray orthogonal basis) with a wavefunction of the form

$$\psi_n = \sum_n d_n^{\frac{1}{2}} |\phi_n) \tag{24}$$

formally gives rise to this signal:

$$C(t) = (\psi_0|e^{-iHt}|\psi_0). \tag{25}$$

Next, we apply the Gaussian filter, and resolve it to a time-dependent integral to obtain (see Ref. [4] for derivation):

$$\mathbf{S}_{ij} = e^{-(E_i - E_j)^2 T^2/4.d0} \int e^{-i(E_i + E_j)t/2} e^{-t^2/4T^2} C(t) dt. \tag{26}$$

A similar equation applies for **H**:

$$\mathbf{H}_{ij} = \frac{1}{2} e^{-(E_i - E_j)^2 T^2/4.d0} \int (E_i + E_j + i\frac{t}{T^2}) e^{i(E_i + E_j)t/2} e^{-t^2/4T^2} C(t) dt. \tag{27}$$

The important feature about this equation is that it does not involve H explicitly, only $C(t)$. Thus, it can be applied to extract frequencies and damping coefficients from a general signal, in spite of the fact that H is unknown. The signal not be quantum in origin - a correlation function of a classical origin [5] applies equally well.

We emphasize the difference between this approach and other methods for extracting signals. Thus, the simplest approach for extracting frequencies from a signal is from the peaks of the power spectrum:

$$P(E) = |\int g(t)e^{iEt}C(t)dt|^2 \tag{28}$$

where $g(t)$ is a damping function. The difficulty with this approach is first that long propagation times (slowly decaying $g(t)$) are required. Further, overlapping resonances cannot be resolved.

An alterate is furnished by a host of methods (Prony's approach, MUSIC, maximum entropy [22, 23, 24]) which extract the full set of frequencies by solving linear algebra equations. Here however we note that for systems with a very large set of frequencies, the required algebra can be formidable. As in the quantum dynamics problem, filter-diagonalization bridges the gap between the single-propagation and the large-matrix diagonalization methods, retaining the advantage of each approach.

We have applied the frequency extraction approach in various contexts. In quantum dynamics studies, it was used [25] for extracting Floquet frequencies in time-dependent studies within the t,t' formalism [26]. We have also tested in on random signals containing 10000 frequencies. Additionally, it was used for a classical molecular dynamics simulation of Ar clusters [5]. Wells has applied the code for extracting overlapping resonance frequencies in laser-molecule reactions [13].

Resonance Functions and Scattering

Another application of Filter Diagonalization is as an aid for reactions involving resonances. Thus, the most fundamental quantity in scattering is the energy resolved scattering function,

$$\Psi(E) = i \int e^{i(E-H)t} \Psi_0 dt, \tag{29}$$

where H is assumed to contain the proper absorbing boundary condition. Straightforward application of approaches to extract $\Psi(E)$ (using either the time-dependent or time-independent version in Eq. 29) would therefore require much propagation time.

A key for solving this difficulty was outlined in Ref. [2] and improved and exemplified in recent work with G. D. Kroes [6]. The key observation is that even in the presence of narrow resonances, most of the wavefunction is damped after a short "direct-reaction" time. Thus, we divide the effort into two parts. We propagate and find the resonances which are long-lived, and then use them to describe the behavior of the wavefunction at long times. (The relevant formulae are discussed in greater detail in [6].)

The first part is the evaluation of the resonance wavefunction. These are extracted, e.g., by applying Filter-Diagonalization with a weight function

$$g(t) = e^{-(t-t_0)^2/2T^2}, \tag{30}$$

where t_0 is a time by which most resonances have decayed, and T is typically taken as $t_0/4$. This time-dependent weight function, sampling late times, is completely equivalent to a Gaussian energy filter

$$f(E - H) = e^{i(E-H)t_0} e^{-(H-E)^2 T^2/2.d0}. \tag{31}$$

The Filter-Diagonalization matrices take then the form

$$\mathbf{S}_{ij} = \exp(-(E_i - E_j)^2 T^2/4.d0)(\psi_0|e^{2i(E-H)t_0} e^{-(H-\frac{E_i+E_j}{2})^2/T^2.d0}|\psi_0), \tag{32}$$

with a similar expression for $\mathbf{H}$. The matrix elements in Eq. 32 can be evaluated using any desired propagation scheme; the final result again reads:

$$\mathbf{S}_{ij} = \exp(-(E_i - E_j)^2 T^2/4.d0) \sum_n \tilde{a}_n((E_i + E_j)/2) R_n, \tag{33}$$

where $\tilde{a}$ are numerical coefficients; when a Chebyshev propagator is used

$$\tilde{a}_n(E) = (2 - \delta_{n0})(-i)^n \int e^{i(E-\bar{H})t} J_n(t\Delta H) e^{-(t-2t_0)^2/4T^2} dt. \tag{34}$$

where $\bar{H}$ and ΔH are the Chebyshev parameters [19] and we introduced the Bessel polynomials. We have outlined one approach for calculating the $\tilde{a}_n$ coefficients in Ref. [4]; an alternate, very efficient algorithm has been recently found [7] so that the these coefficients take negligible computation time. Again, similar equations follow for **H**.

One difficulty in this calculation is that in scattering, due to the need to absorb the wavefunction, an absorbing potential is inserted into H, making the Chebyshev propagation unstable. There are various ways to overcome this difficulty; the simplest is to use damped Chebyshev propagations, as developed by Mandelshtam and Taylor [8]. We have used this approach successfully [6] to extract resonance states in a multidimensional H_2+Cu scattering. Other approaches are possible, including breaking the time-integration into several distinct parts [1] or using generalized Chebyshev Hamiltonians [14], as well as Newton Propagations [21].

Once we have the **S** and **H** matrices, we diagonalize the resulting eigenvalue equation (Eq. 13) and use them to find the resonance wavefunctions, ϕ_n. These are then used to shorten the propagation times, as follows. We rewrite the expression for $\Psi(E)$ in terms of a time-dependent expression as a sum of contributions up to t_0 and from t_0, which readily yields (with $\tau = t - t_0$)

$$\Psi(E) = \int_0^{t_0} e^{i(E-H)t}\Psi_0 dt + e^{iEt_0}\int_0^{\infty} e^{i(E-H)\tau}\Psi_0(t_0), \tag{35}$$

or in the equivalent time-independent form:

$$\frac{1}{E-H}\Psi_0 = \frac{1 - e^{i(E-H)t_0}}{H-E}\Psi_0 + \frac{1}{E-H}e^{i(E-H)t_0}\Psi_0 \tag{36}$$

The physical assumption that the non-resonant scattering vanishes after t_0 implies that the second term in the right hand side of the preceding equations can be evaluated analytically as a sum over the few resonance states,

$$\frac{1}{E-H}e^{i(E-H)t_0}\Psi_0 = \sum_n d_n^{\frac{1}{2}}\frac{e^{i(E-\epsilon_n)t_0}}{E-\epsilon_n}|\phi_n). \tag{37}$$

We further note that typically one does not require the full scattering wavefunction at all grid points but only at a few grid points (where the flux or flux-amplitude is evaluated). This then implies that the resonance states need to be extracted only at a few grid points, avoiding storage difficulties.

As an example of the achievements of this formalism, we show in Figure 2 how Filter-Diagonalization enabled the reduction of the total propagation time by a factor of 8, for an H_2+Cu model [6].

Discussion and Conclusion

Filter Diagonalization emerges as a general approach for extracting high-energy eigenvalues and eigenfunctions from matrices and operators, or, equivalently, extracting frequencies from a time-dependent signal. We outlined the use of the method and have shown that it only requires a small number of residues (or a short segment of the correlation function) for extracting the energy spectrum, part by part. The method is stable and allows for the extraction of overlapping resonances.

The success of the method stems from the fact that in constructing the small matrices to be diagonalized, the residues are effectively "filtered", i.e., summed at the desired energy range. Thus, one restricts the sampling to a small portion of the spectrum.

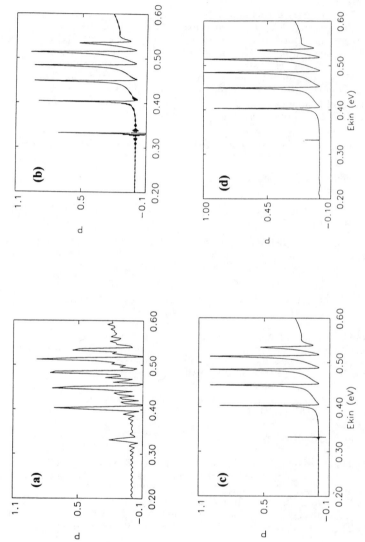

Figure 2. A demonstration of the application of Filter-Diagonalization within an $H_2 + Cu$ scattering model [6]. Parts (a-c) show the result of direct propagation to 0.6, 2.4, and 19.6 psec; Part (d) shows and with extraction of the resonance states the total propagation time can be reduced to $t_0 = 0.6$ psec (with reduction of the total effort by a factor of 8).

Technically, new developments in Filter-Diagonalization are still emerging. For example, most recently Mandelshtam and Taylor have derived an alternate equations to (14,15), also appropriate for signal processing but using a $1/(H - E)$ filter rather than a Gaussian Filter. The use of Lancosz techniques after a Filter-Diagonalization stage can also be anticipated. We also note that the method may prove a useful tool for extracting frequencies from real-time path integral simulations, as it does not require long-time portions of the full wavefunction.

Finally, we expect that the method would have impact in areas where the dynamics is only approximately of an eigenvalue form, i.e., in regions where the strict frequency sum of Eq. 25 is only approximately valid. These include, for example, approximate quantum dynamics calculations (e.g., TDSCF [27] or more complicated approaches, as pursued recently by Manthe and Matzkies [11]) where the method would enable extraction of approximate eigenvalues (analogous to the classical normal mode calculation in Ref. [5]). The method also allows the study of the transition, in classical mechanics, from instantenous normal-mode ($T \to 0$) to long-time averaged modes [28].

Acknowledgements

I gratefuly acknoweldge discsussions with R. Baer, R. Kosloff, V. A. Mandelshtam and H. S. Taylor. Many of the ideas reporeted here were developed in discussions with my students J. Pang and M. Wall, and through a frutiful collaboration with Dr. G. D. Kroes of Leiden University. Financial support of the NSF, the NSF Early Career Award program and the Sloan Foundation is acknowledged.

References

[1] D. Neuhauser, J. Chem. Phys.**93**, 2611 (1990).

[2] D. Neuhauser, J. Chem. Phys.**95**, 4927 (1991).

[3] D. Neuhauser, J. Chem. Phys.**100**, 5076 (1994).

[4] M. R. Wall and D. Neuhauser, J. Chem. Phys.**102**, 8011 (1995).

[5] J. W. Pang and D. Neuhauser, Chem. Phys. Lett.**252**, 173 (1996).

[6] G. D. Kroes, J. W. Pang and M. R. Wall, submitted for publication (1996).

[7] D. Neuhauser, J. W. Pang and M. R. Wall, to be submitted.

[8] V. A. Mandelshtam, T. P. Grozdanov and H. S. Taylor, J. Chem. Phys.**103**, 10074 (1995).

[9] V. A. Mandelshtam, H. S. Taylor and W. H. Miller, J. Chem. Phys.**105**, 496 (1996).

[10] V. A. Mandelshtam and H. S. Taylor, J. Chem. Phys., submitted.

[11] U. Manthe and F. Matzkies, Chem. Phys. Lett.**252**, 71 (1996).

[12] M.J. Bramley, J.W. Tromp, T. Carrington Jr. and B.T. Sutcliffe, J. Chem. Phys. **98**, 10104 (1993).

[13] J. C. Wells, submitted for publication, 1996.

[14] D.J. Kouri, W. Zhu, G. Parker, and D.K. Hoffman, Chem. Phys. Lett. **238**, 395 (1995).

[15] S.K. Gray, J.Chem.Phys.**96**, 6543 (1992).

[16] R. Q. Chen and H. Guo, J. Chem. Phys.**105**, 1311 (1996).

[17] R. E. Wyatt, Phys. Rev. E **51**, 3643 (1995).

[18] N. Moiseyev, P. R. Certain and F. Weinhold, Mol. Phys. **36**, 1613 (1978).

[19] R. Kosloff, J. Phys. Chem. **92**, 2087 (1988).

[20] J.K. Cullum and R.A. Willoughby, *Lanczos Algorithms for Large Symmetric Eigenvalue Computations*, birkhäuser, Boston, 1985.

[21] M. Berman, R. Kosloff and H. Tal-Ezer, J. Phys. A **25**, 1283 (1992). S. M. Auerbach and C. Leforestier, Comp. Phys. Comm. **78**, 55-66 (1993).

[22] R. Roy, B. G. Sumpter, G. A. Pfeffer, S. K. Gray and D. W. Noid, Phys. Rep. **205**, 109 (1991).

[23] S. K. Gray, J. Chem. Phys.**96**, 6543 (1992).

[24] F. Remacle, R. D. Levine and J. L. Kinsey, Chem. Phys. Lett.**205**, 267 (1993). R.D. Levine, J. Phys. A **13**, 91 (1980).

[25] J. W.. Pang, D. Neuhauser and N. Moiseyev, J. Chem. Phys., submitted (1996).

[26] U. Peskin and N. Moiseyev, J. Chem. Phys.**99**, 4590 (1993).

[27] B. Vekhter, M. A. Ratner and R. B. Gerber, J. Chem. Phys.**99**, 7916 (1993).

[28] R. Stratt, private communication.

Chapter 3

Spectral Analysis of the HO$_2$ Molecule

Jörg Main[1,3], Christof Jung[2], and Howard S. Taylor[1]

[1]Department of Chemistry, University of Southern California, Los Angeles, CA 90089
[2]UNAM Instituto de Matematicas, Unidad Cuernavaca, Av. Universidad s/n, 62191 Cuernavaca, Mexico

We present a scaling technique to analyse quantum spectra, i.e. to obtain from quantum calculations detailed information about the underlying important classical motions and the statistical properties of level spacings. The method can be applied to systems without a classical scaling property as, e.g., the rovibrational motion of molecules. A demonstration on the conventionally unassignable vibrational spectrum of the HO$_2$ radical reveals remnants of classical broken tori embedded in the chaotic phase space and leads to a new assignment of spectral patterns in terms of classical Fermi resonances between the local mode motions. The nearest neighbor distribution of level spacings undergoes a transition from mixed phase space behavior at low energies to the Wigner distribution characteristic for chaotic sytems at energies near the dissociation threshold.

Quantum spectra of classically regular systems are interpreted in terms of the underlying classical dynamics that influence the quantum spectra, e.g., in molecules often the normal or local mode vibrational motions that take place on stable tori in phase space *(1)*. Since many systems, especially upon excitation show classically chaotic motion, it is of fundamental importance to perform an analogous analysis for chaotic systems, i.e. to extract information about the important classical motions directly from quantum spectra and to assign spectral patterns in terms of broken tori or unstable periodic orbits *(2)*. This is a nontrivial problem especially if the underlying classical dynamics of the quantum system changes significantly with energy. Exceptions are systems possessing a classical scaling property, i.e. the classical phase space structure is the same for all values of an appropriate scaling parameter which is usually some power of an external field strength or, for Hamiltonians with homogenous potentials, the energy and can be controlled in the laboratory. Examples are the three-body Coulomb system *(3)*

---

[3]Permanent address: Institut für Theoretische Physik I, Ruhr-Universität Bochum, D–44780 Bochum, Germany

or atoms in magnetic fields *(4)* where information about the underlying classical dynamics can be extracted from quantum spectra by application of *scaled-energy spectroscopy (5, 6)*. In contrast, e.g., molecular vibrational Hamiltonians are prototypical examples of classically non-scaling systems.

Here we present an extended version of the scaling technique which does not allow scaled-energy spectroscopy in the lab but can be applied in theoretical quantum calculations to provide information on the classical dynamics, periodic motions, and statistical properties of spectra at any given energy. Such information can usually not be obtained from a single eigenstate ψ_n at energy $E = E_n$ but only from many states at energies around E. Analysing these states automatically implies a smearing over the energy. The aim of the scaling technique is to obtain additional information by creating a family of quantum systems all of which have the same underlying classical dynamics at energy E. The eigenstates of these systems can be analysed without any smearing in energy to create a diagram which highlights at any given energy, certain values of the periods. Classical trajectory calculations, by finding periodic orbits with matching periods, then reveal those motions and phase space structures that influence both the regular and irregular quantum spectra. The importance of the method is for chaotic dynamics as regular systems are well understood *(1, 7)*.

The Concept of the Scaling Technique

Consider a multidimensional system given by the Hamitonian $H = T + V(\mathbf{q})$ with a general non-homogenous potential $V(\mathbf{q})$. For the kinetic energy we assume for simplicity $T = \mathbf{p}^2/2m$ in the following, generalizations of the scaling technique to more complex representations of kinetic energy will be straightforward. The basic idea is to enlarge formally the parameter space of the potential $V(\mathbf{q})$ by introducing an additional parameter z. The dependence of the potential on z is defined by

$$V(\mathbf{q}; z) = V(\mathbf{q}/z; z = 1) = V(\mathbf{q}/z) = V(\tilde{\mathbf{q}}) \qquad (1)$$

where $\tilde{\mathbf{q}} = \mathbf{q}/z$ are scaled coordinates, i.e. z is a scaling parameter describing a blow up or shrinking of the potential, with $z = 1$ corresponding to the true physical situation. It follows that $V(\mathbf{q}; z)$ is the value of the original potential at $\mathbf{q}/z$. By considering the family of systems given by the Lagrangian

$$L = m\dot{\mathbf{q}}^2/2 - V(\mathbf{q}; z) = mz^2\dot{\tilde{\mathbf{q}}}^2/2 - V(\tilde{\mathbf{q}}) \qquad (2)$$

it is easy to show that $\tilde{\mathbf{p}} = z\mathbf{p}$ is the canonical conjugate momentum to $\tilde{\mathbf{q}}$. We can now consider the parametric family of Hamiltonians

$$H = \frac{1}{2m}\mathbf{p}^2 + V(\mathbf{q}/z) = \frac{1}{z^2}\frac{\tilde{\mathbf{p}}^2}{2m} + V(\tilde{\mathbf{q}}) = \frac{1}{z^2}T + V(\tilde{\mathbf{q}}) \quad . \qquad (3)$$

The fact that the transformations of $\mathbf{p}$ and $\mathbf{q}$ are linear and canonical means that the Schrödinger quantisation rules can be applied directly to the new variables $\tilde{\mathbf{p}}$ and $\tilde{\mathbf{q}}$. This also means that the classical motion and particularly the shape of classical (periodic) orbits are not affected by a consistent scaling. Orbits just blow up and shrink in the same way as the potential.

Considering now the quantum mechanics, obviously the systems have different quantum spectra for various z values. Each quantum Hamiltonian $H(z) = \tilde{p}^2/2mz^2 + V(\tilde{q})$ has an effective $\hbar$, $\hbar_{\text{eff}} = \hbar/z$ (or alternatively an effective mass $m_{\text{eff}} = mz^2$), i.e. the Hamiltonians $H(z)$ describe a family of quantum systems with the same underlying classical dynamics. The semiclassical and classical limits are reached as $z \to \infty$ ($\hbar_{\text{eff}} \to 0$). Note that so far the scaling parameter z is not a dynamical variable.

Scaled (E, z) Diagram of the HO₂ Molecule

As an example to illustrate our scaling technique on a real system we study the vibrational spectrum of the hydroperoxyl radical, HO₂ which shows a classically chaotic dynamics. Using the potential surface of Pastrana et al. *(8)* 361 bound states have been calculated recently at the physical value $z = 1$ *(9, 10)*. Only about the 32 lowest states out of 361 bound states can be assigned *(11)* in the conventional regular spectra sense *(1)*, i.e. methods of fitting regular spectra fail at higher energies and wavefunctions show no regular patterns *(9)*. For the quantum calculation of eigenstates at various z values we use a DVR grid representation of basis functions and the computational method of filter diagonalization *(10)*. A technical detail worth noting is that in a first step we diagonalize the Hamiltonian at a fixed value $z = 2$ with matrix dimension > 100000 and then in a second step we use about 2600 eigenfunctions at $z = 2$ as a small basis set to calculate eigenvalues at arbitrary z values. Details of the numerical procedure will be given elsewhere. A part of the resultant (E, z)-diagram is shown in Fig. 1a. For graphical purpose we plot in Fig. 1b $(E - E_b)z$ vs. z where $E_b = -2.38$ eV is the bottom energy of the HO₂ potential surface. In the latter presentation the avoided crossings between levels become more pronounced and indicate the chaotic dynamics of the molecule.

Semiclassical Analysis of the Scaled Spectra

The decisive step for the scaling technique is now to analyse the eigenstates in Fig. 1 not along a line of constant z but along lines of constant energy E (dashed lines in Fig. 1b). By projecting all intersections of the lovi of levels with a line $E = $ const on the z axis we obtain eigenvalues $z_i(E)$, and from this the density of states $\varrho^E(z) = \sum_i \delta(z - z_i^E)$ as a function of z. This is equivalent to considering the scaling parameter z as a new dynamical variable and rewriting Schrödinger's equation

$$\left[\frac{1}{z^2}T + V(\hat{q})\right]|\psi\rangle = E|\psi\rangle \tag{4}$$

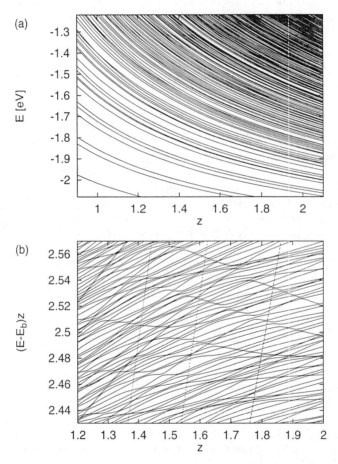

Figure 1: Scaled spectra for the HO_2 molecule. (a) Energy E versus scale parameter z. (b) $(E - E_b)z$ versus z. The dashed lines mark lines of constant energy at (from the left) $E = -0.6$ eV, $E = -0.8$ eV, and $E = -1.0$ eV.

for a particular fixed value of the energy E as a Hermitian eigenvalue equation for $\lambda = z^{-2}$,

$$T^{-1/2}[E - V(\tilde{q})]T^{-1/2}|\Phi\rangle = \frac{1}{z^2}|\Phi\rangle \; ; \; |\Phi\rangle = T^{1/2}|\Psi\rangle \tag{5}$$

where $T = \tilde{p}^2/2m$ is the operator of scaled kinetic energy. The eigenvalue equation (5), which when diagonalized obviously also gives Fig. 1, yields the values of z for which E is an energy eigenvalue. The new Hamiltonian on the left can be used to formally set up the Greens function and density of states

$$
\begin{aligned}
G^E(\lambda) &= \left[\lambda - T^{-1/2}(E - V)T^{-1/2}\right]^{-1} \tag{6} \\
\varrho^E(z) &= \frac{2}{\pi z^3} \, \text{Im Tr } G^E(\lambda) \\
&= -\frac{2}{z^3} \sum_i \delta(\lambda - \lambda_i^E) \\
&= \sum_i \delta(z - z_i^E) \quad . \tag{7}
\end{aligned}
$$

For the interpretation of the quantally calculated $\varrho^E(z)$ we can now formally compare to semiclassical theories, i.e. the Berry-Tabor formula for semiclassical torus quantisation *(12)* or here Gutzwiller's periodic orbit theory for the semiclassical density of states of classically chaotic systems *(2)*. Gutzwiller's trace formula gives the semiclassical approximation to Eq. 7 as

$$\varrho^E(z) = \varrho_0^E(z) + \sum_{\text{p.o. } k} A_k \sin\left(z\tilde{S}_k - \frac{\pi}{2}\mu_k\right) \tag{8}$$

where $\varrho_0^E(z)$ is the mean level density, A_k is the amplitude related to the stability matrix of periodic orbit k, μ_k the Maslov index, and $\tilde{S} = S/z$ the scaled classical action which can also be proven to be the canonically conjugate variable to z. Information about the classical dynamics can now be obtained *from the quantum spectra* by a Fourier transform of $\varrho^E(z)$. From Eq. 8 follows that each peak in the Fourier tansformed action spectra can be identifed either with an individual periodic orbit or with a braid of orbits with similar shapes and periods related to a broken resonance torus.

For the HO₂ molecule the Fourier transform action spectra along lines of constant energy are presented in Fig. 2 in overlay form in the region -2.0 eV $< E < -0.2$ eV. We performed extensive classical calculations and the quantum recurrence spectra are superimposed on the classical bifurcation diagram (dashed lines in Fig. 2) obtained from a search for periodic orbits. The total number of periodic orbits in this threedimensional system is incredibly high and in the classical bifurcation diagram orbits are shown only which exist in longer energy ranges (i.e. do not undergo close in energy bifurcations). Details of the classical calculations will be given elsewhere. The quantum spectra show very sharp and detailed structures. The sharpness of peaks in the Fourier transform indicates

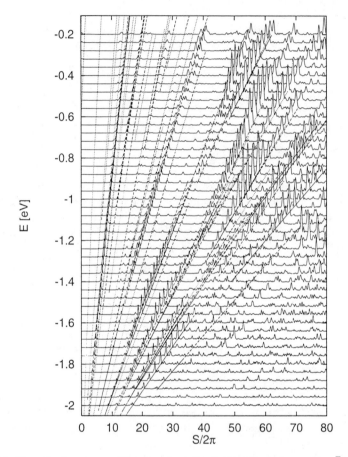

Figure 2: Magnitude square of the Fourier transformed scaled spectra $\varrho^E(z)$ of the HO_2 molecule. The action spectra are superimposed on the classical bifurcation diagram (dashed lines).

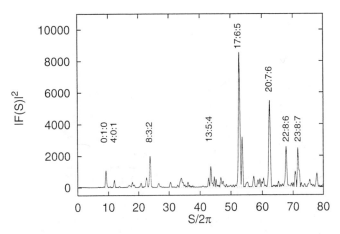

Figure 3: Fourier transform density of states at constant energy $E = -1.0$ eV. Some peaks are assigned by frequency ratios of classical trajectories on broken resonance tori.

the importance of a particular orbit or braid of orbits along the whole range of z values (classical to quantum). The quantally obtained peak heights are related to the amplitudes A_k of periodic orbits in Gutzwiller's trace formula (Eq. 8). A direct semiclassical calculation of these peaks for a chaotic 3D system is a presently unachievable task requiring the finding of *all* periodic orbits up to a certain period. Indeed, a careful examination of Fig. 2 shows that not all quantum structures are explained by those orbits that we have computationally identified. On the other hand orbits not highlightened in Fig. 2, which are needed to converge the periodic orbit sum (Eq. 8), do not make major qualitative contributions to the quantum dynamics. The selected periodic orbit structures (which might be isolated orbits or broken resonance tori if families of similar periodic orbits exist under one peak) are the dynamics which determine the quantum spectra up to intermediate resolution. These structures are markers for regions in phase space where the dynamical potential causes trapping which is seen by quantum mechanics *(13)*.

The periods and amplitudes of the important structures highlightened in Fig. 2 depend on the energy. For example there is a weak resonance structure around $E = -1.2$ eV, $\tilde{S}/2\pi = 45$ whose intensity increases at higher energies to maximum strength around $E = -1.0$ eV, $\tilde{S}/2\pi = 52$ and then decreases again. Here highlighted periodic orbits show that such structures can be explained by classical trajectories moving on broken resonance tori. The individual peaks being associated with recurrent motion have rational frequency ratios $\nu_1 : \nu_2 : \nu_3$ between

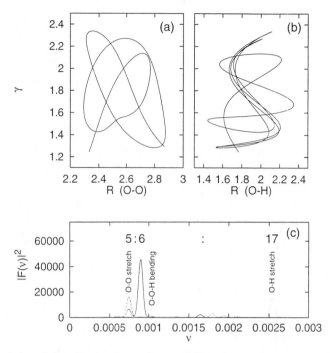

Figure 4: (a) and (b): Projections of a periodic orbit at energy $E = -1.0$ eV in local mode coordinates R_{OO}, R_{OH}, and bending angle γ. (c): Fourier transform of local mode motions showing a frequency ratio of $17 : 6 : 5$.

the three local mode motions, viz. the O-H stretch, the O-O-H bending, and the O-O stretch. Interestingly most of the stronger peaks in Fig. 2 can be identified and assigned in this way, i.e. they represent classical Fermi resonances between the local modes. Fig. 3 shows as an example the assignments of the important structures in the scaled action spectrum at constant energy $E = -1.0$ eV. (If one or two local mode motions have nearly zero amplitude we assign these modes formally by frequencies $\nu = 0$.) The strongest peak at $\tilde{S}/2\pi = 52.5$ belongs to a frequency ratio $\nu_1 : \nu_2 : \nu_3 = 17 : 6 : 5$. This peak can be related to a family of similar periodic orbits localized in phase space, one member is shown in Fig. 4a,b in projections of the local mode coordinates. The Fourier transform of the local mode motions (Fig. 4c) clearly reveals the frequency ratio mentioned above. In fact in place of the labor intensive periodic orbit search used to identify the dynamics under the peaks, time segments of a long trajectory can be Fourier analysed to see if the frequencies of the three local mode motions are in rational ratio. If they are the trajectory segment lies close to important periodic orbits and can be used to find them rather easily.

Vibrogram Analysis of Spectra. A method which supports similar analysis is a windowed Fourier transform (Gabor transform) of quantum spectra resulting in vibrograms being a function of energy and time *(14)*. This analysis has been applied to various classically non-scaling systems *(15, 16)* and has the advantage that it can be directly applied to experimental spectra. However, the resolution of vibrograms in energy and time is fundamentally restricted by Heisenberg's uncertainty principle *(15)*. To improve the resolution and to relate structures to various classical orbits, $\hbar$ has to be reduced in theoretical calculations *(16)* which eliminates the experimental connection. In contrast the scaling technique introduced here works at single energies and resolution is not restricted by the uncertainty principle. Results obtained for the HO$_2$ molecule go far beyond a vibrogram analysis which will be presented elsewhere but where diagrams are not defined enough to correlate its broad peaks to specific longer periodic orbits.

Statistical Analysis of Nearest Neighbor Spacings

Another advantage of the scaling method is its application in a statistical analysis of eigenvalues. Statistical properties often depend strongly on energy but if the density of states is low short range energy intervals do not contain a sufficient number of states to carry out a statistical analysis. Here additional monoenergetic quantum information for such an analysis can be generated by application of the scaling technique.

It is well established that the nearest neighbor spacing distribution of integrable quantum systems after unfolding of the spectra to unit mean level spacing $\langle s \rangle = 1$ is given by a Poisson distribution

$$P_{\text{Poisson}}(s) = e^{-s} \,, \tag{9}$$

while quantum systems with a fully chaotic (ergodic) underlying classical dynamics are characterized by the Wigner distribution

$$P_{\text{Wigner}}(s) = \frac{\pi s}{2} e^{-\pi s^2 / 4} \tag{10}$$

obtained from random matrix theory *(17)*. In systems with a mixed regular-chaotic classical dynamics the nearest neighbor spacing distribution can be phenomenologically described by the Brody distribution *(18)*

$$P_{\text{Brody}}(s; q) = (q + 1)\beta s^q e^{-\beta s^{q+1}} \tag{11}$$

where

$$\beta = \Gamma\left(\frac{q+2}{q+1}\right)^{q+1} \tag{12}$$

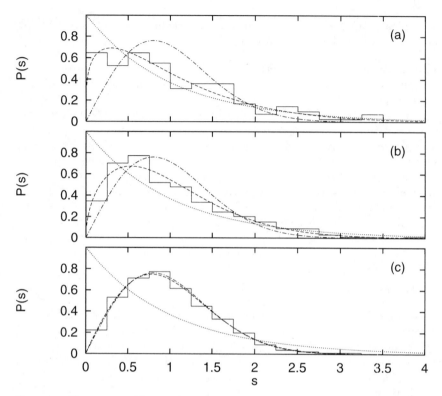

Figure 5: Nearest-neighbor-spacing distribution of eigenvalues at constant energies (a) $E = -1.80$ eV, (b) $E = -1.40$ eV, and (c) $E = -0.20$ eV. Dotted lines: Poisson distribution; dashed-dotted lines: Wigner distribution; dashed lines: Brody distribution with (a) $q = 0.25$, (b) $q = 0.50$, and (c) $q = 0.95$.

and q is a parameter which interpolates between the Poisson distribution ($q = 0$) and the Wigner distribution ($q = 1$) and is roughly related to the percentage of chatic phase space volume of the underlying classical system.

For the HO_2 molecule the nearest neighbor spacing distribution of the 361 bound states has been calculated and fits well to a Brody distribution with $q = 0.92$ (9). But this analysis does not give any information about statistical properties in various energy regions. The additional quantum information in the scaled (E, z) diagram of Fig. 1a can now be used to study the energy dependence of nearest neighbor spacing distributions by analyzing the (unfolded) level spacings $\Delta z_i^E = z_{i+1}^E - z_i^E$ along lines of constant energy E. Results are presented at three different energies in Fig. 5. At low energy $E = -1.80$ eV the spacing distribution can be fitted by a Brody distribution with $q = 0.25$ which indicates a mixed regular-chaotic behavior of the underlying classical dynamics. The Brody parameter q grows with increasing energy and the nearest neighbor spacing distribution at $E = -0.20$ eV is close to the Wigner statistics characteristic for classically chaotic systems. The transition from almost regular to chaotic behavior can be explained by unharmonicities of the potential surface which increase with energy and cause a break down of more and more torus structures in the classical phase space.

Conclusion

The technique introduced here is a powerful tool to obtain information about the important classical dynamics, torus structures, and periodic orbits that underlie the quantum motion directly from quantum calculations. It also provides additional monoenergetic quantum information for a detailed statistical analysis of spectral properties. Application to the HO_2 molecule reveals remnants of broken tori in classical phase space related to Fermi resonances between the local mode motions and shows a transition from mixed regular-chaotic to completely chaotic dynamics in the nearest neighbor spacing distributions.

Acknowledgments.   We acknowledge stimulating discussions with V. Mandelshtam, M. Kaluža, G. Ezra, and M. Davis and are grateful to V. Mandelshtam for supplying his program for filter diagonalization. This work was supported by DOE Grant number DE-FG03-94ER14458. J. M. is grateful to Alexander von Humboldt-Stiftung for a Feodor-Lynen scholarship, and J. M. and C. J. thank Prof. Howard Taylor for his kind hospitality at the University of Southern California.

Literature Cited

1. Herzberg, G. *Molecular Spectra and Molecular Structure*, Van Nostrand, Toronto, 1945, 1966; Vols. II and III.

2. Gutzwiller, M. C. *Chaos in Classical and Quantum Mechanics*, Springer, New York, 1990.

3. Wintgen, D., Richter, K., and Tanner, G. *Chaos 2* **1992**, 19.

4. Friedrich, H. and Wintgen, D. *Phys. Rep. 183* **1989**, 37.

5. Holle, A., Main, J., Wiebusch, G., Rottke, H., and Welge, K. H. *Phys. Rev. Lett. 61* **1988**, 161.

6. Main, J., Wiebusch, G., Welge, K. H., Shaw, J. and Delos, J. B. *Phys. Rev. A 49* **1994**, 847.

7. Kellman, M. E. in *Molecular Dynamics and Spectroscopy by Stimulated Emission Pumping*, Eds. Dai, H.-L. and Field, R. W., World Scientific, Singapore, 1995.

8. Pastrana, M. R., Quintales, L. A. M., Brandao, J. and Varandas, A. J. C. *J. Phys. Chem. 94* **1990**, 8073.

9. Dobbyn, A. J., Stumpf, M., Keller, H.-M. and Schinke, R. *J. Chem. Phys. 103* **1995**, 9947.

10. Mandelshtam, V. A., Grozdanov, T. P., and Taylor, H. S. *J. Chem. Phys. 103* **1995**, 10074.

11. Zhang, D. H. and Zhang, J. H. *J. Chem. Phys. 101* **1994**, 3671.

12. Berry, M. V. and Tabor, M. *Proc. R. Soc. London, Ser. A 349* **1976**, 101.

13. Hose, G., Taylor, H. S., and Bai, Yi. Yan *J. Chem. Phys. 80* **1984**, 4363.

14. Johnson, B. R. and Kinsley, J. *J. Chem. Phys. 91* **1989**, 7638.

15. Hirai, K., Heller, E. J., and Gaspard, P. *J. Chem. Phys. 103* **1995**, 5970.

16. Ezra, G. S. *J. Chem. Phys. 104* **1996**, 26.

17. Haake, F. *Quantum Signatures of Chaos*, Springer, New York, 1991.

18. Brody, T. A., Flores, J., French, J. B., Mello, P. A., Pandey, A., and Wong, S. S. M. *Rev. Mod. Phys. 53* **1981**, 385.

Chapter 4

The Multiresonant Hamiltonian Model and Polyad Quantum Numbers for Highly Excited Vibrational States

William F. Polik and J. Ruud van Ommen

Department of Chemistry, Hope College, 35 East 12th Street, Holland, MI 49423

The multi-resonant Hamiltonian model is a spectroscopic Hamiltonian capable of fitting highly excited vibrational states in polyatomic molecules with a minimum number of parameters. It treats nonresonant interactions among harmonic oscillator product basis states using second-order perturbation theory and resonant interactions explicitly using harmonic oscillator matrix elements. The pure vibrational spectrum of formaldehyde (H_2CO) is analyzed up to 10000 cm^{-1} with several multi-resonant Hamiltonian models, resulting in fits with less than 3 cm^{-1} standard deviation. The identified vibrational resonances imply three polyad quantum numbers for formaldehyde: $N_{oop} = v_4$, $N_{vib} = v_1 + v_4 + v_5 + v_6$, and $N_{res} = 2v_1 + v_2 + v_3 + v_4 + 2v_5 + v_6$. The goodness of these vibrational polyad quantum numbers is discussed, and generalizations to other chemical systems are offered.

A fundamental understanding of chemical reactivity requires a detailed understanding of the reactant molecules. Molecular structure has been studied in great detail for many molecules at or near the equilibrium geometry. However, far fewer studies have been carried out on molecules in excited vibrational states. For a reaction with an activation energy barrier, only molecules in excited vibrational states can sample phase space far from the equilibrium geometry and ultimately undergo chemical reaction. While molecules reacting on a single potential energy surface possess both rotational and vibrational energy, the significant geometry change required to reach the transition state geometry typically requires large amounts of vibrational energy. Thus, a model for chemical reactivity must account for the nature of highly excited vibrational states in reactant molecules.

Spectroscopic characterization of excited vibrational states offer direct insight into the nature of energetic molecules. At low energies in small molecules, interactions between specific vibrational states can cause level shifting. Such interactions are characterized as spectroscopic perturbations and are typically analyzed by using nondegenerate perturbation theory or diagonalizing small matrices. At higher energies, the onset of extensive state mixing can lead to the apparent destruction of the regular patterns of energy levels which existed at lower energies. At still higher energies or in larger molecules, fractionation of energy levels may be observed due to mixing of the spectroscopically active zero-order state with unspecified background states. Ultimately, complete state mixing leads to a statistical description of energy levels. The mixing of vibrational states is commonly referred to as Intramolecular Vibrational Redistribution (IVR) (*1*). The extent of IVR is directly measurable in a spectrum, assuming that the evidence for state mixing is not obscured by another effect, *e.g.*, rotational congestion.

Accurate experimental measurements of excited state properties also assist the theoretical description of excited molecules. The dynamics of a chemical reaction is governed by the underlying potential energy surface (PES). While quantum chemical calculations are generally reliable for obtaining the correct PES topology and are becoming increasingly accurate, they have not yet achieved "spectroscopic accuracy." Thus, one of the best ways to obtain an accurate PES is to start with a calculated PES, which has the correct functional form and good initial values for the potential parameters, and refine this surface using accurate experimental parameters. A second way in which experiments assist theory is by serving as a rigorous test for quantum chemistry methodologies. If general methods can be developed that accurately calculate experimental results for several prototypical systems, then those methods can be more trusted to make accurate predictions for similar, unmeasured chemical systems. The determination of high quality data sets for excited molecular states is therefore essential to the continued development of quantum chemistry methodologies.

In this paper, several methods for obtaining spectra of excited vibrational levels are reviewed and contrasted. One method for obtaining vibrational spectra which are free from rotational congestion is described, and results are presented for formaldehyde. The multi-resonant Hamiltonian model is developed and used to fit the assigned formaldehyde spectrum. The power of this model is that it accounts for the apparent destruction of regular spectroscopic sequences with a minimum number of parameters, each of which has a simple physical interpretation. An analysis is performed on the resonances used in the Hamiltonian model, revealing conserved dynamical quantities known as polyad quantum numbers. Conclusions are drawn regarding the dynamics of formaldehyde, and generalizations to other molecular systems are offered.

Spectroscopy of Excited Vibrational States

Spectra of excited states in molecules are more difficult to obtain than spectra of states near equilibrium. The most straightforward method of accessing excited states is **high overtone absorption spectroscopy**. In this method, a laser is scanned through the energy region of interest and absorption is directly or indirectly monitored. A particularly sensitive implementation is intracavity optoacoutistic spectroscopy, in which the sample cell is placed inside a dye laser cavity, the excitation source is chopped, and absorbance is monitored with an acoustic microphone at the chopping frequency through a lock-in amplifier (*2*). Because the transitions being measured are vibrational overtones, which are forbidden in the harmonic oscillator approximation, they usually have extremely weak absorbances. The absorbance strength of a transition depends on the anharmonicity and the quantum number change of the vibration. High overtone spectroscopy is therefore best suited for studying C-H or O-H stretching vibrations, as stretches are typically more anharmonic than bends and fewer quanta of excitation are required to reach a given energy due to the relatively high frequency of the hydrogen stretch. High overtone spectroscopy suffers from broad spectroscopic linewidths, typically 50 cm^{-1} or greater, due to inhomogeneous broadening from rotational congestion at room temperature. Such broad linewidths can prevent detection of weaker transitions or nearby states, as well as obscure the homogeneous linewidth which is the measure of the IVR rate. The recently developed method of infrared laser assisted photofragment spectroscopy (IRLAPS) allows infrared absorption spectra to be recorded in a supersonic expansion, thereby reducing the problems associated with rotational congestion (*3*).

A tremendous breakthrough in the study of excited states came with the development of **Stimulated Emission Pumping** (SEP) (*4*). In this method, a fixed "pump" laser populates a single rovibronic level in an excited electronic state, and fluorescence from the laser populated level is monitored. A scanning "dump" laser stimulates emission from the level populated by the pump laser to the excited states of interest. Transitions to excited states are detected as dips in the fluorescence from the laser populated level. Since SEP involves an electronic transition, transition intensities depend on Franck-Condon factors. SEP spectroscopy therefore complements high overtone spectroscopy by accessing different excited states. SEP has two main advantages over high overtone spectroscopy. SEP is a double-resonance method which allows the use of rotational selection rules to simplify a spectrum or to confirm assignments. Also, spectroscopic linewidths are typically laser or Doppler limited, resulting in linewidths on the order of 0.1 cm^{-1} with pulsed lasers. The chief limitation of SEP is that signals are detected as small changes on top of a large, potentially noisy, fluorescence background. In practice, the dynamic range of SEP is limited by saturation effects or upward transitions from high "dump" laser intensities at one end and by the extent to which fluctuations in the background fluorescence level can be reduced at the other end. A variation of SEP which circumvents the problem of a fluctuating background is degenerate four-wave mixing spectroscopy (SEP-DFWM) which offers significantly increased sensitivity at the expense of signal linearity (*5*).

Dispersed Fluorescence (DF) spectroscopy has enjoyed a recent resurgence as a method for obtaining excited state spectra (*6*). In this method, a "pump" laser populates a single rovibronic level in an excited electronic state, just as with SEP. Instead of using a second laser to stimulate emission down to excited vibrational states, however, a monochromator is used to disperse the fluorescence. Transitions to excited states are detected as peaks in the fluorescence intensity spectrum. As with SEP, DF spectroscopic intensities depend on Franck-Condon factors and rotational selection rules can be used to simplify a spectrum. The principal advantages of DF over SEP are that DF is a zero background technique, is not subject to saturation effects, and can be recorded much more quickly. Spectroscopic transitions observed by DF are not obscured by background fluorescence fluctuations, although in practice one is ultimately limited by detector readout noise. Intensities are much more reliable in DF because there is no danger of saturation. A 10000 cm^{-1} DF spectrum can be recorded in a day, whereas a comparable SEP spectrum would require months or years of effort along with multiple dye and optics changes. The principle disadvantage of DF relative to SEP is resolution, which is typically monochromator limited resulting in linewidths of several cm^{-1}. DF and SEP methods are complementary in that DF permits rapid acquisition of medium resolution survey spectra while SEP is useful for higher resolution studies of more limited frequency ranges.

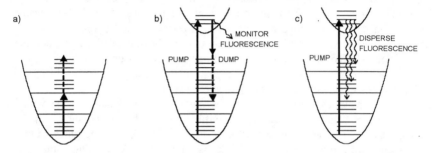

Figure 1. Techniques for recording spectra of excited vibrational states. a) Overtone spectroscopy. b) Stimulated Emission Pumping (SEP). c) Dispersed Fluorescence (DF).

Pure Vibrational Spectroscopy of Formaldehyde

Formaldehyde (H_2CO) is one of the simplest and most studied polyatomic molecules. Excited vibrational levels in formaldehyde have been studied by all three aforementioned methods: overtone, SEP, and DF spectroscopy. Prior to 1984, 24 vibrational states had been assigned primarily by analysis of Fourier transform infrared (FTIR) overtone and combination band spectra. In 1984 Reisner *et al.* assigned 57 new vibrational states using SEP spectroscopy and obtained the first complete set of harmonic and anharmonic spectroscopic constants, ω_i° and x_{ij}, for a tetra-atomic molecule (7). In 1996 Bouwens *et al.* assigned 198 new vibrational states using DF spectroscopy in a supersonic expansion, bringing the total number of assigned vibrational states in S_0 formaldehyde to 279 (6). Bouwens *et al.* also corrected several misassignments in the literature and determined an improved set of vibrational spectroscopic constants. Formaldehyde has also been the subject of many theoretical studies. The most accurate *ab initio* calculation of the PES to date is the 1993 study by Martin, Lee, and Taylor (8). Burleigh, McCoy, and Sibert have used the data of Bouwens *et al.* to refine this surface to 7,600 cm^{-1}, resulting in an average mean deviation between experimental and calculated energies of 1.5 cm^{-1} (9).

The tremendous increase in the number of assigned states for formaldehyde was due to the recently developed technique of **pure vibrational spectroscopy**, in which selection rules are used to eliminate rotational congestion entirely (6). Formaldehyde belongs to the C_{2v} point group, and the S_0 and S_1 electronic states have

A_1 and A_2 electronic symmetry, respectively. Vibrational states for which both v_4 and v_5+v_6 are even have A_1 vibrational symmetry, states for which both v_4 and v_5+v_6 are odd have A_2 vibrational symmetry, states for which v_4 is odd and v_5+v_6 is even have B_1 vibrational symmetry, and states for which v_4 is even and v_5+v_6 is odd have B_2 vibrational symmetry. Evaluation of the electric dipole transition integral by symmetry arguments reveals that in the $S_1 \leftrightarrow S_0$ electronic transition of formaldehyde, transitions between vibrational states where Δv_4=odd and $\Delta(v_5+v_6)$=odd are A-type, transitions between states where Δv_4=odd and $\Delta(v_5+v_6)$=even are B-type, and transitions between states where Δv_4=even and $\Delta(v_5+v_6)$=odd are C-type. Transitions between states where Δv_4=even and $\Delta(v_5+v_6)$=even are electric dipole forbidden, but they are magnetic dipole allowed and are A-type. Analysis of the rotational selection rules for ΔJ, ΔK_a, and ΔK_c (*10*) reveals that from the $J_{K_a K_c}$=0_{00} state of an S_1 vibronic state, only a single rovibronic transition is allowed to each S_0 vibrational state. For example, the 0_{00} rotational level of 4^1 formaldehyde can fluoresce to only the 1_{11}, 1_{10}, and 1_{01} rotational levels of A_1, A_2, and B_2 levels, respectively. Such spectra are pure vibrational spectra in that they are completely free from rotational congestion. *Every line in the spectrum arises from a different vibrational state!*

Figure 2 presents the pure vibrational spectrum of H_2CO recorded from the 0_{00} rotational level of the 4^1 S_1 vibronic state (*6*). The signal to noise ratio of this spectrum is 21000 to 1, and the FWHM resolution in the central portion of the spectrum is 4 cm^{-1}. Highly congested features in this spectrum were recorded again with 1.2 cm^{-1} resolution. Vibrational states with energies up to 14000 cm^{-1} are observed in the DF spectrum. Similar spectra have been recorded from 4^0, $3^1 4^1$, and 5^1 S_1 H_2CO. The vibrational lines in these spectra were assigned by considering their frequencies, relative intensities within a spectrum, and relative intensities among spectra. A table with the complete set of assignments and expanded figures of each spectrum annotated with assignments are presented by Bouwens *et al.* (*6*).

Multi-Resonant Hamiltonian Model

The model used by Reisner *et al.* (*7*) and by Bouwens *et al.* (*6*) to fit H_2CO energy levels was a traditional Dunham-type power expansion of the energy of a state in terms of its vibrational quantum numbers

$$E\left(v_1,\ldots,v_{3N-6}\right) = hc\sum_i \omega_i^\circ v_i + hc\sum_{i \geq j} x_{ij} v_i v_j \tag{1}$$

where ω_i° and x_{ij} are harmonic and anharmonic vibrational spectroscopic constants. This model is appropriate in the limit of low vibrational interactions. As noted in Bouwens *et al.*, however, it begins to fail when states become appreciably mixed. Such mixing is to be expected for the case of highly excited states, as regions of the potential far from the harmonic minimum are being sampled. The approach taken by Bouwens *et al.* was to determine vibrational spectroscopic constants by selectively fitting only states that are both low in energy and not appreciably mixed. In this paper, a method is presented which explicitly incorporates the effects of state mixing and therefore allows every assigned state to be fit and used in the determination of molecular constants.

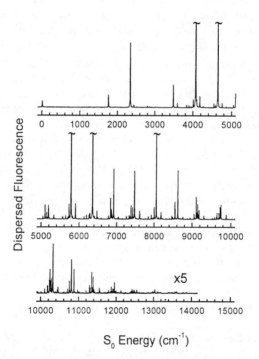

Figure 2. Dispersed fluorescence spectrum of S_0 H_2CO from the 0_{00} rotational level of 4^1 S_1 H_2CO. The horizontal scale is adjusted to read in S_0 rovibrational energy.

A potential energy surface (PES) can be written as a power series expansion in $3N-6$ dimensionless vibrational normal coordinates q_i

$$V(q_1, \ldots, q_{3N-6}) = \frac{1}{2} hc \sum_i \omega_i q_i^2 + hc \sum_{i \geq j \geq k} k_{ijk} q_i q_j q_k + hc \sum_{i \geq j \geq k \geq l} k_{ijkl} q_i q_j q_k q_l + \ldots \quad (2)$$

where ω_i, k_{ijk}, and k_{ijkl} are harmonic, cubic, and quartic potential constants, respectively. The effect of anharmonic potential constants on the vibrational energy spectrum can be approximated using second-order perturbation theory (*11*). The resulting expression for the vibrational energy levels is equation (1), thereby justifying the use of a power series in vibrational quantum numbers to fit vibrational energy levels in the perturbative limit. Explicit formulas for ω_i° in terms of ω_i and x_{ij} and for x_{ij} in terms of k_{ijk} and k_{ijkl} may be found in advanced spectroscopy texts (*10-13*). Note that for a given level of perturbation theory, there are more potential constants than spectroscopic constants. Thus while one can calculate spectroscopic constants from potential constants, it is not generally possible to determine all the potential constants from a spectrum.

It is apparent from the form of equation (1) that the effect of anharmonicity in the perturbative limit is to compress or expand progressions of equally spaced harmonic energy levels. While this effect is observed in virtually every vibrational spectrum, occasionally complete destruction of the quasi-regular pattern is observed. Such deviations are due to strong interactions among states which lead to extensive state mixing. A strong interaction among states is called a **resonance**. Resonances typically arise when two nearly degenerate vibrational basis states can interact through a nonzero potential constant. For example, if $\omega_i + \omega_j \approx \omega_k$ then k_{ijk} can resonantly mix harmonic oscillator product basis states $|\ldots, v_i, v_j, v_k, \ldots\rangle$ and $|\ldots, v_i+1, v_j+1, v_k-1, \ldots\rangle$. Since one quantum of mode k is being exchanged with one quantum of mode i and one quantum of mode j in this case, the resonance is denoted by $k_{ij,k}$. It is not surprising that perturbation theory can fail to describe excited vibrational states accurately, for the assumptions of perturbation theory break down when the perturbation matrix element is large relative to the energy difference between the basis states it connects

$$\frac{\hat{H}'}{\Delta E^\circ} = \frac{k_{ijk} \langle v_i, v_j, v_k | q_i q_j q_k | v_i', v_j', v_k' \rangle}{E^\circ(\ldots, v_i, v_j, v_k, \ldots) - E^\circ(\ldots, v_i', v_j', v_k', \ldots)} > 1 \quad (3)$$

Two approaches for dealing with strongly interacting states are to go to higher-order perturbation theory or to account for every interaction explicitly using matrix mechanics. The problem with both of these approaches is that the resulting model has many more parameters than can be readily determined from spectroscopic data. While such approaches may be appropriate for calculating energy levels from a theoretical set of potential constants, they are inappropriate for extracting constants from experimental data.

The multi-resonant Hamiltonian model is a hybrid approach that blends perturbation theory with matrix mechanics. In doing so, it uses the minimum number of parameters to fit the observed spectrum and it gives each parameter a simple physical interpretation. The multi-resonant Hamiltonian model treats all nonresonant interactions by second-order perturbation theory and all resonant interactions explicitly. A matrix is formed using harmonic oscillator products as basis states. The diagonal matrix elements (in cm^{-1} units) of the multi-resonant Hamiltonian are

$$\langle v_1,\ldots,v_{3N\text{-}6}|H|v_1,\ldots,v_{3N\text{-}6}\rangle = \sum_i \omega_i^\circ v_i + \sum_{i\leq j} x_{ij} v_i v_j \qquad (4)$$

Off-diagonal matrix elements are of the form

$$\langle \ldots,v_i,v_j,v_k,\ldots|H|\ldots,v_i+1,v_j+1,v_k-1,\ldots\rangle$$
$$= k_{ij,k}\langle v_i|q_i|v_i+1\rangle\langle v_j|q_j|v_j+1\rangle\langle v_k|q_k|v_k-1\rangle \qquad (5)$$
$$= k_{ij,k}\left(\frac{v_i+1}{2}\right)^{1/2}\left(\frac{v_j+1}{2}\right)^{1/2}\left(\frac{v_k}{2}\right)^{1/2}$$

where $k_{ij,k}$ represents a resonance between modes i and j and mode k. Similar expressions may be derived for other types of resonances using harmonic oscillator matrix elements (11). In the multi-resonant Hamiltonian model, the diagonal constants ω_i° and x_{ij} incorporate the effects of nonresonant perturbations and the off-diagonal constants $k_{ij,k}$ and $k_{ij,kl}$ represent resonances which are too large to be treated by perturbation theory. The scaling of harmonic oscillator matrix elements with vibrational quantum numbers allows the multi-resonant Hamiltonian with relatively few resonance parameters to account for vibrational state mixing throughout the entire spectrum.

Computationally, the multi-resonant Hamiltonian model is implemented by starting with a seed basis state and generating all resonantly connected basis states by recursively applying the appropriate harmonic oscillator selection rules. The restrictive nature of these rules severely limits the number of interacting states. Figure 3 illustrates the result of generating all basis states connected by two particular resonances to a seed basis state. Once all connected states have been determined, matrix elements are evaluated and the resulting matrix is diagonalized to yield energy eigenvalues and wavefunction eigenvectors. When fitting experimental data, the Marquardt nonlinear least squares algorithm is an effective method for adjusting model parameters to minimize the deviations between predicted and observed energy levels (14). The Hellman-Feynman theorem is the most efficient way to calculate the required first partial derivatives of energy with respect to parameter values (15). Difficulties can arise when matching calculated positions to observed positions using basis state labels for assignments, as a basis state may be mixed by resonances into many eigenstates. In this work, the eigenstate with the largest component of the basis state assignment was selected. Changes in eigenstate composition during the fitting process and the use of unique basis state labels for highly mixed states were accommodated by weighting eigenstates with their proximity to the observed state. Specifically, the match was made by selecting the eigenstate with the largest coefficient squared of assigned state weighted by the inverse of energy difference between the calculated eigenvalue and the observed position.

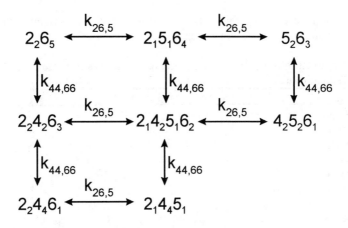

Figure 3. States connected to 2_26_5 by the resonances $k_{26,5}$ and $k_{44,66}$. Restrictive harmonic oscillator selection rules and the requirement that $v_i \geq 0$ severely limit the number of connected states.

The only seemingly arbitrary aspect of the multi-resonant Hamiltonian model is the selection of resonances to be treated explicitly. Two procedures are suggested here for making this decision. First, if theoretical or experimental estimates of the potential constants are available, then a systematic search for "important" resonances can be undertaken. The quantity $\hat{H}'/\Delta E°$ in equation (3) is evaluated for all pairs of energy levels, and values larger than a predetermined threshold are flagged for explicit treatment as resonances. Second, if an experimental spectrum is being fit with the multi-resonant Hamiltonian model, then the quality of fit can be monitored as a function of resonance inclusion. If including a particular resonance significantly improves the fit, then it is retained as an "important" resonance. If a resonance does not appreciably affect the fit, then it is not treated explicitly and its effect is incorporated into the perturbative spectroscopic constants.

A search for important resonances in H_2CO was performed by systematically evaluating $\hat{H}'/\Delta E°$ using the spectroscopic constants of Martin, Lee, and Taylor (8) and cubic and quartic force constants of Green et al. (16). The relative importance of these resonances in describing the observed spectrum of Bouwens et al. (6) was determined by calculating the standard deviation of the fit of the spectral data to a multi-resonant Hamiltonian model as a function of each included resonance. The results of this analysis are given in Table I. Most of the deviation is accounted for by the inclusion of three resonances, $k_{26,5}$, $k_{36,5}$, and $k_{11,55}$. For fits of states with energies up to 7000 cm^{-1}, including $k_{44,66}$ resulted in significant improvement while including additional resonances lead to only marginal improvements. For fits of states with energies up to 10000 cm^{-1}, $k_{25,35}$ was also needed to improve the fit but additional resonances resulted in negligible improvements. By including only these resonances, the standard deviation of the fit is less than 3 cm^{-1}, which approaches the experimental accuracy of 1 cm^{-1}. The remaining small discrepancy likely arises from the approximation of accounting for every nonresonant potential constant with a finite number of anharmonic spectroscopic constants.

Table I. Dependence of fit standard deviation on resonances included in H_2CO multi-resonant Hamiltonian

	Fit 1	Fit 2	Fit 3	Fit 4	Fit 5	Fit 6	Fit 7
$k_{26,5}$		x	x	x	x	x	x
$k_{36,5}$		x	x	x	x	x	x
$k_{11,55}$		x	x	x	x	x	x
$k_{44,66}$			x	x	x	x	x
$k_{25,35}$				x	x	x	x
$k_{26,36}$				x	x	x	x
$k_{1,44}$						x	x
$k_{1,66}$							x
0–7000 cm^{-1} fit std. Dev. (cm^{-1})	11.19	3.16	2.64	2.57	2.39	2.39	2.23
0–10000 cm^{-1} fit std. Dev. (cm^{-1})	23.39	4.34	3.34	2.80	2.76	2.62	2.57

The parameter values for the fits in Table I are given in Table II. The ω_i° and x_{ij} are spectroscopic constants, from which the potential constants ω_i can be calculated. The $k_{ij,k}$ and $k_{ij,kl}$ fit parameters are estimates of the potential constants k_{ijk} and k_{ijkl} in equation (2). Thus, the entire set of harmonic potential constants is determined, but only the small subset of cubic and quartic potential constants which are involved in resonances are determined. One must be cautious in comparing these values directly to *ab initio* calculations. First, the value obtained for a particular parameter depends on which specific resonances were included and excluded in the fit. For example, if a model excluded an important resonance, then the effect of that resonance would be inappropriately incorporated into the other parameters. The variation of a parameter among fits to models with increasingly weaker resonances therefore offers an estimate of the how well it is determined. In this case, the values of most parameters has stabilized to within a few percent or better as additional resonances are included. Second, the effects of higher order resonances are directly incorporated into the value of each resonance parameter being determined. For example, the fitted value of $k_{ij,k}$ includes the effect of potential constant k_{ijk} as well as constants k_{ijkll} for all values of l. One must assume that such higher order contributions are small relative to the value being determined. Third, the potential constants determined here correspond to a restrictive summation over vibrational coordinates, whereas many theoretical calculations report constants for a nonrestrictive summation (*13*). Finally, the sign of the potential constant is not determined by the resonance fit parameter value. The sign could be chosen to agree with the sign of an *ab initio* calculation, or sometimes it can be determined by additional constraints among the fit parameters.

Table II. H_2CO multi-resonant Hamiltonian fit parameters for 0–10000 cm^{-1} fit (all values are cm^{-1})

Parameter	Fit 1	Fit 2	Fit 3	Fit 4	Fit 5	Fit 6	Fit 7
ω_1°	2818.9	2812.3	2813.7	2817.4	2818.0	2815.5	2809.9
ω_2°	1759.9	1755.6	1755.8	1755.7	1755.8	1756.0	1756.0
ω_3°	1479.5	1499.1	1499.3	1499.2	1498.8	1499.1	1499.4
ω_4°	1169.8	1170.5	1169.8	1169.7	1169.7	1167.3	1167.3
ω_5°	2859.6	2882.8	2879.2	2873.8	2875.9	2876.0	2877.7
ω_6°	1260.6	1254.8	1251.5	1251.9	1251.8	1251.9	1249.8
x_{11}	-40.1	-29.8	-30.7	-34.4	-35.0	-35.9	-33.8
x_{12}	1.3	0.9	0.7	0.4	0.4	0.3	0.5
x_{13}	-35.5	-27.2	-27.7	-26.9	-26.8	-27.2	-27.6
x_{14}	-8.5	-9.3	-9.5	-9.4	-9.4	-15.0	-14.9
x_{15}	-127.3	-133.9	-130.8	-131.5	-133.4	-136.1	-137.0
x_{16}	-3.2	-12.2	-12.8	-13.2	-13.0	-13.4	-17.6
x_{22}	-10.6	-10.1	-10.1	-9.9	-9.9	-9.9	-9.9
x_{23}	-4.8	-7.0	-7.1	-8.0	-7.8	-7.9	-8.1
x_{24}	-7.4	-6.8	-6.8	-7.0	-7.1	-7.2	-7.3
x_{25}	-23.4	5.5	5.2	1.7	1.4	1.2	0.9
x_{26}	-0.8	-15.5	-14.2	-11.6	-12.2	-12.2	-12.1
x_{33}	10.8	0.3	0.4	0.6	0.6	0.5	0.5
x_{34}	-2.7	-1.9	-1.5	-1.4	-1.4	-1.1	-1.2
x_{35}	-3.8	-17.3	-17.2	-23.9	-23.9	-23.4	-21.1
x_{36}	-23.3	-12.9	-12.4	-8.7	-8.0	-8.6	-9.2
x_{44}	-3.0	-3.2	-2.9	-2.8	-2.8	-0.9	-0.9
x_{45}	-13.5	-22.2	-21.8	-22.0	-22.2	-20.6	-20.2
x_{46}	1.7	3.1	5.2	5.3	5.5	5.6	5.5
x_{55}	-14.3	-38.2	-36.3	-34.7	-35.5	-36.0	-37.6
x_{56}	-9.8	-0.2	-3.0	-13.5	-14.4	-13.9	-12.2
x_{66}	-5.2	-2.8	-2.1	-2.2	-2.3	-2.2	-0.7
$|k_{26,5}|$		148.6	146.7	138.6	143.7	143.6	145.1
$|k_{36,5}|$		129.3	129.6	135.1	130.9	130.7	129.3
$|k_{11,55}|$		140.5	137.4	129.3	127.8	132.4	138.2
$|k_{44,66}|$			21.6	23.3	24.2	23.4	18.1
$|k_{25,35}|$				18.5	23.1	23.1	23.2
$|k_{26,36}|$					5.7	5.7	6.9
$|k_{1,44}|$						78.0	79.4
$|k_{1,66}|$							55.0

When a single potential constant is the dominant contribution to spectroscopic constants, relationships among these constants can develop. Mills and Robiette demonstrated that the large diagonal anharmonicity of stretches, particularly those involving a hydrogen, leads to relationship between x_{ij} and $k_{ii,jj}$ for molecules containing symmetrically equivalent sets of stretching vibrations (*17*). This Darling-Dennison x-k relation for formaldehyde is

$$2x_{11} = \frac{1}{2}x_{15} = 2x_{55} = \frac{1}{2}k_{11,55} = x_m(CH) \tag{6}$$

where $x_m(CH)$ is the Morse anharmonicity of the CH stretch. The values predicted by equation (6) are compared to the observed values from Fit 7 in Table III, and the Darling-Dennison x-k relation is seen to hold very well for formaldehyde.

Table III. Darling-Dennison x-k relation for formaldehyde (all values are cm^{-1}).

Parameter	Calculated[a]	Fit[b]	Observed (Fit 7)
x_{11}	−32.2	−34.5	−33.8
x_{15}	−128.6	−138.0	−137.0
x_{55}	−32.2	−34.5	−37.6
$k_{11,55}$	−128.6	−138.0	−138.2

[a]Assuming $x_m(CH) = -64.3$ cm^{-1} from CH radical (18)
[b]Assuming a best fit value of $x_m(CH) = -69.0$ cm^{-1}

The multi-resonant Hamiltonian is a model which describes the essential physical features of vibrational state mixing with a minimum number of parameters. The ω_i° and x_{ij} describe the smooth variation of vibrational basis state energy with quantum number. The $k_{ij,k}$ and $k_{ij,kl}$ parameters describe resonances which lead to state mixing. Since the off-diagonal resonance matrix elements scale according to harmonic oscillator matrix elements, a relatively small set of resonances can be applied over a wide quantum number and energy range. The parameters ω_i° and x_{ij} determine which resonances are important by controlling the near degeneracy among basis states. They also determine how basis states tune in energy relative to one another as a function of quantum number, and hence whether a resonance is local or persistent throughout the spectrum (19). Thus, the multi-resonant Hamiltonian model accounts for extensive mixing of vibrational states over wide energy ranges with a limited set of parameters.

Polyad Quantum Numbers

Since resonances lead to extensive mixing of basis states, a natural question to ask is whether any conserved quantities remain. The ability of the multi-resonant Hamiltonian to block diagonalize the Hamiltonian matrix into finite-sized submatrices suggests that conserved quantities do exist. Each block-diagonal submatrix involves basis states which are mixed among themselves, but are negligibly mixed with other basis states. A set of such basis states is called a **polyad**. The new quantum numbers that arise to describe each polyad are called **polyad quantum numbers**. In this section, a method for determining polyad quantum numbers (20,21) is reviewed and then applied using the previously determined "important" resonances of formaldehyde.

Vibrational quantum states may be treated as a $m = 3N-6$ dimensional vector space, with each dimension representing one quantum number. In the case of formaldehyde, $m = 6$ and the vibrational quantum numbers give rise to a 6-dimensional vector space, (v_1, v_2, v_3, v_4, v_5, v_6), where $v_i \geq 0$. Resonances cause mixing between harmonic oscillator basis states which differ by a number of quanta in particular modes as prescribed by harmonic oscillator selection rules. Thus, resonances may be treated as vectors which represent exchange of vibrational quanta. For example, the $k_{26,5}$ resonance in formaldehyde mixes two states which differ by +1 quanta in modes 2 and 6 and −1 quanta in mode 5, or vice-versa, and by zero in all other modes. This resonance is represented by the vector $(0,1,0,0,-1,1)$. Similarly, $k_{44,66}$ is represented by $(0,0,0,2,0,-2)$. The resonance vectors span a **resonance vector subspace** of dimension n. The dimensionality n is less than or equal to the number of resonance vectors, depending on their linear independence. The subspace orthogonal to the resonance subspace is called the **polyad vector subspace** and is of dimension $m-n$. The polyad vector subspace is spanned by $m-n$ linearly independent basis vectors, which are termed polyad vectors. While the number of polyad vectors is determined by the dimensionality of the polyad vector subspace, the polyad vectors themselves are not uniquely determined and need not even be orthogonal. Any set of linearly independent basis vectors forms a valid set of polyad vectors. However, polyad vectors are usually chosen to consist of small, positive integers with a simple physical interpretation. Figure 4 describes the vector subspaces which are involved in determining polyad quantum numbers.

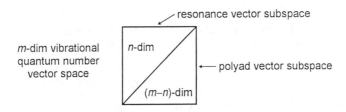

Figure 4. The resonance vector subspace and polyad vector subspace are orthogonal subspaces which comprise the quantum number vector space.

Since the polyad vectors are linearly independent of the resonance subspace, they represent polyad quantum numbers which are unmixed by resonance interactions. Interactions may exist between states with the same polyad quantum numbers, but states with different polyad quantum numbers are not mixed by any of the "important" resonances. Thus, the Hamiltonian matrix is block-diagonal in polyad quantum numbers, which therefore serve as useful state labels. Polyad quantum numbers are the conserved quantities which remain after extensive vibrational mixing by resonances. Note that additional conserved quantities can arise from other restrictions on the problem, *e.g.*, molecular symmetry results in a vibrational symmetry quantum number in addition to polyad quantum numbers.

As a simple example, consider a 2-dimensional system characterized by the two quantum numbers v_1 and v_2 with $\omega_1 \approx 2\omega_2$, as in Figure 5. A nonzero value for $k_{1,22}$ mixes modes 1 and 2. In vector language, the vibrational quantum number vector space is (v_1,v_2) and the resonance under consideration is $(1,-2)$. In this case, v_1 and v_2

are each no longer good quantum numbers. For example, $k_{1,22}$ mixes the following states

$$(3,0) \leftrightarrow (2,2) \leftrightarrow (1,4) \leftrightarrow (0,6)$$

to form a polyad by exchanging one quantum of mode 1 for two quanta of mode 2. However, one resonance reduces the number of conserved quantities by at most one, and a polyad vector analysis identifies the remaining conserved quantities. In this case, the resonance subspace is the 1-dimensional subspace spanned by $(1,-2)$. The orthogonal space is of dimension $2-1=1$ and is spanned by $(2,1)$. Figure 6 describes the vector subspaces involved in this example. The polyad quantum number corresponding to $(2,1)$ is $N=2v_1+v_2$. In the example of the specific mixed states above, $N=6$ for each of them, verifying that the polyad quantum number N is conserved within a polyad.

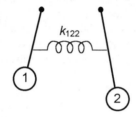

Figure 5. Harmonic oscillators coupled by $k_{1,22}$.

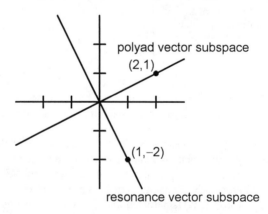

polyad vector subspace
(2,1)

(1,-2)

resonance vector subspace

Figure 6. Vector subspaces arising from $k_{1,22}$.

Higher dimensional systems can be analyzed using linear algebra algorithms on a computer. Gram-Schmidt (G-S) orthogonalization is the conventional method for searching for linear dependencies among resonance vectors and for constructing basis vectors for the polyad vector subspace. However, a more computationally robust method for determining the resonance and polyad vector subspaces is Single Value Decomposition (SVD) (22). SVD is a powerful numerical method for dealing with matrices that are singular or numerically close to singular. SVD uniquely decomposes any $m \times n$ matrix $\mathbf{A}$ with $m \geq n$ into the product of an $m \times n$ column-orthonormal matrix $\mathbf{U}$, an $n \times n$ diagonal matrix $\mathbf{W}$, and the transpose of an $n \times n$ orthonormal matrix V

$$\left[\left(\begin{array}{c} \\ \mathbf{A}_1 \\ \\ \end{array} \right) \cdots \left(\begin{array}{c} \\ \mathbf{A}_n \\ \\ \end{array} \right) \right] = \left[\left(\begin{array}{c} \\ \mathbf{U}_1 \\ \\ \end{array} \right) \cdots \left(\begin{array}{c} \\ \mathbf{U}_n \\ \\ \end{array} \right) \right] \left[\begin{array}{ccc} w_1 & & \\ & \ddots & \\ & & w_n \end{array} \right] \left[\quad V \quad \right]^{\mathrm{T}} \tag{7}$$

In this application of SVD, $\mathbf{A}$ consists of n m-dimensional column vectors which represent resonances. The m-dimensional column vectors $\mathbf{U}$ which correspond to nonzero w_i span the vector space generated by the column vectors of $\mathbf{A}$. Values of w_i which are zero, or small relative to the largest w_i, indicate linear dependence. SVD can be used to test for linear independence of resonance vectors by checking whether $\mathbf{W}$ contains any diagonal values of zero. Once the resonance subspace has been defined by n linearly independent vectors, a basis for the orthogonal $(m-n)$-dimensional polyad subspace can be determined by constructing the $\mathbf{A}$ matrix from linearly independent resonance vectors and then squaring it up with columns of zeroes. The column vectors of $\mathbf{U}$ corresponding to values of w_i equal to zero are the polyad vectors. In both the G-S and SVD methods, linear combinations of the generated polyad vectors are then taken in order to yield polyad vectors which consist of small, positive integers.

From the multi-resonant Hamiltonian fitting in Table I, it was demonstrated that the three most important resonances in S_0 H_2CO are $k_{36,5}$, $k_{26,5}$, and $k_{11,55}$. These resonances are represented by the vectors $(0,0,1,0,-1,1)$, $(0,1,0,0,-1,1)$, and $(2,0,0,0,-2,0)$, respectively. The linearly independent subspace is spanned by the vectors $(0,0,0,1,0,0)$, $(1,0,0,1,1,1)$, and $(2,1,1,1,2,1)$ which correspond to the following polyad quantum numbers

$$\begin{aligned} N_{\mathrm{oop}} &= v_4 \\ N_{\mathrm{vib}} &= v_1 + v_4 + v_5 + v_6 \\ N_{\mathrm{res}} &= 2v_1 + v_2 + v_3 + v_4 + 2v_5 + v_6 \end{aligned} \tag{8}$$

These particular vectors were chosen because the corresponding polyad quantum numbers have simple physical interpretations and there is no need to change polyad quantum number definitions as additional resonances are considered. N_{oop} represents the number of out-of-plane bending quanta. Its existence as a polyad number implies that out-of-plane vibrations do not couple with in-plane vibrations in formaldehyde. Of course, Coriolis-coupling is well-known to mix modes 4 and 6 in formaldehyde (*23*); however, the multi-resonant Hamiltonian developed here only includes pure vibrational resonances and hence does not treat the case of Coriolis coupling. N_{vib} represents a particular sum of vibrational quanta in formaldehyde. Its particular form is chosen so that when N_{oop} is lost as a polyad quantum number, N_{vib} remains a good polyad number. Finally, N_{res} represents the energy of the polyad. Examination of the vibrational frequencies of formaldehyde reveals that modes 1 and 5 are approximately 2800 cm^{-1} and the other modes range between 1100 cm^{-1} and 1800 cm^{-1}, averaging 1400 cm^{-1}. The apparently unusual form of N_{res} arises from the frequency ratio of vibrational modes, and its value approximately equals the energy of the polyad divided by 1400 cm^{-1}. There is little coupling between basis states with different N_{res} because they tend to be of different energy. In order for two states of different N_{res} to be nearly degenerate, they must differ from each other by many quanta, making the interaction appear in very high order and hence quite weak. It should be noted that in addition to these three polyad quantum numbers, vibrational symmetry is a conserved quantity among vibrational basis states and takes on the values A_1, A_2, B_1, and B_2 in formaldehyde.

Table I indicates that the resonance $k_{44,66}$ is the next most important resonance. This resonance causes the loss of N_{oop} as a polyad quantum number. Eventually, inclusion of $k_{1,44}$ or $k_{1,66}$ causes loss of N_{vib}, although the relative weakness of these resonances suggests that N_{vib} would only be partially lost. The only polyad number which remains good for every resonance considered in this analysis is N_{res}. In addition, vibrational symmetry is a rigorously good quantum number for all vibrational resonances.

It is very interesting to compare the formaldehyde polyad numbers with those for acetylene (HCCH). Analysis of all known resonances in acetylene yields 3 polyad numbers: l, the total vibrational angular momentum quantum number; N_s, the total stretch quantum number; and N_{res}, the energy quantum number (19,21,24,25). The existence of N_{oop} in formaldehyde and l in acetylene suggests that certain classes of vibrational motion can be so different in frequency or nature from other vibrations in the molecule that they do not couple well to the rest of the molecule. The existence of an N_{res} in both molecules suggests that an energy polyad number may be general to most small molecules, for N_{res} arises as a consequence of vibrational frequency ratios rather than as a consequence of specific anharmonic vibrational potential constants. The existence of such a general polyad number would serve as an extremely useful organizing feature for vibrational energy levels, e.g., it would block-diagonalize vibrational energy level calculations. It would also have important dynamical considerations by reducing the amount of phase space accessible to a molecule.

Conclusions

The study of highly excited vibrational states is essential to gain a fundamental understanding of chemical reactivity. Excited vibrational states sample anharmonic portions of the potential energy surface, and harmonic oscillator basis states therefore experience level shifts and mixing. These effects can be accounted for by the multi-resonant Hamiltonian model, a method which combines perturbation theory with matrix mechanics. The multi-resonant Hamiltonian accounts for perturbative energy shifts to basis states by means of diagonal parameters $\omega_i{}^\circ$ and x_{ij}. Resonances which lead to mixing of harmonic oscillator basis states are treated explicitly by including resonance parameters $k_{ij,k}$ and $k_{ij,kl}$ in off-diagonal matrix elements. The scaling of these harmonic oscillator matrix elements with vibrational quantum numbers results in a method which accounts for extensive vibrational mixing over a wide energy range with a minimum number of parameters. The resulting fit parameters provide good estimates of the potential constants ω_i, k_{ijk} and k_{ijkl}.

A set of states which are strongly coupled among themselves, but are not coupled to other states, is called a polyad. Harmonic oscillator selection rules strictly limit which basis states can interact with each other, leading to block-diagonal submatrices within the multi-resonant Hamiltonian matrix. Analysis of the resonances by a straightforward linear algebra procedure known as polyad vector analysis yields polyad quantum numbers which characterize each block-diagonal submatrix. Polyad quantum numbers are new quantum numbers which remain conserved in the presence of extensive vibrational mixing, and hence are useful for organizing excited vibrational energy levels and for describing the dynamics of excited states.

The technique of pure vibrational spectroscopy, which takes advantage of rotational selection rules to eliminate rotational congestion, has permitted acquisition of the excited vibrational state spectra of formaldehyde with unprecedented detail for a tetra-atomic molecule. The extensive vibrational mixing which is observed throughout the spectrum is successfully fit with a multi-resonant Hamiltonian model containing a limited number of important resonances. By including only five resonances, every assigned state up to $10000\ cm^{-1}$ of vibrational energy is fit with a standard deviation of only $2.8\ cm^{-1}$. Including additional resonances leads to only slight improvements in the fit. Polyad quantum numbers based on these resonances are identified for formaldehyde. While polyad quantum numbers are typically specific to the resonances which are important in a molecule's vibrational energy level structure, an energy polyad quantum number is identified which may generally applicable to small molecules.

Acknowledgments

We gratefully acknowledge Rychard J. Bouwens for assigning the vibrational states of formaldehyde. In addition, we express thanks to Dr. Robert W. Field, Dr. Anne B. McCoy, Dr. Ian M. Mills, and Dr. Edwin L. Sibert III for valuable discussions. This research was supported by National Science Foundation grant CHE-9157713 and a 1996 Summer Research Grant Award of the Royal Netherlands Chemical Society sponsored by the Council of Undergraduate Research.

Literature Cited

1. Freed, K.F.; Nitzan, A. In *Energy Storage and Redistribution in Molecules*; Hinze, J., Ed.; Plenum: New York, 1983.
2. Demtröder, W. *Laser Spectroscopy*; Springer-Verlag: Berlin, 1982.
3. Boyarkin, O.V.; Settle, R.D.F.; Rizzo, T.R. *Ber. Bunsenges. Phys. Chem.* **1995**, *99*, 504.
4. Kittrell, C.; Abramson, E.; Kinsey, J.L.; McDonald, S.A.; Reisner, D.E.; Field, R.W.; Katayama, D.H. *J. Chem. Phys.* **1981**, *75*, 2058.
5. Zhang, Q.; Kandel, S.A.; Wasserman, T.A.W.; Vaccaro, P.H. *J. Chem. Phys.* **1992**, *96*, 1640.

6. Bouwens, R.J.; Hammerschmidt, J.A.; Grzeskowiak, M.M.; Stegink, T.A.; Yorba, P.M.; Polik, W.F. *J. Chem. Phys.* **1996**, *104*, 460.

7. Reisner, D.E.; Field, R.W.; Kinsey, J.L.; Dai, H.-L. *J Chem. Phys.* **1984**, *80*, 5968.

8. Martin, J.M.L.; Lee, T.J.; Taylor, P.R. *J. Mol. Spectrosc.* **1993**, *160*, 105.

9. Burleigh, D.C.; McCoy, A.B.; Sibert III, E.L. *J. Chem. Phys.* **1996**, *104*, 480.

10. Herzberg, G. *Electronic Spectra and Electronic Structure of Polyatomic Molecules*; Van Nostrand Reinhold: New York, 1966.

11. Califano, S. *Vibrational States*; Wiley: London, 1976.

12. Papousek, D.; Aliev, M.R. *Molecular Vibrational-Rotational Spectra*; Elsevier: Amsterdam, 1982.

13. Mills, I.M. In *Modern Spectroscopy: Modern Research*; Rao, K.N.; Mathews, C.W., Eds.; Academic: New York, 1972.

14. Bevington, D. *Data Reduction and Error Analysis for the Physical Sciences*; McGraw-Hill: New York, 1969.

15. Lefebvre-Brion, H.; Field, R.W. *Perturbations in the Spectra of Diatomic Molecules*; Academic: Orlando, 1986.

16. Green Jr., W.H.; Willetts, A.; Jayatilaka, D.; Handy, N.C. *Chem. Phys. Lett.* **1990**, *169*, 127.

17. Mills, I.M.; Robiette, A.G. *Mol. Phys.* **1985**, *56*, 743.

18. Herzberg, G. *Spectra of Diatomic Molecules*; Van Nostrand Reinhold: New York, 1950.

19. Field, R.W.; Coy, S.L.; Solina, S.A.B. *Progress of Theoretical Physics Supplement No. 116* **1994**, 143.

20. Fried, L.E.; Ezra, G.S. *J. Chem. Phys.* **1987**, *86*, 6270.

21. Kellman, M.E. *J. Chem. Phys.* **1990**, *93*, 6630.

22. Press, W.H.; Flannery, P.; Teukolsky, S.A.; Vetterling, W.T. *Numerical Recipes*; Cambridge University Press: Cambridge, 1986.

23. Clouthier, D.J.; Ramsay, D.A. *Ann. Rev. Phys. Chem.* **1983**, *34*, 31.

24. Solina, S.A.B.; O'Brien, J.P.; Field, R.W.; Polik, W.F. *Ber. Bunsenges. Phys. Chem.* **1995**, *99*, 555.

25. Solina, S.A.B.; O'Brien, J.P.; Field, R.W.; Polik, W.F. *J. Phys. Chem.* **1996**, *100*, 7797.

Intramolecular Dynamics: Unimolecular Decay

Chapter 5

Time Scales and Mechanisms of Intramolecular Energy Redistribution

David S. Perry

Department of Chemistry, University of Akron, Akron, OH 44325–3601

The frequency-resolved spectra of CF_3H, methanol, CF_2Cl_2, ethanol, and 1-butyne reveal a wide range of timescales and a variety of mechanisms for intramolecular vibrational and rotational energy redistribution. The fastest timescale ($\tau \lesssim 100$ fs) is the result of strong anharmonic or harmonic low-order resonances. The longest directly measurable timescale (10 ps $< \tau <$ 300 ps) is evidenced by the width of the narrowest features in the methanol overtone spectra and by the clump widths in the eigenstate-resolved spectra of the ethanol fundamentals. The latter spectra indicate the simultaneous action of anharmonic, z-type Coriolis, and x/y-type Coriolis mechanisms. There are even longer timescales that cannot be determined directly from the spectrum of a bright state but can be deduced by modelling the data. For example, random matrix simulations indicate that the time for the completion of K-relaxation in ethanol is on the order of a nanosecond.

Since energy randomization was recognized in the RRK theory (1,2) as a crucial step in unimolecular reactions, it has been assumed to be a fast process. The assumption that intramolecular vibrational redistribution (IVR) is fast compared to the timescale of unimolecular reaction remains the key assumption in the modern - and very successful - RRKM theory of unimolecular reactions (3). It is now evident that IVR is not a single process but multiple processes occurring on a wide range of timescales. In this paper, timescales ranging from 10^{-13} to 10^{-9} seconds and the phenomenology and mechanisms associated with each are discussed.

Most of the quantitative information currently available about the rate and mechanism of IVR derives from frequency-resolved spectra. It is therefore essential to understand the relationship between such spectra and the dynamics of energy redistribution. The zeroth order state that is responsible for the oscillator strength in a particular part of the spectrum is called the bright state ϕ_s (Figure 1). In the work discussed below, the bright state is typically a hydride stretch in the three micron region of the infrared or one of the overtones of a stretching vibration. If the vibrational

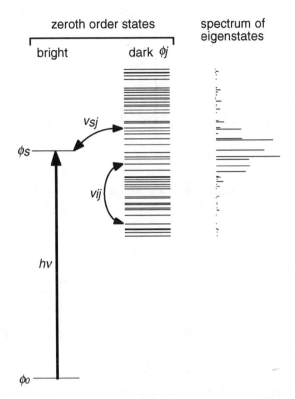

Figure 1. Energy level scheme for IVR. The energies and intensities shown here as an example are from one of the synthetic spectra obtained during the random matrix analysis of the ethanol v_{14} CH stretch.

density of states is high enough, the bright state may be coupled to other zeroth order states ϕ_j, called bath states, by anharmonic or Coriolis matrix elements v_{sj}. In favorable cases, the bath states are dark states in that they carry no oscillator strength; that is, there is only a single bright state. Likewise the dark states may also be coupled together with matrix elements v_{ij}. The result of these interactions is a set of molecular eigenstates, each of which has a portion of the bright state character and appears in the spectrum with a corresponding intensity.

The experimental data discussed here, e.g., Figure 2, are all frequency-resolved spectra in which a laser frequency is scanned through the spectrum of the eigenstates. In the highest resolution experiments, the eigenstates are resolved and appear as discrete features in the spectrum. As the laser is scanned through a particular eigenstate, a stationary state of the system with mixed bright-state/dark-state character is prepared directly. Therefore, no IVR will result from this experiment.

Nonetheless, the frequency-resolved spectra contain the information needed to calculate the result of an hypothetical time-resolved experiment. A coherent excitation of a "clump" of eigenstates, such as those shown in Figure 1, by a sufficiently short light pulse would result in the preparation of the bright state. Since the bright state is a superposition of eigenstates, it is a nonstationary state and would evolve in time. The

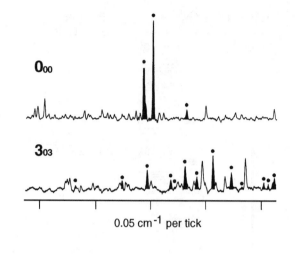

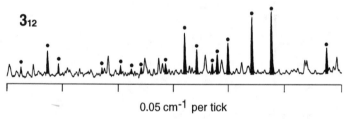

Figure 2. Sections of the slit-jet spectrum of the ν_{14} CH stretch fundamental band of ethanol. The highlighted lines in each section are assigned to the indicated rotational levels. (Data from ref. 7.)

dephasing of the bright state results in the movement of vibrational amplitude from the bright state coordinate (e.g., hydride stretch) into the other vibrational degrees of freedom of the molecule. The time-dependent survival probability of the bright state - which in principle could be monitored by infrared fluorescence - can be calculated from the energies E_i and intensities I_i of the eigenstates: (4-6)

$$\left|\left\langle \Psi(t=0) \middle| \Psi(t) \right\rangle\right|^2 = \frac{\sum_i \sum_j I_i I_j \cos\left[\left(E_i - E_j\right)t/\hbar\right]}{\left(\sum_j I_j\right)^2} \tag{1}$$

Survival probabilities exhibit an initial decay followed by quantum beats which are seen as oscillations around the long-time average ϕ_d. The quantity ϕ_d is also called the dilution factor. The initial decay reflects the rate of energy transfer out of the bright state, and the IVR lifetime τ_{IVR} of the bright state can be obtained as the $1/e$ decay time relative to the long-time average (6). Values of τ_{IVR} can be calculated for each rotational level in a band or decay curves can be averaged for representative groups of bright state rotational levels. For a clump of eigenstates with a Lorentzian envelope, the IVR lifetime is given by $\tau_{IVR} = 1/2\pi\Delta\nu$ where $\Delta\nu$ is the full-width-at-half-maximum of the clump.

At higher energy or in larger molecules, the density of interacting states is too high to permit resolution of individual eigenstates and only an unresolved contour is observed. In this case, it is useful to employ some sort of state-selection to avoid erroneously interpreting the inhomogeneous structure from the thermal distribution of rotational quantum numbers in terms of vibrational couplings and IVR. State selection is also desirable in eigenstate-resolved experiments because it is often difficult and certainly tedious to assign the bright state quantum numbers of each feature in a fully resolved spectrum.

There are additional conditions for the meaningful interpretation of spectra in terms of IVR: (i) The nature of the bright state corresponding to a certain section of spectrum must be understood. Even in the spectral regions of the hydride stretch fundamentals and their overtones where the source of the oscillator strength is reasonably well understood, bright state preparation may also involve some excitation of other internal coordinates. (ii) Because the information needed in equation 1 lies in the intensities as well as in the frequencies, a useful experimental method must provide a measure of the linear transition strength at each frequency. The data from the author's laboratory are direct absorption spectra (8). Below, reference will also be made to infrared-laser-assisted photodissociation spectroscopy (IRLAPS) (9) which has been shown to be linear and also to optothermal detection (10) which is equivalent to an absorption signal in the cases in question.

The objective of this paper is to give the reader a perspective on the timescales involved in IVR and on the phenomena which occur on those timescales. Examples are taken from the author's work with additional references as needed to provide a perspective on the field.

Strong Low-Order Resonances: Ultrafast IVR

The strongest interactions give rise to the largest spectral splittings and therefore to the fastest energy redistribution. The 1:2 CH stretch:bend resonance in CF_3H gives rise to splittings of about 200 cm^{-1} in the overtone spectrum (11,12). These splittings are greater than the width of the rotational contour and are easily observed without the benefit of state-selection. By application of equation 1 and the procedure of ref. 6, the IVR lifetimes of the CH overtones v=3-6 are found to be 26±2 fs. The frequencies and relative intensities in the CF_3H overtone spectrum are well described by a tridiagonal Hamiltonian and the stretch-bend parameter k_{sbb}=100 cm^{-1} (12).

Another strong interaction is the 1:1 CH:OH resonance in the v_{OH}=5 region of the overtone spectrum of methanol (9,13). Here the spectrum comprises two features of comparable intensity split by 50 cm^{-1}. These splittings were also large enough to be observed without state selection, but only under jet-cooled conditions. The assignment of the splitting was derived from isotopic substitution and the systematic behavior through the overtones v=4,5,6. Halonen's internal coordinate model (L. Halonen, submitted for publication) supports the assignment and identifies the principal interaction as harmonic coupling between the OH stretch and the stretch of the CH trans to it. In the normal mode version of Halonen's model, the interaction is dominated by the (anharmonic) $K_{11;12}$ Darling-Dennison constant where 1 and 2 represent the v_1 OH and v_2 CH fundamentals respectively. The timescale is about 100 fs which is somewhat longer than in CF_3H.

A third example is the local-local coupling between CF stretches in CF_2Cl_2 overtone spectra (14). Here the bright states are the local CF stretch and its overtones. At each overtone, a clump of several features is seen in gas phase photoacoustic spectra with an average spacing of 58 cm^{-1}. The IVR lifetime is 108 fs at the CF fundamental and decreases to 72 fs by the fourth overtone. In the local mode picture, the harmonic local-local interaction parameter is 29 cm^{-1} and contains contributions from both the F- and G-matrix elements. The features within each clump can be identified as normal mode overtones and combinations. Therefore, if one were to choose one of these normal mode feature states as the bright state, the relevant IVR lifetime would be much longer and would correspond to the homogeneous width of the feature, which is obscured in the photoacoustic spectra by the inhomogeneous rotational contour.

The above are examples of strong anharmonic and harmonic vibrational interactions that lead to large splittings and energy redistribution on a subpicosecond timescale. However only a very limited volume of phase space is coupled to the bright state on this timescale. In each case, the vibrational amplitude is constrained to remain in the specific modes directly involved in the strong interaction until additional (and weaker) couplings come into play on a longer timescale. The pattern of widely spaced levels produced by such strong low-order interactions is sometimes called a polyad. For the purposes of this paper, we will refer to the timescale of the ultrafast IVR resulting from the strongest vibrational resonances as τ_1.

There are only few known kinds of interactions of such strength and these only occur where the relevant zeroth order states are in resonance. Thus there are many molecular systems where the fastest observed IVR process is orders of magnitude slower than indicated here.

In the methanol $v_{OH}=5$ region, one of the two features that result from the CH-OH interaction is further split into several features spread over about 20 cm^{-1}. (See Figures 5 and 6 of ref. 13.) The full-width-at-half-maximum of this secondary polyad is about 5 cm^{-1} and corresponds to a 1 ps timescale which we call τ_2. The zeroth-order states responsible have not yet been identified, but additional low-order anharmonic interactions are probably involved. These additional interactions serve to extend the coupling to new degrees of freedom, but the volume of phase space explored is still limited. The timescale τ_2 is not fundamentally different from τ_1, but the weaker interactions involved result in slower energy redistribution and are rather more difficult to assign. The complexity of the dispersed fluorescence and stimulated emission pumping spectra of acetylene is also the result of multiple low-order anharmonic interactions (15-17). Kellman's polyad treatment of the acetylene spectra (18) includes conserved quantum numbers which reflects the limited volume of phase space explored by these interactions.

Vibrational Randomization: Slower IVR

The vibrational randomization stage of IVR is illustrated by the ethanol CH fundamental spectra shown in Figure 2 (7) and by similar 1-butyne spectra (5) recorded in the author's laboratory. The clumps of eigenstates corresponding to each bright state are spread over a few tenths of a cm^{-1} which implies IVR lifetimes of 59 ps and 270 ps for ethanol and 1-butyne respectively. In these spectra, the density of recorded lines is equal to or greater than the calculated density of vibrational states per symmetry species. A careful analysis of two CH fundamental bands of 1-butyne (4,5,19) shows that where Coriolis coupling is unimportant, the experimental level density is in good agreement with the appropriate calculated vibrational density (20). For ethanol, where Coriolis coupling of x-, y-, and z-types is active, the rotational quantum numbers K_a and K_c are not conserved and the experimental level density is larger than the calculated vibrational level density. This implies that the energy redistribution samples the whole vibrational phase space that is quantum mechanically accessible within the interaction width Δv. In this context, these spectra demonstrate vibrational energy randomization. We will label the timescale corresponding to the onset of randomization as τ_3. From a spectral point of view, τ_3 is the last easily discernible timescale before reaching the timescale corresponding to the density of states. τ_3 is the longest timescale revealed directly by the spectrum of a bright state. The rate at which some of the other degrees of freedom reach equilibrium may be much slower than even the longest of the timescales evident from the bright state decay.

The τ_3 timescales for ethanol and 1-butyne are about three orders of magnitude longer than the τ_1 timescales discussed above. For methanol and CF_3H, τ_3 is represented by the width of the narrowest features in the overtone spectra and is on the order of 10 ps. Analysis of the large splittings (τ_1 and τ_2) involves identification of the nature of the zeroth order states involved, but this identification is much more difficult for the finer splittings corresponding to τ_3. Theoretical calculations of transition frequencies are incapable of assigning the zeroth order dark states responsible for each of the large number features separated by hundredths of a cm^{-1} or less that may appear in a spectrum. Furthermore, there is evidence that the zeroth order bath states are substantially mixed with each other. This means that each eigenstate cannot be identified with any particular zeroth order dark state but is a complicated linear combination involving a great many zeroth order dark states in any of the usual vibrational basis sets. Accordingly, one is led to develop statistical measures to characterize the eigenstate spectra and models that incorporate statistical ideas. Useful statistical measures of an eigenstate spectrum include the IVR lifetime τ_{IVR} (6), the interaction width Δv (20), the dilution factor ϕ_d (20,21), and the effective density of coupled levels (20,22).

τ_3 is the timescale most readily studied by eigenstate-resolved infrared spectroscopy, and therefore a substantial body of reliable data is available (6,23,24). A number of generally applicable concepts have been proposed to explain the observed timescales which range from about 10 ps to longer than 1 ns. We will not review all of the emerging concepts (6,23,25-27) here, but one example (24) will serve as an illustration. Figure 3 is a plot of the IVR lifetime of hydride stretches in several molecules where the excited bond is adjacent to bond about which internal rotation is possible. The data show a clear correlation of τ_{IVR} with the torsional barrier height. The more flexible molecules, which have a lower torsional barrier, exhibit faster IVR. The acceleration of IVR in flexible molecules can be rationalized in terms of the extreme anharmonicity of low-barrier torsional motion that enables direct high order coupling to adjacent bond coordinates. The density of low order resonances has been a useful concept in accounting for IVR rates in more rigid systems (28-30).

The ethanol spectra (Figure 2) illustrate the contributions of multiple coupling mechanisms to τ_3. The splitting at $J=0$ can only arise from anharmonic interactions. The two principal features at $J=0$ are the A and E states that arise from torsional tunneling, but they would be superimposed in this figure in the absence of anharmonic perturbations. The increase in the number of coupled states at $J=3$, $K=0$ is evidence for x/y-type Coriolis coupling to the bath. A further increase in the number of coupled states and in the interaction width for $K=1$ arises from z-type Coriolis coupling of the bright state to the bath. The observed density of coupled levels, which is greater than the vibrational density but less than the rotation-vibration density, provides a measure of the x/y-type Coriolis coupling among the bath states. The effects of these coupling mechanisms have been modelled and the strength of each estimated by means of random matrix calculations.(31,32)

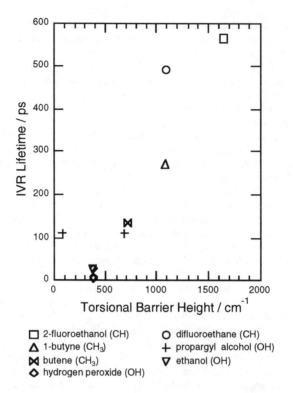

Figure 3. IVR lifetimes for hydride stretch vibrations which are adjacent to a torsional bond with the indicated barrier to internal rotation. (Reproduced with permission from ref. 24. Copyright 1995 VCH Verlagsgesellschaft mbH, D-69451 Weinheim.)

K-Relaxation and Long-Time Behavior

Although τ_3 is the longest relaxation timescale that can be derived from the spectrum through the use of equation 1, many of the zeroth order bath states will require longer to achieve their long-time average probabilities. Timescales on which such "acceptor" states are populated are referred to as τ_4. A class of acceptor states about which indirect information exists is the rovibrational states that have different K-rotational quantum numbers from the bright state. The question of K-relaxation has an impact on how theories of unimolecular reaction are implemented (33). Physically, K-relaxation can be regarded as a reorientation of the molecule relative to the (conserved) total angular momentum vector.

In ethanol, Coriolis coupling is stronger than anharmonic coupling and the decay of the bright state represents the first stage of K-relaxation. The random matrix calculations mentioned above treated the vibrational parts of the problem statistically but the rotational aspects were treated dynamically. The Coriolis interactions were constrained to obey the known rotational dependence of the Coriolis effect and were matched to the rotational dependence of the experimental measures of IVR. Four root-mean-square average coupling parameters were fitted by the calculation. The parameters specified the anharmonic, Coriolis z-type and Coriolis x/y-type coupling of the bright state to the bath and also the Coriolis x/y-type coupling among the bath states. The Coriolis z-type coupling among the bath states was set equal to the value found for 1-butyne.(31) Therefore, the random matrix calculation constitutes a complete model of the collisionless rotational relaxation processes following the CH stretch excitation.

Time profiles, derived from these calculations, for several ethanol K states (32) are shown in Figure 4. These curves represent the total probability of each K state ensemble-averaged over 32 random matrices. The ensemble averaging has substantially reduced the amplitude of the quantum beats so that the average approach to the long-time limit is easily visible. A longer time is needed to populate the more distant K states and about a nanosecond is required for $K=4$ to reach its long-time limit.

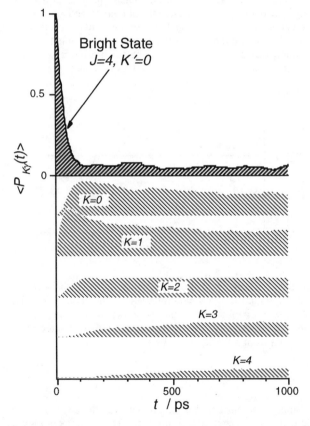

Figure 4.   Upper: Calculated survival probability of the $J=4$, $K'=0$ bright state rotational level in the ν_{14} CH stretch band of ethanol.   Below: Calculated time dependent probability of zeroth-order bath states with the indicated K-rotational quantum numbers.   All curves are ensemble averages of 32 random matrix calculations.   The asymmetry of the molecule splits levels with $K>0$ into states ($\gamma=0,1$) with different rotational symmetry. Only $\gamma=1$ states are represented in this figure. (Data from ref. 32.)

The Coriolis interactions in ethanol are relatively strong which has made rotational effects evident in the spectrum and has enabled extensive modelling with random matrix calculations. In other systems, such as 1-butyne, where the rotational effects are much weaker, one can only suppose that the time required to populate distant values of K is much longer.

The calculated distribution of K-states at long times (Figure 5) remains noticeably different from a statistical distribution. One can see from Figure 5 that the K-states most distant from the bright state have less than half of the final probability of those adjacent to the bright state. The nonstatistical distribution arises from the finite rotation-vibration density of states in the vicinity of the bright state ($\rho \approx 100$ per cm^{-1}) which mandates a maximum relaxation time constant of $\tau \approx \rho/2\pi \approx 0.5$ ns. This is approximately the relaxation time constant evident from the bottom panel of Figure 4 and at the end of 1 ns the distribution has essentially reached the final distribution shown in Figure 5. In a single spectrum, either real or simulated, quantum beats would continue to times much longer than 1 ns. However, in ensemble-averaged simulation represented in Figure 4, the quantum beats are averaged out so that only relaxation processes are seen. If the density of states were higher, relaxation could continue to longer times until a statistical distribution was attained. Therefore, in the present case, the nonstatistical distribution is "frozen in" within 1 ns by the finite density of states.

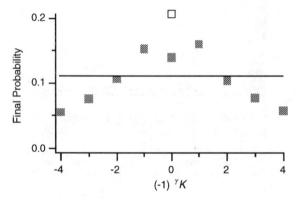

Figure 5. Calculated average probability at long times of the various K-states following the excitation of the $J=4$ $K'=0$ bright state. The open symbol includes the residual bright state probability, but the solid symbols represent probability attributable to the bath states only. For reference, the statistical outcome is indicated by a solid line. (Data from ref. 32.)

Summary

The IVR timescales considered in this paper, which range from 26 fs up to a nanosecond, have been categorized as τ_1 through τ_4. The phenomenology and coupling mechanisms found for each timescale have been explored. The definitions of the different timescales have been left as qualitative distinctions because a continuum of behavior is possible. Nonetheless, we believe that the these distinctions provide a useful context for the evaluation of new IVR data and for the consideration of the role of IVR in other phenomena.

Acknowledgements

Support for this work was provided by the Division of Chemical Sciences, Office of Basic Energy Sciences, Office of Energy Research, United States Department of Energy under grant DE-FG02-90ER14151. This support does not constitute endorsement by DOE of the views expressed in this article.

Literature Cited

1. Rice, O. K.; Ramsperger, H. C., *J. Am. Chem. Soc.* **1927**, *49*, 1617.
2. Kassel, L. S., *J. Phys. Chem.* **1928**, *32*, 225.
3. Baer, T.; Hase, W. L. *Unimolecular Reaction Dynamics Theory and Experiments*; Oxford; New York; 1996; pp. 171.
4. McIlroy, A.; Nesbitt, D. J., *J. Chem. Phys.* **1990**, *92*, 2229.
5. Bethardy, G. A.; Perry, D. S., *J. Chem. Phys.* **1993**, *98*, 6651.
6. Bethardy, G. A.; Wang, X. L.; Perry, D. S., *Can. J. Chem.* **1994**, *72*, 652.
7. Bethardy, G. A.; Perry, D. S., *J. Chem. Phys.* **1993**, *99*, 9400.
8. Perry, D. S., in *Laser Techniques in Chemistry*, Myers, A. B.; Rizzo, T. R., Eds.; Wiley; New York; 1995; Vol. XXIII, 71.
9. Settle, R. D. F.; Rizzo, T. R., *J. Chem. Phys.* **1992**, *97*, 2823.
10. Pate, B. H.; Lehmann, K. K.; Scoles, G., *J. Chem. Phys.* **1991**, *95*, 3891.
11. Dubal, H.-R.; Quack, M., *J. Chem. Phys.* **1984**, *81*, 3779.
12. Segall, J.; Zare, R. N.; Dubal, H. R.; Lewerenz, M.; Quack, M., *J. Chem. Phys.* **1987**, *86*, 634.
13. Lubich, L.; Boyarkin, O. V.; Settle, R. D. F.; Perry, D. S.; Rizzo, T. R., *Faraday Disc. Chem. Soc.* **1995**, *102*, 167.
14. Brabham, D. E.; Perry, D. S., *Chem. Phys. Lett.* **1984**, *103*, 487.
15. Yamanouchi, K.; Ikeda, N.; Tsuchiya, S.; Jonas, D. M.; Lundberg, J. K.; Adamson, G. W.; Field, R. W., *J. Chem. Phys.* **1991**, *95*, 6330.
16. Jonas, D. M.; Solina, S. A. B.; Rajaram, B.; Silbey, R. J.; Yamanouchi, K.; Tsuchiya, S., *J. Chem. Phys.* **1993**, *99*, 7350.
17. Solina, S. A. B.; O'Brien, J. P.; Field, R. W.; Polik, W. F., *J. Phys. Chem.* **1996**, *100*, 7797.
18. Kellman, M. E.; Chen, G., *J. Chem. Phys.* **1991**, *95*, 8671.
19. de Souza, A. M.; Kaur, D.; Perry, D. S., *J. Chem. Phys.* **1988**, *88*, 4569.
20. Perry, D. S., *J. Chem. Phys.* **1993**, *98*, 6665.
21. Stewart, G. M.; McDonald, J. D., *J. Chem. Phys.* **1983**, *78*, 3907.
22. McIlroy, A.; Nesbitt, D. J.; Kerstel, E. R. T.; Pate, B. H.; Lehmann, K. K.; Scoles, G., *J. Chem. Phys.* **1994**, *100*, 2596.
23. Lehmann, K. K.; Scoles, G.; Pate, B. H., in *Ann. Rev. Phys. Chem.*, Eds.; Annual Reviews, Inc.; Palo Alto; 1994; Vol. 45, 241.
24. Perry, D. S.; Bethardy, G. A.; Go, J., *Ber. Bunsenges. Phys. Chem.* **1995**, *99*, 530.
25. Nesbitt, D. J.; Field, R. W., *J. Phys. Chem.* **1996**, *100*, 12735.
26. Gambogi, J. E.; Timmermans, J. H.; Lehmann, K. K.; Scoles, G., *J. Chem. Phys.* **1993**, *99*, 9314.
27. McIlroy, A.; Nesbitt, D. J., *J. Chem. Phys.* **1994**, *101*, 3421.
28. Gambogi, J. E.; Lehmann, K. K.; Pate, B. H.; Scoles, G.; Yang, X. M., *J. Chem. Phys.* **1993**, *98*, 1748.

29. Gambogi, J. E.; L'Esperance, R. P.; Lehmann, K. K.; Pate, B. H.; Scoles, G., *J. Chem. Phys.* **1993**, *98*, 1116.
30. Stuchebrukhov, A. A.; Marcus, R. A., *J. Chem. Phys.* **1993**, *98*, 6044.
31. Go, J.; Perry, D. S., *J. Chem. Phys.* **1995**, *103*, 5194.
32. Perry, D. S.; Bethardy, G. A.; Davis, M. J.; Go, J., *Faraday Disc. Chem. Soc.* **1995**, *102*, 215.
33. Baer, T.; Hase, W. L. *Unimolecular Reaction Dynamics Theory and Experiments*; Oxford; New York; 1996; pp. 235.

Chapter 6

Intramolecular Dynamics Diffusion Theory: Nonstatistical Unimolecular Reaction Rates

Dmitrii V. Shalashilin[1] and Donald L. Thompson[2]

Department of Chemistry, Oklahoma State University,
Stillwater, OK 74078-3071

A method based on diffusion theory for calculating nonstatistical unimolecular reaction rates is described. The method, which we refer to as *intramolecular dynamics diffusion theory* (IDDT), uses short-time classical trajectory results to obtain the rate of IVR (intramolecular vibrational energy redistribution) between the reaction coordinate and the "bath" modes of the molecule in a diffusion theory formalism (*i.e.*, a classical master equation) to calculate the reaction rate. This approach, which requires much less computer time than the conventional classical trajectory method, accurately predicts the classical rates of unimolecular reactions.

The microcanonical rate coefficient $k(E,t)$ for a chemical reaction can be written in terms of the flux through the critical surface that separates reactant and products:

$$k^{dyn}(E,t) = \frac{\int_{S^*} f(p,q,t)v_\perp \, dS^*}{\int_V f(p,q,t)d\Gamma};$$

(1)

where the integral in the numerator is over the region of the critical surface S^* and that in the denominator is over the configuration space V of the reactant, $v_\perp$ is the velocity perpendicular to the critical surface, and $f(p,q,t)$ is a distribution function. (We have explicitly indicated the time dependence in the quantities since that is what is of interest to us here.) The initial microcanonical distribution in equation 1 can be written as

[1]On leave from the Institute of Chemical Physics, Russian Academy of Sciences, 117334 Moscow, Russian Federation

[2]Corresponding author

$$f_E(p,q,t=0) = f^{mc}(p,q) = \begin{cases} \dfrac{\delta(H(p,q)-E)}{\displaystyle\int_V \delta(H(p,q)-E)d\Gamma} & \text{reactant region} \\[4mm] 0 & \text{product region} \end{cases} \tag{2}$$

The time evolution of an ensemble with microcanonical initial conditions corresponding to this distribution is given by the Liouville equation:

$$\frac{\partial f}{\partial t} = -\sum_i \left(\dot{p}_i \frac{\partial f}{\partial p_i} + \dot{q}_i \frac{\partial f}{\partial q_i} \right) = \sum_i \left(\frac{\partial H}{\partial q_i} \frac{\partial f}{\partial p_i} - \frac{\partial H}{\partial p_i} \frac{\partial f}{\partial q_i} \right). \tag{3}$$

In practice, Hamilton's equations of motion for the ensemble are numerically integrated because it is difficult to solve the Liouville equation.

The lifetimes of a microcanonical ensemble of reactants computed by using classical trajectories can be fit by

$$\ln \frac{N(t)}{N(t=0)} = -k^{traj}(E)t. \tag{4}$$

The decay time is the reciprocal of the rate coefficient: $\tau^{traj} = 1/k^{traj}(E)$. The number of trajectories is usually not large enough to resolve the time dependence of $k^{dyn}(E,t)$.

For an initial microcanonical distribution, equation 1 gives

$$k^{dyn}(E,t=0) = k^{stat}(E) = \frac{\displaystyle\int_{S^*} \delta(H(p,q)-E)v_\perp dS^*}{\displaystyle\int_V \delta(H(p,q)-E)d\Gamma} \tag{5}$$

for the initial rate; that is, if the initial distribution is microcanonical the initial rate is statistical. If the reaction is not sufficiently fast to disturb the shape of the microcanonical distribution, see equation 2, then the rate will be independent of time and thus remain statistical throughout the decay process. This is the basic assumption of statistical theories. In this case the distribution given by the solution of the Liouville equation for the initial conditions in equation 2 is

$$f(p,q,t) = \frac{N(t)}{N(t=0)} f^{mc}(p,q). \tag{6}$$

Substituting equation 6 into equation 1 leads to the time-independent dynamical rate coefficient, which is equal to the statistical rate coefficient:

$$k^{dyn}(E) = k^{stat}(E). \tag{7}$$

This is the result of the rate of IVR (intramolecular vibrational energy redistribution) being fast compared to that of reaction.

If, on the other hand, reactions disturb the microcanonical distribution, the rate coefficient $k^{dyn}(E,t)$ will be time dependent. Then the trajectory rate coefficient $k^{traj}(E)$ is the time-averaged dynamical rate coefficient $<k^{dyn}(E,t)>$, and differs from the $t = 0$ dynamical rate $k^{dyn}(E,t=0)$ which is equal to the statistical rate $k^{dyn}(E,t=0)$:

$$k^{traj}(E) = <k^{dyn}(E,t)> \neq k^{dyn}(E,t=0) = k^{dyn}(E,t=0) \tag{8}$$

At high energies the reaction rate can be fast enough to deplete the distribution near the dividing surface S^*. The IVR replenishes the phase space points in the vicinity of S^*, and thus the IVR rate determines the rate of chemical reaction.

Classical trajectories are the common, practical means of computing the rates of unimolecular reactions when dynamical effects are important. They can be used to simulate the full-dimensional dynamics of complex reactions in large molecules for realistic potential energy surfaces.(*1*) Statistical rates can likewise be obtained, also for realistic potential energy surfaces, by using Monte Carlo variation transition-state theory (MCVTST).(*2-6*)   These two methods can be applied by using the same microcanonical distribution (obtained by a Monte Carlo procedure) and for exactly the same potential energy surface, thus a comparison of the results for the two calculations clearly show the non-statistical effects.   Of course, classical trajectories can theoretically be used to calculate the statistical rates. However, this is usually not feasible because of the practical problem of integrating the trajectories for sufficiently long times for total energies near threshold, where the dynamics are statistical, for reactions to occur. Since trajectory results are usually calculated for energies much higher than threshold, it is necessary to extrapolate the results to obtain rates at lower energies.   However, since the dynamical behavior can differ for the high and low energies, this is often inaccurate; the reaction dynamics can be non-statistical at high energies when the rates are limited by IVR and statistical at energies in the threshold region (which is appropriate for most experiments) where the IVR is fast compared to reaction.

The results of classical trajectory calculations or experiments are often fit to the RRK (Rice-Ramsperger-Kassel) equation

$$k^{RRK}(E) = \nu\left(1 - \frac{E^*}{E}\right)^{(s-1)}, \tag{9}$$

where ν is a frequency factor, E^* is the energy required for reaction, and s is the number of effective degrees of freedom. The RRK equation, which is based on the assumption of harmonic vibrations, is used simply as a convenient analytical expression for the energy dependence of the unimolecular rate constant.   However, the value of the parameter s obtained by fitting dynamical results to equation 9 indicates the extent that the dynamics are statistical.   Theoretically $s = 3N-6$, but for dynamical results it will be less than that if the dynamics are non-statistical. The theoretical value holds for harmonic systems for which the rate of IVR is faster than the rate of reaction so that the reaction is statistical. Since equation 9 has often been used to fit trajectory results, there are numerous results in the literature.(*7*) Comparisons of the computed s values to the theoretical values (*3N-6*) for various molecules illustrate the problem of using a statistical theory to describe dynamical reactions. The values of *s* computed from classical trajectory results are in better agreement with the theoretical (*3N-6*) values for small molecules, *e.g.*, NH_3 (*7e*) and SiH_2.(*7j*)   The agreement is good for the dissociation of silicon (*7f*) and argon (*7a*) clusters, for which the IVR is rapid because there are no significant "bottlenecks" in the phase space since all of the bonds are identical. However, for larger, more complex molecules with several different types of bonds, the agreement is not very good, *e.g.*, RDX (hexahydro-1,3,5-1,3,5-triazine).(*7o*)

Figure 1 illustrates the basic behavior of the rates. We have sketched the generally observed behavior of the statistical (solid line) and dynamical (dashed line) rate coefficients. At high energies the dynamical rate is lower than the statistical rate.

Bunker and Hase,(*8-11*) in the early 1970's, discussed the basic differences in statistical and non-statistical behaviors of unimolecular reactions. They used the term *intrinsic non-RRKM* (Rice-Ramsperger-Kassel-Marcus) to describe the behavior of reactions when the rate of IVR is slow relative to the reaction rate; that is, when the IVR determines the unimolecular reaction rate. This behavior is sometimes attributed to "bottlenecks" in the phase space that restricts the system in regions not connected to the reaction coordinate. Intrinsic non-RRKM behavior is to be to distinguished from the non-statistical behavior observed in the low-pressure limit where activation of the molecules by collisional energy transfer is rate determining.

We describe here a method based on diffusion theory (the classical version of a master equation) that can be used, with some input of short-time trajectory results, to predict the dynamical rates of unimolecular reactions. This method provides the means to determine dynamical rate coefficients at a considerable savings in computer time, and the rates can be determined at energies where it is not feasible to calculate them by the standard classical trajectory methods.   This approach also provides a rigorous formalism for interpreting intrinsic non-RRK behavior.

In this approach, which we refer to as *intramolecular dynamics diffusion theory* (IDDT), we assume that the system is comprised of a set of active modes strongly coupled to the reaction coordinate with the remaining molecular modes serving as an energy reservoir.   It is the flow of energy between these two sets of modes that controls the reaction, and the rate can thus be determined from the time evolution of the distribution of energy between them.

We have recently reported studies in which we have explored the accuracy of this method.   In the first study,(*12*) we considered simple N-N bond rupture in dimethylnitramine (DMNA) to illustrate the basic behaviors of the reaction dynamics for the statistical and nonstatistical regimes. We formulated a simple, but realistic potential-energy surface which we used in classical trajectory simulations to obtain dynamical rate coefficients and in MCVTST to calculate statistical rate coefficients as functions of energy.   The observed behavior of the statistical and dynamical rates is that illustrated in Fig. 1 (also see Figure 5 in reference 12). In that initial study, we used a qualitative estimate of the IVR rate to illustrate that the IDDT formalism predicts the general behavior of the rate as a function of energy.   In our second study,(*13*) we selected a system for which identification of the two sets of modes is clear and used it to demonstrate how short-time trajectory simulations can be used to determine the time evolution of a microcanonical distribution which is needed in the IDDT formalism to make accurate predictions of reaction rates. Because the rate of IVR can be determined from dynamics calculated on the time scale of a few femtoseconds, the method is considerably less expensive than direct classical trajectory simulations of the reactions.

In the present Article we present the formulation of the IDDT approach and show that it is applicable to the unimolecular dissociation of a molecule (DMNA) where the reaction coordinate modes are not clearly separable from the remaining (bath) modes.

Theory

General equations for harmonic modes: The motivation for using diffusion theory is the practical problem of calculating rates of unimolecular reactions in the non-statistical regime. The solution of the Liouville equation for a polyatomic molecule is not feasible, and the accurate determination of the time dependence of the dynamical rate by classical trajectory simulations requires that very large ensembles be integrated for times corresponding to the reaction time. The IDDT approach offers an accurate, tractable alternative.

The general equations are obtained by replacing the Liouville equation by a kinetic Fokker-Planck equation, which describes the dynamics as diffusion of phase space points. We will apply it in a manner somewhat akin to that used in the *diffusion theory of chemical reaction* (DTCR),(*14*) which describes the kinetics of the thermal excitation of molecules. The present application requires a change in the physical meaning of the parameters, however, the formalism remains the same.

We assume that a molecule with $s = n$ vibrational degrees of freedom can be treated as a set of harmonic oscillators which can be divided into two subsystems of n_1 and n_2 modes ($n_1 + n_2 = n$). Subsystem 1 includes the modes that make up the reaction coordinate, while subsystem 2 comprises a set of "bath" modes.

If the distribution in the entire phase space is microcanonical, then the distribution of energy in the first subsystem is easily obtained. The probability that subsystem 1 has energy E_1 is

$$F^{eq}(E_1) = \frac{\rho_1(E_1)\rho_2(E - E_1)}{\rho(E)} = \frac{E_1^{n_1-1}\left(E - E_1\right)^{n_2-1}}{E^{n-1}}\frac{[(n-1)!]}{\left[(n_1-1)!(n_2-1)!\right]} \; ; \qquad (10)$$

where

$$\rho_1(E_1) = \frac{E_1^{n_1-1}}{(n_1-1)!\prod_{i=1}^{n_1}\hbar\omega_i}, \qquad (11)$$

$$\rho_2(E - E_1) = \frac{(E - E_1)^{n_2-1}}{(n_2-1)!\prod_{i=1}^{n_2}\hbar\omega_i}, \qquad (12)$$

$$\rho(E) = \frac{E^{n-1}}{(n-1)!\prod_{i=1}^{n}\hbar\omega_i}; \qquad (13)$$

ρ_1, ρ_2, and ρ are, respectively, the densities of states for subsystems 1 and 2, and the entire system; and E is the total energy. The distribution in equation 10 has recently been analyzed by Oref.(15) The function $F^{eq}(E_1)$ can be obtained from $f^{mc}(p,q)$, equation 2, by averaging:

$$F^{eq}(E_1) = \int f^{mc}(p,q)\delta(H_1(p_1,q_1) - E_1)dpdq, \qquad (14)$$

where $H_1(p_1,q_1)$ is the Hamiltonian of subsystem 1.

Considering the molecular modes to be separated into an active mode plus a set of bath modes, the equilibrium distribution is

$$F^{eq}(E_1) = \frac{(E - E_1)^{n-2}}{E^{n-1}}(n-1) . \qquad (15)$$

If n is large and $E_1 << E$ this distribution is Boltzmann-like:

$$F^{eq}(E_1) = \exp\left\{\ln\left[(1 - E_1/E)^{n-2}\right]\right\}\left[(n-1)/E\right] \cong \exp\left(-\beta^{eff} E_1\right)\beta^{eff} . \qquad (16)$$

Thus the average energy per mode corresponds to an effective temperature: $E/n = 1/\beta^{eff} = (kT^{eff})$. The distributions in equations 15 and 16 are essentially the same. We will use the approximate form in equation 16 in the development presented here. Also, we consider the case $n_1 = 1$, however, we note that generalization to $n_1 > 1$ is not difficult.

The equation for $F(E_1,t)$ must have two properties. First, it should satisfy the principle of detailed balance and it must have the form of equation 15 or 16 at equilibrium. The first condition leads to the divergent form of the equation:

$$\frac{\partial F(E_1,t)}{\partial t} = -\frac{\partial}{\partial E_1} j(E_1,t), \qquad (17)$$

where $j(E_1,t)$ is the diffusion current, or more specifically, the flux along the E_1 axis. Equation 17 conserves the normalization of the distribution,

$$\int F(E_1,t) dE_1 = \text{constant} \qquad (18)$$

The flux $j(E_1)$ is a functional of $F(E_1,t)$, that is,

$$j(E_1,t) = \int W(E_1,\tilde{E}_1) F(\tilde{E}_1,t)d\tilde{E}_1 . \qquad (19)$$

The only assumption in the development of the Fokker-Planck equation is that the kernel W is "local," that is, only energies $\tilde{E}_1$ close to E_1 contribute to the integral in equation 19; higher order terms can thus be neglected. Thus, $F(\tilde{E}_1, t)$ can be expanded to first order in $\Delta E_1 = (\tilde{E}_1 - E_1)$ around E_1, that is,

$$F(\tilde{E}_1, t) \cong F(E_1, t) + \frac{\partial F(E_1, t)}{\partial E_1}(\tilde{E}_1 - E_1). \tag{20}$$

Then, equation 19 is

$$j(E_1, t) = \int W(E_1, E_1 + \Delta E_1) \, F(E_1 + \Delta E_1, t) \, d\Delta E_1$$

$$\cong F(E_1, t) \int W(E_1, E_1 + \Delta E_1) \, d\Delta E_1 + \frac{\partial F(E_1, t)}{\partial E_1} \int W(E_1, E_1 + \Delta E_1) \, \Delta E_1 \, d\Delta E_1$$

$$= A(E_1)F(E_1, t) + B(E_1)\frac{\partial F(E_1, t)}{\partial E_1}. \tag{21}$$

The requirement that the distribution have the correct behavior at equilibrium means that the flux must be zero at equilibrium. This leads to the relationship $A(E_1)/B(E_1) = -(\partial F^{eq}(E_1)/\partial E_1)/F^{eq}(E_1)$ between the coefficients A and B. Introducing the notation $b(E_1) = -A(E_1)$, as in ref. 14, we obtain

$$j(E_1, t) = -b(E_1)\left[F(E_1, t) - \left(\frac{F^{eq}(E_1)}{\dfrac{\partial F^{eq}(E_1)}{\partial E_1}} \right) \frac{\partial F(E_1, t)}{\partial E_1} \right] \cong$$

$$\cong -b(E_1)\left[F(E_1, t) + \beta^{eff} \frac{\partial F(E_1, t)}{\partial E_1} \right] \tag{22}$$

for the diffusional flux.

Thus, the kinetic equation

$$\frac{\partial F(E_1, t)}{\partial t} = -\frac{\partial}{\partial E_1}\left\{ -b(E_1)\left[F(E_1, t) + \beta^{eff}\frac{\partial F(E_1, t)}{\partial E_1} \right] \right\}, \tag{23}$$

instead of the Liouville equation, can be solved. This describes the relaxation of the distribution F to the equilibrium distribution F^{eq}. Another term must be introduced to take into account the effects of chemical reaction on the distribution $F(E_1, t)$, thus the final equation, which is formally equivalent to that of DTCR,(*14*) is

$$\frac{\partial F(E_1, t)}{\partial t} = -\frac{\partial}{\partial E_1}\left\{ -b(E_1)\left[F(E_1, t) + \frac{1}{\beta^{eff}}\frac{\partial F(E_1, t)}{\partial E_1} \right] \right\} - \frac{F(E_1, t)}{\tau(E_1)}. \tag{24}$$

For $n_1 = 1$ the expression for the reaction rate constant and reaction time τ is

$$k(E_1) = \frac{1}{\tau(E_1)} = v\,\theta\!\left(E_1 - E^*\right) = v \begin{cases} 1, & E_1 - E^* > 0 \\ 0, & E_1 - E^* < 0 \end{cases}. \tag{25}$$

In a fashion similar to what is done in other applications (14,16-18) the parameter $b(E_1)$ is taken to be proportional to the mean square of the energy transferred to the reaction coordinate modes (subsystem 1) from the bath modes per unit time, that is,

$$b(E_1) = \frac{\beta^{eff}}{2}\frac{\partial}{\partial t} < \Delta E_1^2 >. \tag{26}$$

In the present case, $b(E_1)$ characterizes the rate of energy exchange between the reaction coordinate mode and the remaining modes of the molecule.

In some cases, such as a harmonic system with high vibrational temperature ($\beta^{eff} = 0$), the diffusion equation, and the parameter b, can be derived directly from Hamilton's equations.(16,17)

Nonstatistical behavior is observed when reaction is fast compared to IVR, that is,

$$\tau_{rxn} \ll \tau_{IVR}. \tag{27}$$

In this case the second term in equation 24 can be replaced by a boundary condition. Assuming that reaction takes place instantly when the energy in the active mode equals the activation energy E^* effectively introduces a boundary condition at $E_1 = E^*$, that is,

$$F^{ST}\!\left(E^*\right) = 0. \tag{28}$$

The initial distribution $F(E_1,t)$ evolves into a steady-state (SS) distribution F^{SS}, equation 18, which is determined by the condition of stationary flux,(14)

$$-b(E_1)\left[F^{SS} + (\beta^{eff})^{-1}\frac{dF^{SS}}{dE_1} \right] = j = \text{constant}; \tag{29}$$

where constant j means that the flux does not depend on the energy (but it can be time dependent). The evolution of the initial distribution $F(E_1,t{=}0)$ to the steady-state distribution F^{SS} is illustrated in Figure 2 (also, see Figure 7 in reference 12), where we have sketched the behavior for the limit of fast reaction. The solution of equation 29 which satisfies the boundary condition equation 28 is

$$F^{SS} = j(t)\exp(-\beta^{eff} E_1) \int_{E_1}^{E^*} \frac{\exp(\beta^{eff}\,\tilde{E}_1)}{b(\tilde{E}_1)}\,\beta^{eff}\,d\tilde{E}_1. \tag{30}$$

Substituting this into equation 24 and integrating over energy, gives

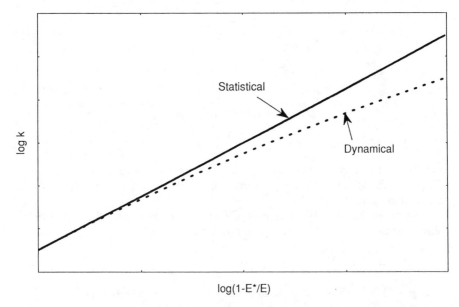

Figure 1. An illustration of the divergence of statistical (solid line) and dynamical (dashed curve) unimolecular rate coefficients as functions of energy. The rate is statistical at energies approaching threshold, but the rate becomes IVR-limited as the energy increases and the dynamical rate is lower than the statistical rate.

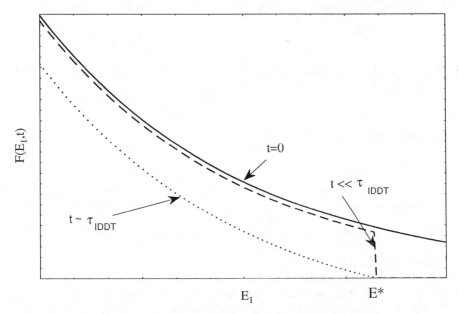

Figure 2. An illustration of the evolution of a microcanonical distribution in the limit of fast reaction. The three curves correspond to three times: the solid curve is for $t = 0$, the dashed curve for a time short relative to τ_{IVR}, and the dashed curve for $t \sim \tau_{IVR}$. This behavior is discussed in refs. 12 and 13.

$$\frac{dj}{dt} = -\frac{j}{\tau_{IDDR}} \quad ; \tag{31}$$

where

$$\tau_{IDDT} = \int_0^{E^*} \exp(-\beta^{eff} E_1) \left\{ \int_{E_1}^{E^*} \frac{\exp(\beta^{eff} \tilde{E}_1)}{b(\tilde{E}_1)} \beta^{eff} d\tilde{E}_1 \right\} dE_1. \tag{32}$$

thus, the reaction time is determined by the parameter b which depends on the rate of energy transfer between the bath and reaction coordinate. It is the only nontrivial parameter in the IDDT formalism.

Anharmonic Modes. We have presented elsewhere the equations for the case of anharmonic modes.(13) The equations are written in terms of

$$g(E_1,t) = F(E_1,t)/\rho_1(E_1), \tag{33}$$

where $\rho_1(E_1)$ is the density of states in subsystem 1. Equation 24 becomes

$$\frac{\partial g(E_1,t)}{\partial t} = -\frac{1}{\rho_1(E_1)} \frac{\partial}{\partial E_1} \left\{ -b(E_1)\rho_1(E_1) \left[g(E_1,t) + \beta^{eff} \frac{\partial g(E_1,t)}{\partial E_1} \right] \right\} - \frac{g(E_1,t)}{\tau(E_1)}. \tag{34}$$

In the limit of fast reaction times, using equation 24 in equation 34, we obtain

$$\tau_{IDDT} = 2 \int_0^{E^*} \rho_1(E_1) \exp(-\beta^{eff} E_1) \left\{ \int_{E_1}^{E^*} \frac{\exp(\beta^{eff} \tilde{E}_1)}{\rho_1(\tilde{E}_1) \frac{\partial \langle \Delta E^2 \rangle}{\partial t}} d\tilde{E}_1 \right\} dE_1. \tag{35}$$

The reaction rate coefficient is

$$k = 1/\tau_{IDDT}. \tag{36}$$

It is determined by the rate of IVR, which is defined quantitatively by $\frac{\partial}{\partial t} \langle \Delta E_1^2 \rangle$, equation 24. This is the significant parameter in the integral in equation 35.

Practical Application of IDDT

Equation 35 is the practical result of the theoretical formulation, since it provides a means of determining the rate of a unimolecular reaction from the rate of energy transfer to the reaction coordinate. In an earlier study,(13) we showed how classical trajectories can be used to calculate the rate of change of the $\langle \Delta E_1^2 \rangle$, i.e., $\frac{\partial}{\partial t} \langle \Delta E_1^2 \rangle$, and thus the IDDT rate of reaction. This can be done by using short-time (on the order of femtoseconds) classical trajectory simulations. Although it is necessary to perform the simulations as a function of the E_1, the computer time needed is insignificant compared to that needed to compute the rate directly using the standard classical trajectory approach.

Results

To demonstrate the accuracy of the IDDT method for predicting the dynamical rates of chemical reactions, we have carried out calculations for the dissociation of dimethylnitramine (DMNA) (the structure is shown in the Figure 3). We use a model of the reaction in which only simple bond fission occurs:

$$(CH_3)_2N\text{-}NO_2 \rightarrow (CH_3)_2N\bullet + \bullet NO_2. \tag{37}$$

The N-N bond length is the reaction coordinate, thus we use the N-N stretch (which does not correspond to a normal mode) as subsystem 1. It is described by the Hamiltonian for a Morse oscillator:

$$H_1(P_{N-N}, R_{N-N}) = \frac{1}{2\mu_{N-N}} P_{N-N}^2 + D\{\exp[-\alpha(R_{N-N} - R_{N-N}^0)] - 1\}^2. \tag{38}$$

We integrated ensembles of 1000 trajectories for 1.2 fs for total energies of 250, 300, and 400 kcal/mol, and computed the rate of energy flow by monitoring the energy in the reaction coordinate, using equation 38, at fixed intervals of ΔE_1. Additional initial conditions were imposed for the energy of the N-N atom pair to be within the interval E_1 to $E_1 + \Delta E_1$. Figure 4 shows how an initially narrow distribution spreads as a function of time. The rate of this spreading is described by $\frac{\partial}{\partial t}\langle \Delta E_1^2 \rangle$. Figure 5 shows the dependence of $\frac{\partial}{\partial t}\langle \Delta E_1^2 \rangle$ on E_1.

To calculate the integral in equation 35 we fit the results in Figure 5 to the function

$$\frac{\partial}{\partial t}\langle \Delta E_1^2 \rangle = 10^{16}\left[a + b\tanh\left(\frac{E_1}{c}\right) \right] \tag{39}$$

The dependence of $\frac{\partial}{\partial t}\langle \Delta E_1^2 \rangle$ on E_1, the energy in the N-N stretch, is shown in Figure 4 total energies $E = 250$, 300, and 400 kcal/mol. The dependence on the total energy is negligible; the parameters $a = 0.07$ kcal^2/s, $b = 0.3$ kcal^2/s, and $c = 10$ kcal/mol provide a reasonable fit for all three energies.

The IDDT results (triangles) are compared in Figure 6 with the dynamical (circles), statistical (MCVTST) (squares), and RRK (straight line) results reported earlier.(*12*) The rates calculated by using IDDT are quite close to the results obtained from a classical trajectory simulation,(*12*) in fact, they are in quantitative agreement. And, as expected, the dynamical results are lower than the statistical rates. The IDDT method accurately predicts the nonstatistical dynamical effects.

It is important to note that the trajectory calculations to determine the IVR rate require very short integration times (on the order of femtoseconds) and thus, even

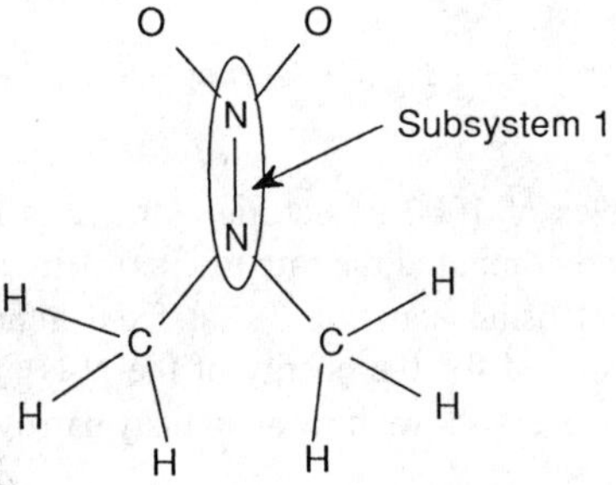

Figure 3. Dimethylnitramine (DMNA). The active mode, subsystem 1, is the N-N atom pair.

though it is necessary to perform calculations for the range of energies from zero to the dissociation energy, the savings in computer time compared to that required for a standard classical trajectory calculation of the rate coefficient is significant.   The same amount of computer time is required to calculate the rates for any total energy, while the cost in computer time for classical trajectory simulations drastically increases as the total energy decreases.   The standard classical trajectory may not even be feasible at energies at the lower end of the dynamical regime, while the IDDT method can be easily used to compute the rates there.   However, at sufficiently low energies where the reaction is slow and statistical, the IDDT approach, which is based on the assumption that reaction is fast compared to IVR, is not valid.   Then, the rate must be calculated by using a statistical theory, *e.g.*, MCVTST.   The lower of the rates calculated by MCVTST and IDDT is to be taken as the actual rate of reaction.   We have discussed elsewhere (*12*) how to use a simple interpolation between the two rates; briefly, one can take the actual rate to be

$$k = \frac{k^{IDDT} k^{MCVTST}}{k^{IDDT} + k^{MCVTST}}.$$ (40)

In another study,(*13*) we applied IDDT to the reaction

$$H_3Si\text{-}SiH_3 \rightarrow 2 \ SiH_3,$$ (41)

for which the Si-Si stretch can be easily identified as the reaction coordinate.   Due to the difference in masses of Si and H the Si-Si stretch is a good normal mode, thus the flow of energy between it and the other molecular modes which comprise the energy reservoir is slow.   It is the kind of situation where it is expected that diffusion theory should provide a good description of the reaction dynamics.   However, the accuracy of the IDDT approach for the decomposition of DMNA, equation 27, is less obvious; the N-N stretch is not an isolated mode.   One of the purposes of the present study was to see if the IDDT approach can be used for reactions where there is strong coupling of the reaction coordinate to the bath modes.

The IDDT is a good approximation to the exact classical mechanics when (*15*)

$$\frac{1}{\nu} \frac{\partial \Delta E^2}{\partial t} (\beta^{eff})^2 \ll 1,$$ (42)

where ν is the frequency of dissociating bond.   The condition in equation 42 is obviously not a strong one.   It is expected to hold for the reaction in equation 41 where the Si-Si bond corresponds to an isolated, normal mode, however, we find that it also is valid for the reaction in equation 37 where the reaction coordinate does not correspond to a normal mode.   Also, the condition in equation 41 is violated as the energy approaches the dissociation energy D where $\nu\rightarrow0$, however, we find that this is not a problem in the integration in of equation 35.

A pragmatic conclusion of this study is that one can, in general, simply use the stretching coordinate of the breaking bond as the active mode (i.e., subsystem 1 in our notation) in general.   It is not necessary to apply IDDT only in cases where the reaction coordinate is an isolated mode. The success in the present application suggests that the IDDT method may be generally applicable.

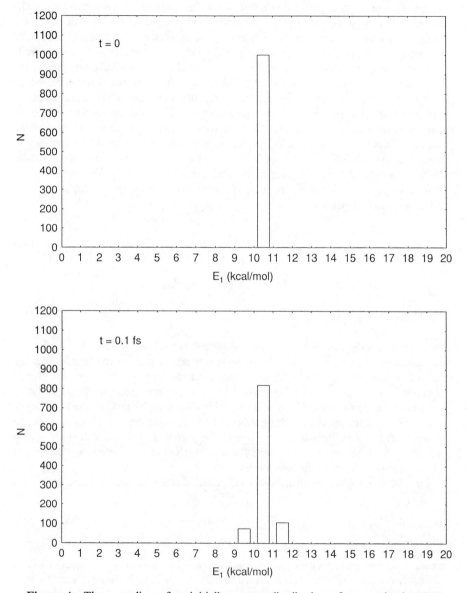

Figure 4. The spreading of an initially narrow distribution of energy in the N-N stretch for an ensemble of 1000 trajectories. The total energy for all the trajectories was E = 400 kcal/mol. The initial conditions for the trajectories were obtained from a Metropolis Monte Carlo random walk with the additional condition that 10 kcal/mol$<E_1<$11 kcal/mol. The rate of the spreading is characterized by $\frac{\partial}{\partial t}\langle \Delta E_1^2 \rangle$.

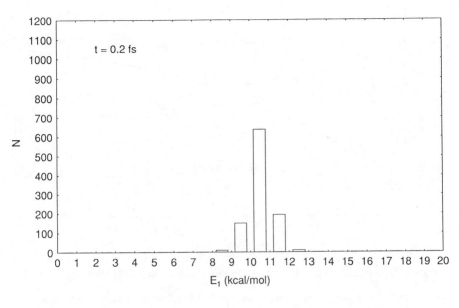

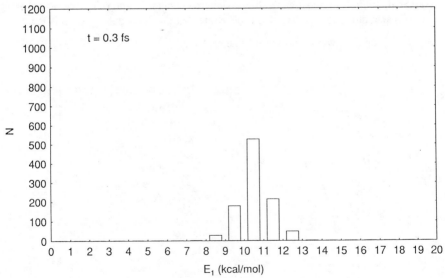

Figure 4. Continued.

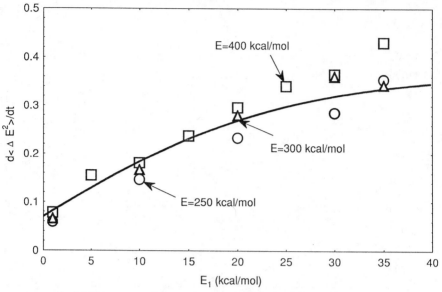

Figure 5. $\frac{\partial}{\partial t}\langle \Delta E_1^2 \rangle$ versus E_1, the energy in the N-N stretching motion, for total energies 400 (squares), 300 (triangles), and 250 kcal/mol (circles). The results for 400 kcal/mol are the most accurate. While the results for the other two energies are less accurate, they are included to illustrate that the dependence of the IVR rate $\frac{\partial}{\partial t}\langle \Delta E_1^2 \rangle$ on E_1 is weak, and thus can be neglected.

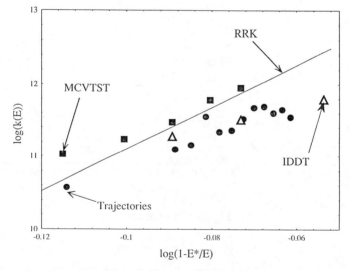

Figure 6. Comparison of the IDDT rates (triangles) with the trajectory (circles), MCVTST rates (squares), and RRK results from reference 13.

Conclusions

We have described a method for calculating nonstatistical rates of unimolecular reactions by using diffusion theory. This method, which we refer to as *intramolecular dynamics diffusion theory* (IDDT), can be used to accurately predict the rates by using the rates of IVR determined by short-time (on the order of femtoseconds) classical trajectory simulations.   The calculations require much less computer time than do standard classical trajectory simulations of unimolecular reactions.

Acknowledgment

This work was supported by the U. S. Army Research Office.

Literature cited

1.   For a recent review, see: Sewell, T. D.; Thompson, D. L. Classical Trajectory Methods for Polyatomic Molecules, *Int. J. Mod. Phys.* B, in press.

2.   Doll, J. D. *J. Chem. Phys.* **1980**, *73*, 2760.

3.   Doll, J. D. *J. Chem. Phys.* **1981**, *74*, 1074.

4.   Viswanathan, R.; Raff, L. M.; Thompson, D. L. *J. Chem. Phys.* **1984**, *81*, 828.

5.   Viswanathan, R.; Raff, L. M.; Thompson, D. L. *J. Chem. Phys.* **1984**, *81*, 3118.

6.   Shalashilin, D. V.; Thompson, D. L., *J. Phys. Chem.*, **1996**, *XX*, XXXX. (in press.)

7.   Some examples are: (a) Brady, J. W.; Doll, J. D.; Thompson, D. L. *J. Chem. Phys.* **1980**, *73*, 2767; (b) Brady, J. W.; Doll, J. D.; Thompson, D. L. *J. Chem. Phys.* **1981**, *74*, 1026; (c) Viswanathan, R.; Raff, L. M.; Thompson, D. L. *J. Chem. Phys.* **1984**, *80*, 4230; (d) NoorBatcha, I.; Raff, L. M.; Thompson, D. L. *J. Chem. Phys.* **1986**, *84*, 4341; (e) Rice, B. M.; Raff, L. M.; D. L. Thompson, *J. Chem. Phys.* **1986**, *85*, 4392; (f) Gai, H.; Thompson, D. L.; Raff, L. M. *J. Chem. Phys.* **1988**, *88*, 156; (g) Agrawal, P. M.; Thompson, D. L.; Raff, L. M. *J .Chem. Phys.* **1988**, *89*, 741; (h) Agrawal, P. M.; Thompson, D. L.; Raff, L. M. *J .Chem. Phys.* **1990**, *92*, 1069; (i) Sewell, T. D.; Thompson, D. L. *J. Chem. Phys.* **1990**, *93*, 4077; (j) Schranz, H. W.; Raff, L. M.; Thompson, D. L. *J. Chem. Phys.* **1991**, *94*, 4219; (k) Schranz, H. W.; Raff, L. M.; Thompson, D. L. *Chem. Phys. Letters* **1991**, *182*, 455, and references therein; (l) Sewell, T. D.; Schranz, H. W.; Thompson, D. L.; Raff, L. M. *J. Chem. Phys.* **1991**, *95*, 8089; (m) Sorescu, D. C.; Thompson, D. L.; Raff, L. M. *J. Chem. Phys.* **1994**, *101*, 3729; (n) Rice, B. M.; Adams, G. F.; Page, M.; Thompson, D. L. *J. Phys. Chem.* **1995**, *99*, 5016; (o) Chambers, C. C.; Thompson, D. L. *J. Phys. Chem.* **1995**, *99*, 15881.

8.   Bunker, D. L.; Hase, W. L. *J. Chem. Phys.* **1973**, *59*, 4673.

9.   Bunker, D. L.; Hase, W. L. *J. Chem. Phys.* **1978**, *69*, 4711.

10. Hase, W. L. In *Potential Energy Surfaces and Dynamical Calculations*, Truhlar, D. G., Ed.; Plenum Press, New York, 1981; p.1..

11. Hase, W. L. In *Dynamics of Molecular Collisions*, Miller, W. H., Ed.; Plenum Press, New York, 1976, Vol. VII B; p.121.

12. Shalashilin, D. V.; Thompson, D. L. *J. Chem. Phys.*, **1996**, *105*, 1833.

13. Shalashilin, D. V.; Thompson, D. L. *J. Chem. Phys.*, submitted.

14. Nikitin, E. E. *Theory of Elementary Atomic and Molecular Processes in Gases*; Clarendon Press, Oxford, 1974.

15. Oref, I. *Chem. Phys.*, **1994**, *187*, 163.
16. Zaslavskii, G. M.; Sagdeev, R. Z. *Introduction into Nonlinear Physics*; Nauka, Moscow, 1988.
17. Lichtenberg, A. J.; Liberman, M. A. *Regular and Chaotic Dynamics*; Springer-Verlag, New York, 1992.
18. Kolmogorov, A. N. *Uspehi Matematicheskih Nauk*, **1938**, *5*, 5 (in Russian).

Chapter 7

Photodissociation of Ozone at 276 nm by Photofragment Imaging and High-Resolution Photofragment Translational Spectroscopy

D. A. Blunt and A. G. Suits

Chemical Sciences Division, Lawrence Berkeley National Laboratory, University of California, Berkeley, CA 94720

The photodissociation of ozone at 276nm is investigated using both state resolved ion imaging and high-resolution photofragment translational spectroscopy. Ion images from both [3+1] and [2+1] resonance enhanced multiphoton ionization of the $O(^1D)$ photofragment are reported. All images show strong evidence of $O(^1D)$ orbital alignment. Photofragment translational spectroscopy time-of-flight spectra are reported for the $O_2(^1\Delta_g)$ photofragment. Total kinetic energy release distributions determined from these spectra are generally consistent with those distributions determined from imaging data. Observed angular distributions are reported for both detection methods, pointing to some unresolved questions for ozone dissociation in this wavelength region.

The power of state resolved ion imaging to reveal sometimes subtle effects in a number of chemical processes is well established (*1*). However, there is often the lack of an overall picture of the process under investigation. Without this it can be difficult to properly interpret and indeed validate the observations. This study combines in one apparatus the imaging technique with the universal detection of high-resolution photofragment translational spectroscopy (PTS) (*2,3*). Photodissociation of ozone in the Hartley band is investigated as a test case to show that results from both detection methods are consistent. In future studies where only one method is employed for a given fragment this proof of consistency will be crucial. For example, it is theoretically possible to measure a fragment's recoil angular distribution with both imaging and PTS, however, the former method is much faster in collection time and simpler in subsequent analysis. The PTS method can have a higher resolving power in the laboratory frame, especially important where there are a number of competing channels leading to the formation of a given fragment (*4*). Additionally, there are a number of species for which ionization or excited-state formation schemes are undetermined. In this case imaging cannot be used. Overall, there is great flexibility afforded in having the two detection methods in one apparatus. We will demonstrate that the two methods are consistent for the most part and where they differ explanation is given.

We chose ozone because the photodissociation dynamics in the Hartley band have been investigated by several groups (*4-7*). At 276nm the primary channels are:

$$O_3 + h\upsilon \rightarrow O_2(^1\Delta_g) + O(^1D)$$

$$\rightarrow O_2(^3\Sigma_g^-) + O(^3P)$$

where 85-90% of fragmentation leads to singlet species (5-7). Total cross-section for the process is large - approximately $5x10^{-16}cm^2$ (8). The dissociation energy for the singlet channel is 386kJ/mol while it is only 101kJ/mol for the triplet case (8). Experimentally the recoil velocities of singlet photofragments are more favorable for study. There is the possibility of resolving vibrational structure in both imaging and PTS time-of-flight spectra whereas with the faster ground state fragments this would be difficult. Vibrational structure is observed in previous PTS studies of ozone near this wavelength and also in the pioneering work of Fairchild et al. (5).

    In our experiment the molecular oxygen fragment is investigated with PTS. Higher background counts at mass-to-charge ratio 16 makes study of the $O(^1D)$ fragment more difficult, instead a newly published scheme for detecting $O(^1D)$ by [3+1] resonance enhanced multiphoton ionization (REMPI) near 276nm is employed (9). This paper presents ion images using this REMPI scheme for the first time. An existing method of detecting the $O(^1D)$ fragment, [2+1] REMPI near 205nm, is also utilized (10). It should be noted that with this discussion of different wavelengths the dissociation step is always with 276nm light.

Experimental

Vacuum Apparatus. Figure 1 illustrates the vacuum apparatus used for these experiments. A complete description will be given elsewhere, however a brief summary is presented here (Blunt, D.; Suits, A.G. unpublished data). The supersonic beam is formed in a rotatable chamber through a pulsed .024" diameter orifice (General Valve Corporation). Pressure in the source is typically 10^{-5}torr with a 860torr helium/ozone mixture. The chamber is pumped by a 600l/s high-throughput magnetically levitated turbomolecular pump (STP-H600C, Seiko Seiki). The mach disk of the expansion is attached to a 1mm diameter skimmer at a distance of 5mm and the beam passes into the main chamber which is held at 10^{-7}torr by a 1000l/s magnetically levitated turbomolecular pump (STP-1000, Seiko Seiki). Quartz windows on the main chamber admit the laser beam(s) on an axis perpendicular to both the molecular beam and detector axes.

Imaging Detector. For ion imaging experiments an electrode assembly consisting of three equispaced mesh plates is mounted directly in front of the source chamber. DC potentials of +700, +380 and 0V are applied to the plates and ions formed by laser/molecular beam interaction are accelerated into a field free tube leading toward the imaging detector. The detector is pumped by a 400l/s magnetically levitated turbomolecular pump (STP-400, Seiko Seiki) and the pressure is typically 10^{-9} torr during operation of the molecular beam.

    Ions are detected on an enhanced 40mm diameter dual microchannel plate (MCP) Chevron-type assembly (3040FM, Galileo). A voltage of -1.9kV DC is applied to the front MCP while the other is held at ground. Electron multiplication takes place and these are accelerated toward a phosphor screen biased at +4.5kV DC. A CCD camera (512x512 16-bit, Princeton Applied Research) records the phosphor image and this is transferred to computer for storage and subsequent analysis. In addition a photomultiplier tube is positioned to allow ion time-of-flight to be recorded. The ion of interest (O^+) is selectively detected by gating the front MCP from -1.4kV to -1.9kV at the O^+ ion flight time.

PTS Detector. For PTS experiments the ion lens assembly is removed and the source rotated to the desired laboratory angle. Photofragments are admitted to the electron impact bombardment ionizer via a 4mm diameter aperture in the main chamber. Ions are formed at 100eV, selectively transmitted through a quadrupole mass spectrometer, and then detected with a standard Daly ion detector (2). The detector region is pumped by two 400l/s magnetically levitated turbomolecular pumps (STP-400, Seiko Seiki) and a third pump of the same type evacuates a differential region between the main chamber and ionization region. With a liquid nitrogen cooled copper shield around the ionizer base pressures of 7×10^{-11} torr are achieved. In addition, a copper plate is mounted in front of the source chamber and maintained at 10K with a heat exchanger (HC-2, Air Products), resulting in low background count rates. The signal from the Daly ion detector is connected to a computer controlled multichannel scaler (Turbo-MCS, EG&G Ortec). Pulsing the nozzle at 40Hz allows laser-on/laser-off shot-by-shot subtraction and this greatly reduces the presence of direct beam signal at small laboratory angles *(4)*.

Ozone Generation. Ozone is generated by discharge through an oxygen flow and is subsequently trapped on silica gel at -80°C. For experiments the trap is warmed to -40°C and 800torr ultra-high purity helium is passed over the gel. All tubing is Teflon with stainless steel fittings. A total nozzle backing pressure of 860torr is achieved. Ozone is seen to remain almost constant during the course of measurements. The velocity distribution of the molecular beam is determined by laser-induced hole burning of the ozone at zero degrees. Least squares fitting of the distribution $f(v) \propto v^2 \exp[-(v-v_o)/\alpha^2]$ gives values of $v_o = 1309$m/s and $\alpha = 105$m/s.

Laser Configuration. A single laser setup is used for both the PTS and single-color imaging experiments. The second harmonic of a seeded Nd:YAG laser (GCR-5, Spectra Physics) pumps a dye laser at 10Hz (PDL-1, Quanta-Ray). With 700mJ/pulse we obtain 200mJ/pulse of 552nm light. This light is doubled with a KD*P crystal in a wavelength extender (WEX-1, Quanta-Ray). Up to 25mJ/pulse of linearly polarized light is available at 276nm. For measurement of vibrational state distributions via PTS, the light is passed through a single fresnel rhomb at 45 degrees to produce circularly polarized light. Where focusing is required (for the [3+1] REMPI) a 25cm focal length fused silica lens is inserted just prior to the beam entering the vacuum apparatus. The dye laser output wavelength is adjusted under computer control to facilitate scanning the REMPI line Doppler width during imaging experiments.

For the [2+1] REMPI measurements (at ~205nm) a two-color scheme is used. The dissociation laser pulse (276nm) is formed via the same scheme described above, however the pump laser is an unseeded Nd:YAG laser (DCR-2A, Quanta-Ray) capable of 500mJ/pulse at 532nm. Up to 10mJ/pulse of 276nm light is available for the dissociation. This is weakly focused at the interaction region with a 47.5cm focal length fused silica lens.

The 205nm light for the [2+1] REMPI is generated using the GCR-5/PDL-1 combination where the dye is Sulphorhodamine 640, giving 190mJ/pulse of 615nm light. This is passed through a double fresnel rhomb and 'UV' half-wave plate before being doubled with a BBO crystal. The remaining 615nm light is mixed with the 307nm light in another BBO crystal to produce 100µJ/pulse of vertically polarized 205nm light. A Pellin-Broca prism separates the 205nm beam and this is then tightly focused at the interaction region with a 10cm focal length fused silica lens.

The purpose of the polarizing elements prior to doubling are to produce enough red light in the correct polarization for both doubling and mixing stages. As in the single laser experiment, dye laser output wavelength is controlled by computer.

Results and Discussion

The primary concern of this paper is to compare the data obtained by ion imaging with PTS time-of-flight spectra. There are two convenient schemes for detecting $O(^1D)$ via resonance enhanced multiphoton ionization (REMPI). Firstly, two resonances near 205nm corresponding to [2+1] processes and, more recently discovered, five resonances near 276nm which are [3+1] processes. We have measured ion images at each of these lines, but for this paper we will restrict our attention to three representative data sets.

PTS. Time-of-flight spectra of the O_2 photofragment are recorded at two laboratory wavelengths (20 and 30 degrees) for each of five dissociation wavelengths near 276nm. In addition, each spectrum is recorded once using circularly polarized light and then again with linearly polarized light. From these measurements the total kinetic energy release and center-of-mass angular recoil distributions are determined. The principal reason for performing the PTS experiments at five different wavelengths is to allow comparison with single-color ion imaging data where the detection is via one of the five $O(^1D)$ [3+1] REMPI lines near 276nm. Comparison with the imaging data is discussed later in this section. A previous investigation of ozone photodissociation in this region of the Hartley band reports rapid variation of the O_2 vibrational state distribution with wavelength, however over the narrow range of wavelengths investigated here it is not possible to draw such a conclusion (7).

Results for two wavelengths using circularly polarized light are presented in Figure 2. In this experimental arrangement the use of circularly polarized light results in isotropic fragment angular recoil direction in the detector-molecular beam plane. Determining the total kinetic energy release distribution, $P(E)$, is simplified without this possible (and likely) anisotropy. The energy available after dissociation for fragment degrees of freedom is no more than 46kJ/mol. Only the lowest three vibrational levels of O_2 are accessible. Peaks corresponding to these levels are clearly resolved in the PTS spectra at 20 degrees. At a wider angle only v = 0 and v = 1 are observed since that angle is outside the maximum velocity circle of v = 2 fragments. Fitting of the data sets is performed by forward convolution of a trial $P(E)$ with experimental parameters (11). This trial distribution is refined until good agreement between experimental and calculated laboratory TOF spectra is achieved.

The partitioning of available energy into the three vibrational levels is determined from these $P(E)$ by calculating the relative contributions of each component to the time-of-flight spectra at 20 degrees. Our findings for all five wavelengths are shown in Table I. Data for each wavelength is collected over a single day. Variation of the v = 1 / v = 0 and v = 2 / v = 0 ratios with wavelength may be in part due to varying initial source conditions each day, yet as noted previously, the ozone signal varied little over the course of a given day. The average value for v = 1 / v = 0 (0.63) is higher than the 0.5 determined by Thelen et al. (7). This discrepancy is possibly due to the different source conditions employed in their experiments. Thelen et al. find a mean flow velocity of 1705m/s and a half-height width of 55m/s, suggesting that the ozone was rotationally cooler than in our setup. With a different set of initial states it is quite possible that a different O_2 vibrational state distribution would result. Average total kinetic energy release for each vibrational level is consistent with previous measurements at 266 and 283nm and is discussed later in this section (5,6).

Table I.   Vibrational State Distributions for the $O_2(^1\Delta_g)$ Photofragment

Wavelength (nm)	v = 1 / v = 0	v = 2 / v = 0
276.53	0.73	0.41
276.55	0.64	0.33
276.66	0.58	0.20
276.70	0.60	0.39
276.75	0.58	0.33

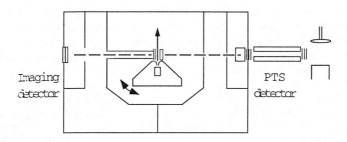

Figure 1. Experimental apparatus.

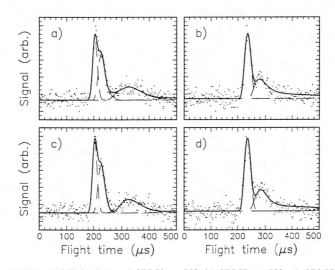

Figure 2. PTS time-of-flight data: a) 276.53nm 20°, b) 276.53nm 30°, c) 276.55nm 20°, d) 276.55nm 30°.

Product recoil angular distributions are determined by convolving the $P(E)$ from above with a trial distribution and fitting this to time-of-flight spectra recorded with linearly polarized light. The trial distribution is of the form

$$I(\theta) = N\left[1 + \beta P_2\left(\cos^2 \theta\right)\right] \qquad (1)$$

where P_2 is the second Legendre polynomial, β is the anisotropy parameter, and N the usual normalization constant. A value of 0.71(0.17) is obtained for β. This value for β is significantly lower than the 1.6(0.2) reported by Thelen et al. (7). The asymptotic limit for β assuming equilibrium geometry for ozone is 1.18 (5). Using this equilibrium geometry may be incorrect since it has been shown that bending motion is important in the ozone decay and thus a different geometry would be a better choice (4). However, since the upper state is more strongly bent than the ground state, the expected asymptotic value for β is lower than the 1.18 determined from the equilibrium geometry, rather than the higher value (1.60) observed by Thelen et al. Another possible interpretation is the role of the parent rotation, cited by Thelen et al. to account for the difference between their result and those of Fairchild et al. in an effusive beam. At any rate this remains one of the interesting unanswered questions.

Ion Imaging. Figure 3 presents the two dimensional projections of the three dimensional ion sphere for three detection wavelengths, 276.53, 276.55 and 205.47nm. This is the first report of images recorded using the [3+1] REMPI scheme near 276nm. Collection times for the [3+1] REMPI images are typically one hour since not only is the ion flux low, but the Doppler width of the transitions must also be scanned. [2+1] REMPI images were collected in just 15 minutes. However, even a one hour collection time is much less than would be required to collect PTS time-of-flight spectra over the same center-of-mass angular range.

There are three obvious features in these images and one perhaps not so clear until the data is transformed into the total kinetic energy release distribution. Three rings are visible in each image. Working outward from the center of the image these correspond to the vibrational levels v = 2, 1, and 0, respectively, of the O_2 fragment. Secondly, the observed angular distributions for the three images are remarkably different. At 276.55nm there is a striking lack of intensity at 0 and 180 degrees for the v = 0 and 1 states. The 205.47nm image shows a truncated distribution when compared to the 276nm images. Finally, there is a variation in the angular distribution with recoil velocity. The first two effects arise from alignment of the $O(^1D)$ orbital, yielding an anisotropic distribution of the REMPI transition dipole moment in the center-of-mass frame. Analysis of this orbital alignment, or v-J vector correlation, will be reported in a future publication. However, recoil velocity information is available even prior to a detailed analysis of the v-J correlation.

To extract the total kinetic energy release distribution a two dimensional slice of the three dimensional ion sphere is reconstructed from the two dimensional projection by an inverse Abel transform (12). Figure 4 presents the total $P(E)$ for two of the images and also that obtained from PTS data. The three vibrational state components are present and importantly the energies of each component are matched between PTS and imaging data. Vector correlation effects in the imaging experiment prevent direct quantitative comparison of the relative contributions of each component. Figure 5 illustrates the variation of observed angular distribution between imaging and PTS data sets. The angular distribution for PTS data is calculated using $\beta = 0.71$ and equation 1.

Figure 3. Ion images from REMPI of O(^{1}D) photofragment. From left to right, 276.53, 276.55, and 205.47nm.

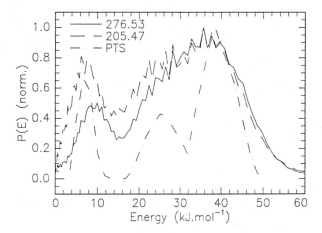

Figure 4. Total kinetic energy release distributions with detection at indicated wavelength.

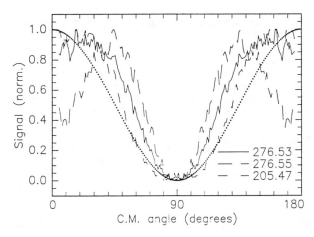

Figure 5. Photofragment recoil angular distributions with detection at indicated wavelength (calculated plot is based on PTS data - see text).

Conclusion

Whilst there are still unresolved questions about the dissociation dynamics of ozone in this wavelength region, the combined techniques of ion imaging and PTS provide a rich source of information. We have presented PTS time-of-flight spectra that are largely consistent with previous investigations. The first ion images of the $O(^1D)$ fragment detected by [3+1] REMPI near 276nm are reported. In addition, images from detection near 205nm are recorded. All images show strong evidence of orbital alignment. Total kinetic energy release distributions from both detection methods are determined and the positions of the three vibrational state components are equivalent. An angular distribution for the dissociating fragments is calculated and shown to be in general agreement with some previous measurements. Vector correlation effects are observed in all $O(^1D)$ ion images preventing the determination of an angular distribution, however, future analysis of these effects is planned and will allow the angular information to be extracted as well as providing detailed insight into nonadiabatic processes in the dissociation dynamics. Finally, we have shown the advantages of an apparatus possessing these two complementary detection methods.

Acknowledgments

This work was supported by the Director, Office of Energy Research, Office of Basic Energy Sciences, Chemical Sciences Division of the U.S. Department of Energy under Contract No. DE-AC03-76SF00098.

Literature Cited

1.   Houston, P.L. *J. Phys. Chem.* **1996**, 100, 12757.
2.   Lee, Y.T.; McDonald, J.D.; LeBreton, P.R.; Herschbach, D.R. *Rev. Sci. Instrum.* **1969**, 40, 1402.
3.   Wodtke, A.M.; Lee, Y.T. *J. Phys. Chem.* **1985**, 89, 4744.
4.   Stranges, D.; Yang, X.; Chesko, J.D.; Suits, A.G. *J. Chem. Phys.* **1995**, 102(15), 6067.
5.   Fairchild, C.E.; Stone, E.J.; Lawrence, G.M. *J. Chem. Phys.* **1978**, 69, 3632.
6.   Sparks, R.K.; Carlson, L.R.; Shobatake, K.; Kowalczyk, M.L.; Lee, Y.T. *J. Chem. Phys.* **1980**, 72(2), 1401.
7.   Thelen, M.-A.; Gejo, T.; Harrison, J.A.; Huber, J.R. *J. Chem. Phys.* **1995**, 103(18), 7946.
8.   Okabe, H. *Photochemistry of Small Molecules*; Wiley: New York, NY, 1985.
9.   Richter, R.C.; Hynes, A.J. *J. Phys. Chem.* **1996**, 100(20), 8061.
10.   Pratt, S.T.; Dehmer, P.M.; Dehmer, J.L. *Phys. Rev. A* **1991**, 43(1), 282.
11.   Zhao, X.; Nathanson, G.M.; Lee, Y.T. *Acta Physico-Chimica Sinica* **1992**, 8, 1.
12.   Hansen, E.W.; Law, P.-L. *J. Opt. Soc. Am. A* **1985**, 2(4), 510.

Chapter 8

Unimolecular Decay of Azabenzene: A Mechanistic Model Based on Pyrazine's Photoreactive Relaxation Pathways

James D. Chesko[1] and Yuan T. Lee[2]

Department of Chemistry, University of California, and Lawrence Berkeley
National Laboratory, Berkeley, CA 94720

Relaxation processes inducing chemical reactions are effective
dissipation mechanisms for azabenzenes. Photofragment translational
spectroscopy has characterized a rich variety of photodissociation
products whose evolution follows the amount of internal energy
available and the degree of nitrogen substitution. Although a diverse
range of reactivity is observed, a reaction model consistent with orbital
symmetry, radiationless transition theory and energetic constraints
suggests both ground and excited state processes, including
isomerization, ring opening and extensive rearrangement is presented.
The partitioning of translational energy into the primary reaction
product, HCN, results from several mechanisms including a retro-Diels-
Alder concerted elimination and a stepwise, biradical pathway.

Benzene and its nitrogen substituted derivatives, the azabenzenes, have historically
been a fruitful source of studies in spectroscopy, radiative and non-radiative relaxation
processes (1-6). The list of remarkable phenomena observed-- the onset of fast
radiationless processes (channel III behavior (7-8)), the role of geometric isomers in
describing characterized ultraviolet photochemistry, the novel concerted triple
dissociation of highly substituted triazine (9) and tetrazine (10) species, the large
transfer of energy from a diazine during 'supercollisions' (11)--are consequences of
dynamic energy flow through the aromatic system. Although the azabenzenes invite
comparison based upon their structural and spectroscopic similarities (12-13), the
reactive behavior they span may appear too broad to generalize. While benzene
decomposes primarily by loss of hydrogen, the more heavily substituted s-triazine and
s-tetrazine triple dissociate into three fragments such as HCN and molecular nitrogen
(see Figure 1). The diazines form a bridge midway between these two reaction

[1]Current address: Pacific Northwest National Laboratory, Battelle, MS K2–14, 902 Battelle Boulevard,
Richland, WA 99352

[2]Current address: Academica Sinica, Taipei 11529, Taiwan

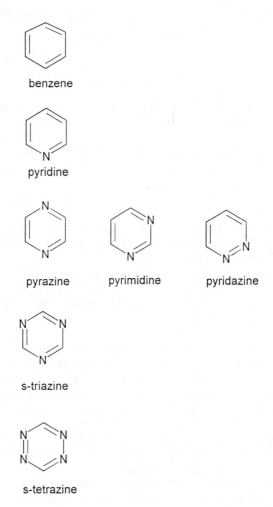

Figure 1. Benzene and the azabenzene (nitrogen derivatives) discussed in this paper.

extremes. Despite the detailed measurements of electronic structure and relaxation which have been performed on 1,4-diazine (pyrazine), little is known about its reactive character beyond a small propensity to isomerize (14) and report of triple dissociation (15) at high photon energies (λ<160 nm).

The unimolecular decay of the diazines was characterized in an attempt to bridge the wide range of behavior observed and evaluate the consequence of each reaction product, with each channel intimating portions of a reaction path followed. The sum of the reaction paths, weighted by their branching ratios, describes the molecule's reaction profile. The excess energy dependence of the paths hinted at the relaxation processes which led to their expression. From the synthesis of these parts emerged a description which proved useful across the full range of azabenzenes studied.

A detailed consideration of the mechanistic details of each path suggested a subtle interplay between accessible valence isomers, concerted versus stepwise reactions, conservation of orbital symmetry, molecular versus radical products and simple bond fission versus complex rearrangements. Ab initio methods confirmed the accessibility of key species along the relaxation pathway. Based upon differences due to induction, energetics and radiationless transition theory, the results were used to predict the reactive trend observed for the other azabenzenes through a range of excitation energies.

Unimolecular Decay Products of Azabenzenes following Ultraviolet Excitation

A dilute mixture (c. 0.1 %) of azabenzene seeded in an inert carrier gas (1.5 bar He) was expanded supersonically through a small nozzle and skimmed twice downstream to form a collimated molecular beam. Hole burning of the parent by ultraviolet photodissociation was performed to characterize the beam velocity distribution (1.7×10^3 ms^{-1} ,3% FWHM) and check for the presence of clusters. An additional test for clusters involved heating the expansion nozzle to 200 °C , although it was found that velocity spread and long term stability of the dilute mixture was superior to the heated nozzle. A molecular beam apparatus described earlier (16) was used in two distinct configurations, the first being for detecting relatively heavy fragments at angles 5 to 45 degrees from the incident beam direction and the second at right angles to an unskimmed expansion. After intersecting the dissociating laser the recoiling fragments with suitable resultant vectors pass through defining apertures, undergo electron impact ionization and quadrupole mass filtration before impacting a Daly detector and inducing a shower of electrons which strike a scintillator and are detected by a photomultiplier tube.

Polarization dependence of the diazine and triazine ultraviolet photoproducts was measured by rotating a stack of 10 quartz plates at Brewster's angle to provide 95% polarized light oriented at various angles between the detection axis and electric field vector. The anisotropy of the fast HCN molecules and H atoms was found to be approximately 8±10%, or isotropic within the uncertainty of the measurement process. This implies that photo-excited complex survives longer than a rotational period and decomposes on the time scale of at least several nanoseconds. The polarization dependence of HCN elimination is similar to that reported for s-triazine (17), but differs from the behavior observed for nitrogen elimination from s-tetrazine (10), which shows a strong anisotropy.

Pyrazine's Photochemical Fate.

Elimination of HCN was the predominant reaction process. Surprisingly, it occurred following up to four distinct pathways (i.e. three different co-products plus the contribution from triple dissociation). Other minor reaction channels included the elimination of both atomic and molecular hydrogen as well as the production of methylene nitrile radical pairs. The intensity dependence of the products was tested over the range of 2 mJ/cm^2 to 2000 mJ/cm^2. With laser intensities over 100 mJ/cm^2 multiphoton effects were readily observable. Figure 2 shows a time-of-flight spectrum for m/e=27 (HCN) and its momentum matched partner C_3H_3N (m/e=53) following 248 nm excitation. Three distinct peaks corresponding to three channels are observed. At higher photon energy (193 nm, 6.5 eV) the two translationally fast peaks at m/e=53 will disappear but a single peak at m/e=52 that momentum matches the fast HCN product persists. Loss of a proton is required to stabilize the ionic species at m/e=53 when its neutral precursor is carrying greater internal energy. At higher laser intensity (> 100 mJ/cm^2) the fast m/e=53 peak will dissociate into translationally fast fragments of HCN and C_2H_2. Because the species at m/e=53 cracks to daughter fragments primarily at m/e=26 we were unable to rule out the possibility of triple dissociation, except to note that all the HCN fragments could be accounted for by two body dissociations.

Concerted HCN Elimination. Accounting for the multiplicity in reaction channels required detailed consideration of the mechanisms involved. The large amount of translational energy released suggested a concerted process with a repulsive exit barrier that would bestow some stability to the C_3H_3N fragment. A retro-Diels-Alder reaction could be responsible, with the π bonds of HCN acting as the dienophile and C_3H_3N playing the role of the diene. Because of the high degree of unsaturation in this diene and the requirement of a cisoid configuration, a cyclic transition state structure would be preferable, for cis-fused cyclic dienes are very reactive in this manner, especially for smaller rings where the bonding distances of the diene ends are convenient. The dipolar character of HCN and inductive (electron withdrawing) effect of the nitrogen will both enhance the reactivity. The process is formally a $[_\sigma 4_s + _\pi 2_s]$ cycloreversion which is thermally allowed by orbital symmetry correlations (see Fig. 3 and (19)). The mechanism of Diels-Alder reactions, namely the simultaneous making (or breaking) of bond pairs, directly addresses the topological difficulty of extracting two atoms (carbon and nitrogen) intrinsic to a ring (i.e. disrupting at least two bonds) in a concerted manner.

Arriving at a configuration conducive to cycloreversion can be achieved by examining the various valence isomers of the aromatic ring and considering how their structure and electronic configuration satisfy demands imposed by the reaction path. Of interest is the cis-bicyclo[2.2.0]-2,5-diene, also known as the 'Dewar' isomer. The conversion of the familiar, planar, aromatic structure to the Dewar form is a photo-allowed $[_\pi 2_s + _\pi 2_s]$ (disrotary) process. Once formed, the Dewar isomer cannot

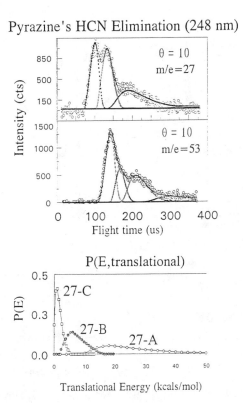

Figure 2. Time-of-flight spectra taken for the photodissociation reaction of pyrazine into HCN and C_3H_3N. Three distinct translational energy distributions are seen and referred to as channels 27-A, 27-B and 27-C in order of decreasing average translational energy release.

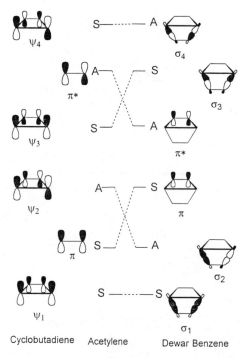

Figure 3. A orbital correlation diagram for a [$_\sigma 4_S$ + $_\pi 2_S$] retro-Diels-Alder cycloreversion, formally symmetry allowed in the ground state. The reverse reaction, a [$_\pi 4_S$ + $_\pi 2_S$] cycloaddition, shares the same correlation by the principle of microscopic reversibility.

readily convert to the ground state form because the thermally allowed conrotatory ring opening results in a highly strained cis,cis,trans triene.

With the orbital symmetry correlations well established, the question of energetic accessibility became an issue. Computer ab initio methods are well-suited to answering this question, so MP2/6-311+G* calculations were run with Gaussian 92 and are summarized in Figure 4. The results confirm that the energy of the Dewar isomers decrease with increasing nitrogen substitution. Furthermore, the opposite trend is observed for the energy of the azete (nitrogen substituted cyclobutadiene) relative to the dissociation products HCN, acetylene and molecular nitrogen. Herein lies the thermodynamic reasons underlying the diverse range of reactive behavior: elimination of the 'first' HCN (or N_2, but not C_2H_2) can occur through a concerted [4+2] elimination, but the remaining fragment must follow a [2+2] or biradical pathway to further dissociate. Thus when HCN is eliminated from pyridine the cyclobutadiene co-fragment must undergo an energetically unfavorable endothermic reaction to 'triple dissociate' into two additional acetylenes. Experimentally, this second dissocation is not seen (22). For pyrazine (or pyrimidine), the monoazete intermediate is still energetically favored over separated acetylene and HCN fragments and the orbital symmetry prohibition to dissociation provides enough of an effective barrier to greatly limit the amount of secondary dissociation. Alternatively, a [1,3] hydrogen migration can occur with ring opening to form the more stable acrylonitrile species. For s-triazine, the diazete intermediate is not energetically favored over two HCN molecules and can further dissociate, perhaps via a $[_\sigma2_s + _\sigma2_a]$ concerted process. This mechanism strongly suggests that s-triazine and perhaps s-tetrazine undergo sequential dissociation steps. Recent experiments on s-triazine (17) have provided substantial evidence that stepwise elimination occurs, contrary to earlier reports (9). The large amount of bending excitation found in the HCN (21) is predicted by the large change in the C-H bond angle in going from Dewar compound to linear HCN. The case for s-tetrazine is weaker due to the absence of acetylene product (even from multiphoton processes), although one may note that the original theoretical work describing the triple dissociation (23) did not consider the possible influence of a Dewar intermediate and only a very weak polarization dependence was observed for the HCN product (10), a possible consequence of either the dissociation geometry or perhaps lifetime of a diazete intermediate. On the other end of the spectrum, elimination of acetylene from benzene is not as favored due to the greater endothermicity of the process and the unsubstituted groups that would be involved in the process.

Stepwise elimination of HCN through a biradical intermediate.   The small amount of translational energy released in the slowest HCN produced is consistent with a stepwise elimination involving a biradical.   One may compare the reaction to elimination of nitrogen from a cyclic azo compound, a facile process in which a biradical intermediate has been suggested (24). As one would predict, the diazine expressing this reaction channel the strongest is pyridazine (see Fig. 1). Production of translationally slow HCN from triazine (9,28) following 193 nm excitation has been observed and would be expected if a biradical intermediate were the initial product following ring opening.

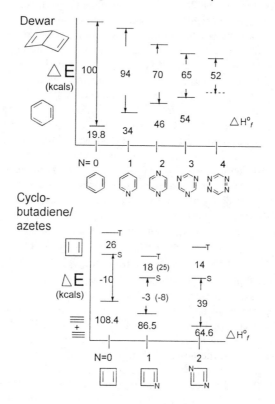

Energies of Relevant Species

Aromatic hydrocarbons taken from J.A. Miller and C.F.
Melius, Combustion and Flame, 91:21-39 (1992)
S-T splittings from J. Michl et al., JACS, 111:6140-6146
Heats of formation from JANAF tables,
Azete energies MP2/6-311+G*
Numbers in parentheses from experiment

Figure 4. Some important valence isomers of benzene. Understanding the photochemistry of benzene has led to a recognition of the central role these species play in transformations such as exchange and isomerization processes. The processes which interconvert these various forms strong reflect the principle of conservation of orbital symmetry.

Simple bond rupture and H atom elimination.   Simple bond fission forming an H atom and a pyrazyl radical became an important pathway, especially at higher excitation energy. With typical pre-exponential (A) factors of 10^{14}, this reaction was expected to dominate at higher internal energies, especially since the strength of the C-H bond is likely to decrease due to nitrogen's inductive effects. While this assertion holds for benzene and to a lesser extent pyridine, for the diazines and especially s-triazine it is not followed. Although it may appear more facile to simply break one bond exocyclic to the aromatic ring system, the reactive pathways followed by the heavily substituted azabenzenes clearly suggest that the perturbation introduced by the nitrogens is drastic with respect to the reactive behavior, even though the spectroscopic transition energies and molecular geometries are quite similar. The heats of formation of the pyrazyl and triazyl radicals can be estimated from the maximum translational energy release of the ejected H atom:

$$\Delta H_f^o(pyrazyl) = \Delta H_f^o(pyrazine) + \hbar\omega - \Delta H_f^o(H,{}^2S_{1/2}) - E_{t,\max} - E_{\text{int}} \qquad (1)$$

The standard enthalpies of formation for pyrazine and atomic hydrogen are well known, as well as the photon energy. If we assume that a negligible amount of internal energy was present in the molecular beam expansion before photoexcitation and that the fastest products recorded had very little internal excitation, a heat of formation for the radical species may be determined This measurement was difficult due to the presence of multiphoton effects, when combined with modest quantum yields and the background of fast HCN (fragmenting to H+), made this value uncertain to about 10 kcal/mol. The results are summarized in Table I. The pyrazyl radical does not survive electron bombardment very well, but its daughter fragments (m/e=52,40,26) could be traced, especially m/e=40 which had very small background levels.

Table I. Energetics of C-H Bond Fission

Species	Ab initio Dissociation Energy (kcals/mol)	Measured Dissociation Energy (kcals/mol)
Benzene	114.6	115.2
Pyridine	107	101±10
Pyrazine	104	98±10
Triazine	102	96±10

Formation of Methylene Nitrile Radical Pairs.   The ring opening process which is usually followed by elimination of HCN can follow a second path, namely dissociation into a pair of methylene nitrile radicals. Time-of-flight of spectra for this fragment are shown in Figure 5. Of interest is the speculation by Woodward and Hoffman (19) that the conversion of the Dewar isomer may occur through a biradical pathway due to symmetry prohibition of a concerted pathway. For example, the bridging bond breaks and is followed by a twisted boat distortion, thus creating two allyl radical systems which then separate. An upper limit for the standard enthalpy of formation can be

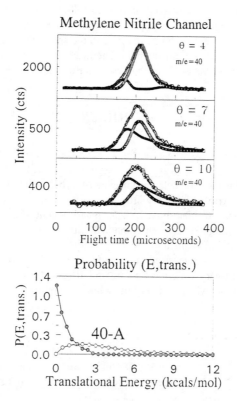

Figure 5. Ab initio results verifying the energetic accessibility of several key reaction intermediates and products in azabenzene unimolecular decay. The number of nitrogens present in each cyclic species markedly influences the propensity to follow reactive (dissociative) pathways.

estimated at 74±2 kcal/mole, about 8 kcal/mole less than that measured for the propargyl radical (25). This process likely occurs in a stepwise manner from the vibrationally excited open chain biradical.

Methyl Radical Elimination. The loss of a methyl radical which has been reported as a minor channel for benzene (26) and pyridine (22) was only found for the diazine pyrimidine, suggesting that three contiguous carbons (CH groups) are necessary to supply the necessary hydrogen atoms. The existence of this channel is strong evidence for the process of ring opening, for the extent of H atom migration (sigmatropic rearrangements) is difficult to rationalize within a closed ring. In the case of pyridine and pyrimidine the 'initiation' step along this pathway, ring opening to a biradical, can be promoted by immediate formation of an energetically stable nitrile group coupled with a hydrogen [1,2] or [1,3] shift. Although this channel is quite minor, it reflects important similarities in the dynamics on the vibrationally hot S_0 manifold.

Comparison of Azabenzene Reactivity as a function of Excitation Energy.

The branching ratios of the various product channels reflect the propensity of the relaxation processes down the available reaction paths. Table II summarizes the variation in product expression at 5.0 and 6.5 eV of excitation energy while Figure 6 describes the various paths. Two general features should be noted. First of all, at the

Table II. Branching Ratios for Unimolecular decay of Azabenzenes following 5.0 and 6.5 eV of photoexcitation

Molecule(ref)	Energy	H	H_2	CH_3	HCN	CH_2CN
Benzene(26)	5.0 eV	0.00	0.96	0.04	0.00	0.00
	6.5	0.80	0.16	0.04	0.00	0.00
Pyridine(27)	5.0	0.05	0.00	0.01	0.80	0.00
(22,27)	6.5	0.75	0.00	0.02	0.15	0.02
Pyrazine(28)	5.0	0.08	0.02	<0.01	0.85	0.05
	6.5	0.24	0.01	<0.01	0.66	0.09
s-triazine(9,17)	5.0	0.00	0.00	0.00	1.00	0.00
	6.5	0.05	0.00	0.00	0.95	0.00
s-tetrazine(10)	5.0	0.00	0.00	0.00	1.00	0.00

higher excitation energy the processes with the higher A factors (such as simple bond rupture) become more pronounced since the excess energy present is considerably greater than the required activation energy and hence the pre-exponential term in the Arrhenius expression will largely control the overall rate. The growing presence of H atom elimination reflects this situation. The observation that elimination of HCN from the hot ground state is very competitive with this process suggests that for the diazines, triazines and tetrazines the bond breaking process of opening the ring has a comparable A factor and smaller activation energy. The decrease in the concerted elimination

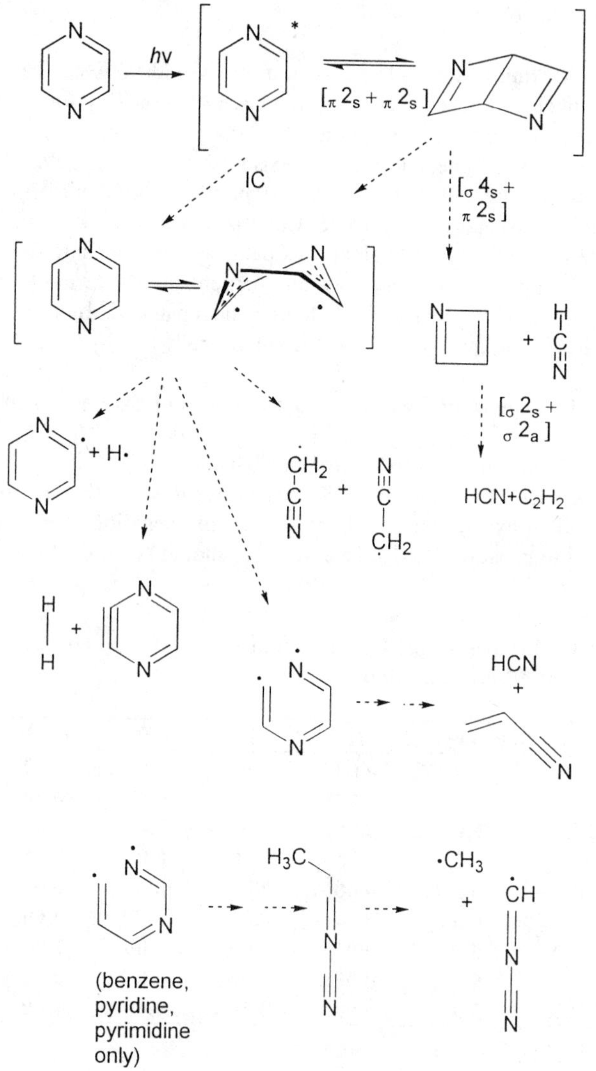

Figure 6. Time-of-flight spectrum taken for the elimination of molecular deuterium from pyrazine. The bottom trace represents the time-of-flight for pyrazyne, the co-product of the reaction. The notable asymmetry in the top trace can be accounted for by discrete quanta of vibrational excitation in the molecular deuterium produced.

pathways (molecular hydrogen elimination, 4+2 HCN cycloreversion) is expected due to the more sterically constrained transition states and lower A factor. Isomerization processes, intersystem crossing and the products generated further down this reaction path will also become less favored relaxation mechanisms as the increase in excess energy drastically increases the rate of internal conversion and stepwise processes such as biradical production followed by migrations and rearrangement. The dramatic shift in translational energy release in s-triazine photodissociation (9) can be explained by appreciating the excess energy dependence of the biradical pathway. Both channels can be observed, but the concerted mechanism dominates following 248 nm excitation while after 193 nm excitation the biradical (translationally slow) products are favored.

Summary.

The unimolecular decay of azabenzenes exhibits a remarkable evolution as the degree of nitrogen substitution and amount of internal energy is varied. On the 'pure aromatic' end of the scale, molecules such as benzene and pyridine primarily photodissociate hydrogen while keeping their cyclic bonding intact. On the heavily substituted end of the triazines and tetrazines the ring integrity is easily compromised, typically yielding three fragments. The intermeditate case of the diazines provided insight into possible mechanisms generating HCN through both concerted and stepwise (biradical) mechanisms and showed continuity along the spectrum of reactivity based upon some fundamental principles of symmetry correlations and radiationless transition theory.

References

1.   Knight, A. E. W., Parmenter, C.S., Chemical Physics, **1976**, 15, 82.
2.   Byrne, J.P., McCoy, E.F., Ross, Aust. J. Chem., **1965**, 18, 1589.
3.   Lower, S.K., El-Sayed, M.A., Chem. Rev. 1966, 66, 199.
4.   Noyes, Jr., W.A., Al Ani, K.E., Chem. Rev.. **1974**, 74, 29.
5.   Yamazaki, I., Murao, T. Yamazaki, T., Yoshihara, K., Chem. Phys. Lett., **1984**, 108, 11.
6.   Avouris, P., Gelbart, W. M., El-Sayed, M. A. Chemical Reviews, **1977**, 77, 793.
7.   Takagi, Y., Sumitani, M., Nakashima, N., O'Connor, D. V., Yoshihara, K., J. Chem. Phys., **1982**, 77, 6337.
8.   Yoshihara, K., O'Connor, D.V., Sumitani, M., Takago, Y., Nakashima, N., "Single Vibrational Level Dependence of Picosecond Fluorescence in the Channel 3 Region of Benzene", in Application of Picosecond Spectroscopy to Chemistry, NATO ASI Series C, Mathematical and Physical Sciences, ed. Eisenthal, K.B., **1984**, 127, 261.
9.   Ondrey, G. S., Bersohn, R., J. Chem. Phys., 1984, 81(10),4517.
10. Zhao, X., Miller, W. B., Hintsa, E. J., Lee, Y. T., J. Chem. Phys., **1989**, 90(10), 5527.
11. Mullin, A. S., Park, J., Chou, J. Z., Flynn, G. W., Weston, Jr., R.E., Chemical Physics, 1993, 175, 53.
12. Innes, K. K., Byrne, J. P., Ross, I.G. , J. Mol. Spectroscopy, **1967**, 22, 125.
13. Innes, K. K., Ross, I. G., Moomaw, W. R., J. Mol. Spectroscopy, **1988**, 132, 492-544.

14. Lablanche-Combier, A., "Photoisomerization of Six-Membered Heterocyclic Molecules," in Photochemistry of Heterocyclic Compounds, ed. Butchardt, O., Vol. 4, p. 208.

15. Innes, K. K., Abstracts, 149th National Meeting of the American Chemical Society, Detroit, Michigan, April 1965.

16. Wodtke, A. M. and Lee, Y. T., J. Phys. Chem., **1985**, 89, 4744.

17. Gejo, T., Harrison, J.A., Huber, J. R., J. Phys. Chem., **1990**, 100, 13941.

18. Coxon, J.M., Halton, B., "Organic Photochemistry", 2nd ed., Cambridge University Press, **1987**.

19. Woodward, R.B., Hoffman, R., "The Conservation of Orbital Symmetry", Verlag Chemie, New York, **1970**.

21 Goates, S.R., Chu, J.O., Flynn, G.W., J Chem. Phys., **1984**, 81, 4521.

22. Prather, K. A. , Lee, Y.T., Israel Journal of Chemistry, **1994**, 34, 43.

23. Scheiner, A C., Scuseria, G.E., Schaefer III, H.F., J. Am. Chem. Soc., **1986**, 108, 8160.

24. Hoffman, R., J. Am.Chem.Soc., **1968**, 90, 1475.

25. Wu, C. H. and Kern, R.D., J. Phys. Chem., **1987**, 91, 2691.

26. Yokoyama, A., Zhao, X., Hintsa, E. J., Continetti, R. E., Lee., Y.T., J.Chem.Phys., **1990**, 92, 4222.

27. Chesko, J. D., Lee, Y.T., unpublished results.

28. Chesko, J. D., Ph.D. Thesis, University of California at Berkeley, **1995**.

INTERMOLECULAR DYNAMICS:
NONREACTIVE ENERGY TRANSFER

Chapter 9

Nonbonding Interactions of the Boron Atom in the Excited 2s²3s ²S Rydberg and 2s2p² ²D Valence States

Paul J. Dagdigian, Xin Yang, and Eunsook Hwang[1]

Department of Chemistry, Johns Hopkins University, Baltimore, MD 21218

Laser fluorescence spectroscopic studies of electronic transitions built upon the B atomic $2s2p^2\ ^2D \leftarrow 2s^22p\ ^2P$ resonance transition to an excited valence state in the BNe, BAr, and BKr van der Waals complexes are presented. Information on the relevant potential energy curves of these electronic states, as well as those built upon the B atomic $2s^23s\ ^2S$ Rydberg state, derived from laser excitation spectra, is also reported. Considerable variation in the interaction energies is found for the excited valence and Rydberg states in the different systems. Some of the excited electronic states do not fluoresce because of predissociation, and their observation was achieved through fluorescence depletion experiments. The contrasting binding energies and decay behavior of the excited states of BNe, BAr, and BKr are discussed in terms of our knowledge of electronic states of these complexes.

Considerable information on non-bonding interactions of ground state and electronically excited metal atoms with rare gas atoms has been derived from spectroscopic studies of van der Waals complexes of these species (1). In our laboratory, we have concentrated on studies involving the boron atom. In these experiments, the complex under study is prepared in a pulsed free jet by 193 nm photolysis of diborane diluted in an appropriate seed gas mixture and is interrogated by laser fluorescence excitation (FE) downstream. The experimental apparatus is thoroughly described in previous publications (2-5).

We have observed FE spectra of the $B^2\Sigma^+ - X^2\Pi$ electronic transitions of the BNe and BAr complexes and characterized the three lowest electronic states [$X^2\Pi$, $A^2\Sigma^+$, $B^2\Sigma^+$] of these species in conjunction with electronic structure calculations by Alexander and co-workers (2, 4, 6, 7). The $X^2\Pi$ and $A^2\Sigma^+$ electronic states correlate with the lowest atomic asymptote $B(2s^22p\ ^2P)$ + Rg [Rg = Ne, Ar], while the $B^2\Sigma^+$ electronic state is associated with $B(2s^23s\ ^2S)$ + Rg. With our experimental observations and these collaborative electronic structure calculations, it was possible to derive potential energy curves for the $B^2\Sigma^+$ state of both BNe and BAr. We find these potential energy curves to be dramatically different. The $BAr(B^2\Sigma^+)$ state

[1]Current address: Molecular Physics Laboratory, SRI International, Menlo Park, CA 94025

possesses a deep inner well $[D_0' \approx 1010 \text{ cm}^{-1}]$ and a shallow outer well, with an intervening barrier, approximately 140 cm^{-1} high (2). By contrast, the BNe($B^2\Sigma^+$) state is purely repulsive, with only a slight inflection on the repulsive wall as evidence of a weak attractive interaction (4). The binding energy for the ground $X^2\Pi$ states of both BNe and BAr were found to be small $[D_0'' = 21 \text{ cm}^{-1}$ and $\geq 75 \text{ cm}^{-1}$, respectively (2, 4)].

We have now extended these studies to BRg electronic transitions correlating with the $2s2p^2 \, ^2D \leftarrow 2s^22p \, ^2P$ boron atomic transition at 208.9 nm. In this case, the upper state is an excited *valence* state. In this paper, we briefly review our published work on the BNe complex and report new results on the BAr and BKr complexes. A more complete account of our recent experiments on BAr is forthcoming (8), while a publication describing our observations on BKr is in preparation.

Table I presents the observed and expected BRg electronic states with energies < 50,000 cm^{-1}. As can be seen in Table I, the approach of a spherical Rg atom to B($2s2p^2 \, ^2D$) leads to 3 molecular electronic states. Since the B–Rg repulsion at short range will be minimized with the B $2p$ electrons in the π vs. σ orbital, we expect the most attractive of these states to be $C^2\Delta$. In Table I, we see that there are several states below 50,000 cm^{-1}, correlating with the $2s2p^2 \, ^4P$ and $2s^23p \, ^2P$ asymptotes, which are not optically accessible through 1-photon transitions. We will find that the presence of these states significantly affects the spectra of the BAr and BKr complexes.

Table I. Boron atomic levels below 50 000 cm^{-1} and the BRg electronic state emanating from the B + Rg asymptotes [a]

Atomic state	Term energy (cm^{-1})	Molecular states[b]
$2s^22p \, ^2P$	0	$X^2\Pi, A^2\Sigma^+$
$2s2p^2 \, ^4P$	28,877	$^4\Sigma^-, \, ^4\Pi$
$2s^23s \, ^2S$	40,040	$B^2\Sigma^+$
$2s2p^2 \, ^2D$	47,857	$C^2\Delta, D^2\Pi, E^2\Sigma^+$
$2s^23p \, ^2P$	48,612	$^2\Pi, \, ^2\Sigma^+$

[a]   Atomic energy levels taken from Ref. (9).
[b]   Molecular states labeled with letters have been observed in excitation or emission.

The BNe Complex

We have observed two progressions near the atomic $^2D \leftarrow \, ^2P$ transition in the FE spectrum of this species (5). These have been assigned as the (v',0) progressions of the $C^2\Delta - X^2\Pi_{1/2}$ and $D^2\Pi - X^2\Pi_{1/2}$ transitions. To the blue of these discrete bands, we observed a continuous excitation, assigned as excitation into the B(^2D) + Ne continuum. From the dissociation threshold, we obtained estimates of dissociation energies for the ground and electronically excited BNe states. In contrast to the purely repulsive nature of the BNe($B^2\Sigma^+$) state, we find that the binding energy of the $C^2\Delta$ state is somewhat larger $[D_0' = 111 \text{ cm}^{-1}]$ than that of the ground state $[D_0'' = 21 \text{ cm}^{-1}]$. Since the $C - X$ transition involves a $2p\pi \leftarrow 2s\sigma$ promotion, this increased binding can be rationalized as resulting from decreased electron repulsion. We were also able to resolve the rotational structure in several $C - X$ bands and found that the average internuclear separation for this state is 2.87 Å. The $D^2\Pi$ state was found to be weakly bound $[D_0 = 8 \text{ cm}^{-1}]$, while no evidence for the transition to the $E^2\Sigma^+$ state was found in the spectrum. The $E^2\Sigma^+$ state is probably purely repulsive, and the $E - X$ transition is most likely buried in the B(^2D) + Ne continuum excitation. Simple molecular orbital arguments (5) predict that the strength of these transitions will be in the order $C - X > D - X > E - X$.

The BAr Complex

Figure 1 presents a survey FE spectrum for this complex in the region of the atomic $^2D \leftarrow {}^2P$ transition. In addition to the atomic lines, we observe a number of BAr molecular bands, as well as a broad unstructured feature to the blue. In analogy with the BNe spectrum, the onset of this continuous transition is assigned to the threshold of the $B(^2D)$ + Ar continuum. With this assignment, we obtain an experimental value for the ground state dissociation energy, $D_0'' = 102.4 \pm 0.3$ cm^{-1}, assuming no barrier in the excited state potential energy curve. In contrast to the BAr($B^2\Sigma^+$) state (2), there is no evidence for barriers in the BAr states built upon $B(2s2p^2\ ^2D)$ + Ar. The wavenumber differences between the $B(^2D)$ + Ar threshold and the transition wavenumbers for the observed bands provide estimates of excited state binding energies. The binding energy of the redmost band [63 cm^{-1}] is significantly less than for BNe($C^2\Delta$, v' = 0) [111 cm^{-1}]. Since it is unlikely that BAr binding energies would be less than the corresponding values for BNe, the BAr $C - X$ bands must lie further to the red than the bands observed by FE (displayed in Figure 1). We have searched to the red of these bands and have found no other bands in this spectral region.

In order to observe the BAr($C^2\Delta$) state, which apparently decays nonradiatively, we have applied the fluorescence depletion (FD) technique, which we recently implemented in our laboratory for the observation of the spectrum of the B···H$_2$ complex in this spectral region (10). The FD technique is a folded variant of optical-optical double resonance in which the two lasers access the same lower level. Transitions to non-fluorescing excited states are observed by monitoring the effect of the depletion laser on fluorescence induced by the probe laser. The FD technique has also been utilized by other groups for the study of van der Waals complexes (11, 12).

FD spectra of BAr were recorded with the probe laser tuned to the (8,0) band of the $B - X$ transition (2). Since ^{11,10}BAr isotopic splittings are resolved in this band system, FD spectra could be recorded separately for these isotopomers. All ground state rotational levels are detected with approximately equal efficiency since individual rotational lines in the $B - X$ (8,0) band are overlapped because of large homogeneous linewidths due to predissociation (2). Survey spectra are displayed in Figure 2. We see a new series of bands, with substantial isotope splittings, to the red of the bands observed by FE. These are assigned to the ^{11,10}BAr $C^2\Delta - X^2\Pi$ transition, and the transition wavenumbers are reported in Table II. It should be noted that the bands and the dissociation continuum observed in the FE spectrum are also seen in the FD spectra.

Table II. Transition wavenumbers T_v, rotational constants B_v', and homogeneous line widths Γ for the ^{11,10}BAr $C^2\Delta - X^2\Pi_{1/2}$ (v',0) bands observed by FD spectroscopy [a]

v'	T_v(^{11}BAr)[b]	T_v(^{10}BAr)[b]	B_v'(^{11}BAr)[c]	Γ[c]
13	47,403.6	47,467.0		
14	47,524.1	47,582.5	0.263(8)	0.15(6)
15	47,627.5	47,679.5	0.249(5)	0.35(6)
16	47,714.4	47,760.1	0.226(5)	0.42(6)
17	47,786.5	47,825.1	0.195(6)	0.15(6)
18	47,843.3			

[a] For an explanation of the vibrational assignments, see text. All spectroscopic constants reported in cm^{-1}.

[b] Estimated uncertainties ± 0.2 cm^{-1}.

[c] Estimated uncertainties reported in units of the least significant quoted digit.

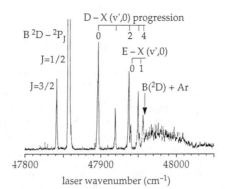

Figure 1. Low-resolution FE spectrum of BAr. The B atomic $^2D - {}^2P_{1/2,3/2}$ transitions are identified. The observed BAr bands are assigned to $D - X$ and $E - X$ (v',0) progressions, as described in the text. The threshold to the $B(^2D) + Ar$ continuum is also marked.

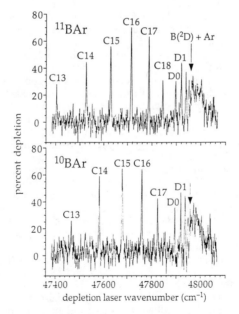

Figure 2. Low-resolution FD spectra of the BAr isotopomers. Bands are assigned to the $C - X$ (v',0) progression, as described in the text, are labeled Cv', while the $D - X$ (0,0) and (1,0) bands are denoted $D0$ and $D1$, respectively. The probe laser was tuned to the $B - X$ (8,0) band of each isotopomer.

With the availability of transition wavenumbers $T_v^{(i)}$ for two isotopomers, it is possible to assign excited state vibrational quantum numbers v'. We have fit these to a quadratic expression in (v' + 1/2):

$$T_v^{(i)} = T_{e0} + \rho_i \, \omega_e' \, (v' + 1/2) - \rho_i^2 \, \omega_e x_e' \, (v' + 1/2)^2 - \delta_i , \qquad (1)$$

where T_{e0} is the separation between the bottom of the $BAr(C^2\Delta)$ potential energy curve and the $^{11}BAr(X^2\Pi_{1/2}, v'' = 0)$ level. The factor ρ_i is a mass scaling factor, which equals unity for ^{11}BAr and $[\mu(^{11}BAr)/\mu(^{10}BAr)]^{1/2} = 1.03828$ for ^{10}BAr. The term δ_i is the isotope shift for $BAr(X^2\Pi_{1/2}, v'' = 0)$ and equals zero for ^{11}BAr and 0.88 cm^{-1} for ^{10}BAr (2). The transition wavenumbers listed in Table II were fit with various guesses for the v' numbering. The assignments given in Table II seem quite certain since the χ^2 of the fit increased by more than an order of magnitude when the v' values were changed by ±1. With our v' assignment, we obtain the following vibrational constants for the C state: $\omega_e' = 336.64 \pm 1.72$ cm^{-1} and $\omega_e x_e' = 7.873 \pm 0.054$ cm^{-1}. The energy separation T_{e0} was found to equal 44,254 ± 10 cm^{-1}. This leads to an estimate of 3705 ± 10 cm^{-1} for the dissociation energy D_e of the $BAr(C^2\Delta)$ state. The quoted uncertainty represents the statistical error and does not include the fact that there is a strong correlation among the fitted parameters in Equation 1, because of the limited range of v' levels observed.

We have also taken high-resolution FD scans over several of the ^{11}BAr $C - X$ bands in order to confirm our electronic assignment of the upper levels and to determine mean internuclear separations through measurement of rotational constants. Figure 3 presents such a scan of the (16,0) band. We find that the rotational structure is well fit with the assumption that the upper state has $^2\Delta$ symmetry. The C state was not detected by FE and hence has a negligible fluorescence quantum yield as a result of a fast nonradiative decay process. Fitting spectra such as that shown in Fig. 3 required linewidths significantly greater than the laser linewidth. Hence, we have fit the spectra with Voigt line profiles, i.e. a convolution of a Lorentzian linewidth Γ and a Gaussian laser linewidth profile. Table II presents estimates of the upper state rotational constants and predissociation linewidths. We see that the rotational constants $B_{v'}$ decrease with increasing v'. We also observe a significant variation of the linewidths with v'. For our best simulations of the experimental spectra, we set the spin-orbit constant A' to zero, as expected from simple molecular orbital arguments about the $C^2\Delta$ state (5).

Since the redmost bands, observed by the FD technique and listed in Table II, have been assigned to the $C^2\Delta - X^2\Pi$ transition, some or all of the bands observed in the FE spectrum displayed in Figure 1 should be assignable to the $D^2\Pi - X^2\Pi$ transition. In Figure 1, we have assigned 5 bands to the $D - X$ (v',0) progression. Since the ground v'' = 0 vibrational level has no nodes in its wave function, we expect a smooth v' variation of the intensities of these bands. Thus, the low intensity of the (1,0) band is puzzling. We have measured the decay lifetimes for the $B(2s2p^2 \, ^2D)$ atomic levels, as well as the excited BAr v' = 0, 1, and 2 levels (8). We find that the lifetimes of the BAr levels [19.7 ± 2.3, 16.1 ± 0.8, and 18.9 ± 2.4 ns, respectively] are all significantly shorter than the atomic lifetime [23.1 ± 1.2 ns (13)]. This suggests that all vibrational levels of $BAr(D^2\Pi)$ are weakly predissociated. Since the lifetime of v' = 1 is the shortest, this implies that this level has the smallest fluorescence quantum yield. This provides an explanation for the reduced intensity of the (1,0) band in Figure 1.

We have taken high-resolution FE scans of the bands in Figure 1 assigned to the $D - X$ transition. While a completely satisfactory fit of the rotational structure of these bands has not yet been obtained, the band contours appear to be consistent with a $^2\Pi - X^2\Pi$ transition. Preliminary values of excited state rotational constants have been derived. Two weak bands in Figure 1 near the dissociation threshold have a different rotational structure and have been assigned to the $E^2\Sigma^+ - X^2\Pi$ band system.

We have constructed potential energy curves for the BAr $C^2\Delta$ and $D^2\Pi$

electronic states which reasonably well reproduce our experimental observations. These are displayed in Figure 4. For the former state, a Morse-spline-van der Waals (MSV) form (*14*, *15*) was employed, while a Morse function was adequate for the latter. We see that the binding energy of the C state is substantial. The equilibrium internuclear separation in this state is much smaller than that of the $X^2\Pi$ state [3.6 Å (*2*)]. The observed v′ levels are those for which the overlap of the outer lobe of these wave functions with the v″ = 0 vibrational wave function is large. By contrast, the binding is relatively weak for the D state, and its equilibrium separation is slightly larger than for the X state.

In contrast to the BNe($C^2\Delta$) state which decays radiatively (*5*), we find that the BAr($C^2\Delta$) state undergoes predominantly nonradiative decay, through predissociation. Figures 5 and 6 compares known and estimated potential energy curves for all the BNe and BAr electronic states below 50,000 cm^{-1}. These potential energy curves come from a combination of electronic structure calculations (*2*, *4*, *6*) and fits to experimental data (*2*, *5*). The sole exceptions are the $^4\Sigma^-$ and $^4\Pi$ states correlating with the B($2s2p^2$ ^4P) + Rg asymptote. The electron configurations of these states can be written as $2s\sigma2p\pi^2$ and $2s\sigma2p\sigma2p\pi$, respectively. The $X^2\Pi$ and $A^2\Sigma^+$ states, which have electron configurations $2s\sigma^22p\pi$ and $2s\sigma^22p\sigma$, respectively, also differ by having an electron in a $2p\pi$ vs. $2p\sigma$ orbital. The former states [$X^2\Pi$ and $^4\Sigma^-$, in the two cases] are expected to exhibit stronger binding because of reduced electron repulsion. Electronic structure calculations indeed find that the BNe and BAr $X^2\Pi$ states have a stronger binding than the corresponding $A^2\Sigma^+$ states (*2*, *4*). In the absence of information about the quartet potential energy curves, we have simply assumed for the construction of Figures 5 and 6 that the curves for the BRg $^4\Sigma^-$ and $^4\Pi$ states are the same for the corresponding $X^2\Pi$ and $A^2\Sigma^+$ states. We actually expect the quartet states to be less repulsive than the doublet states because of reduced repulsion with an electron in a $2p\pi$ vs. $2s\sigma$ orbital. (Compare the electron configurations of the $^4\Sigma^-$ and $X^2\Pi$ states.) Because of the compressed energy scale, shallow attractive wells are not discernible in these figures.

The BNe($C^2\Delta$) state has a small binding energy and relatively large equilibrium internuclear separation. From the potential energy curves shown in Figure 5, we see that the repulsive $^4\Pi$ state passes inside the inner limb of the $C^2\Delta$ state. Thus, this state decays radiatively since it is not crossed by any repulsive curve emanating from a lower asymptote. We see from Figure 5 that the $^4\Pi$ state does cross the $B^2\Sigma^+$ state. This crossing has no effect on the decay of BNe($B^2\Sigma^+$) since its potential energy curve is purely repulsive, and optical excitation to this state is accompanied by rapid dissociation to yield free B($2s^23s$ ^2S) (*4*).

In contrast to the situation with BNe, we see in Figure 6 that the repulsive BAr $^4\Pi$ state does cross the $C^2\Delta$ curve. Simple molecular orbital arguments (*16*) indicate that the one-electron part of the spin-orbit matrix element between these states will be nonzero. Hence, the $C^2\Delta$ state can be predissociated by spin-orbit coupling. Sohlberg and Yarkony (*17*) have carried out an electronic structure calculation of the BAr($C^2\Delta$) state, as well as a theoretical study (*18*) of the nonradiative decay of this state. They indeed find that predissociation is the dominant mode of decay for the $C^2\Delta$ state. The repulsive $^4\Pi$ state also crosses the BAr($B^2\Sigma^+$) state. However, there is no evidence that this state is predissociated (*2*). The one-electron part of the spin-orbit matrix element between the $B^2\Sigma^+$ and $^4\Pi$ states is zero. Most likely, predissociation induced by the weak two-electron part of the spin-orbit matrix element cannot compete with the fast radiative decay of the $B^2\Sigma^+$ state. [The radiative lifetime of the B($2s^23s$ ^2S) state is 4 ns (*13*).]

Our decay lifetime measurements show that the BAr($D^2\Pi$) state undergoes a weak predissociation. We see from Figure 6 that this nonradiative decay process cannot be mediated by direct coupling to a repulsive curve emanating from a lower atomic asymptote. The B($2s^23p$ ^2P) level lies only 755 cm^{-1} (*9*) above B($2s2p^2$ ^2D), and BRg($^2\Pi$, $^2\Sigma^+$) electronic states arise from this asymptote (see Table I). If one of these crosses the $D^2\Pi$ state from above and then crosses the repulsive $^4\Pi$ state, this would provide a mechanism for indirect predissociation (*16*) of the $D^2\Pi$ state.

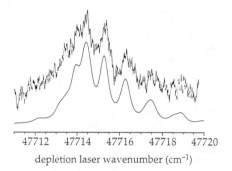

depletion laser wavenumber (cm⁻¹)

Figure 3. High-resolution FD spectrum of the ^{11}BAr $C-X$ (16,0) band. The lower trace is a simulated spectrum using the spectroscopic constants given in Table II, with a rotational temperature of 3.5 K and Gaussian laser bandwidth of 0.15 cm⁻¹.

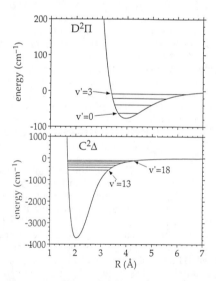

Figure 4. Potential energy curves for the BAr $C^2\Delta$ and $D^2\Pi$ electronic states. The observed vibrational levels are indicated by horizontal lines.

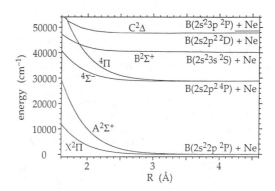

Figure 5. Potential energy curves for BNe states below 50,000 cm^{-1}.

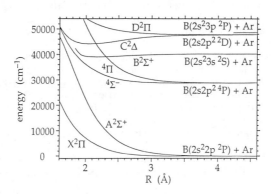

Figure 6. Potential energy curves for BAr states below 50,000 cm^{-1}.

Quantum chemical calculation of the relevant potential energy curves of these highly excited states and their couplings would, unfortunately, be a severe challenge with present computational capabilities.

The BKr Complex

Figure 7 presents a survey FE spectrum for this complex in the region of the atomic $^2D \leftarrow ^2P$ transition. This spectrum is qualitatively similar to the BAr FE spectrum shown in Figure 1. The BKr spectrum displays a series of bands leading to a convergence limit, beyond which appears a broad unstructured feature. We interpret the latter as excitation to the B(^2D) + Kr dissociation continuum. For this complex, the threshold cannot be so precisely determined as for BAr. From the wavenumber difference between this threshold and the B $^2D \leftarrow ^2P_{1/2}$ atomic transition, we estimate the dissociation energy $D_0'' = 160 \pm 2$ cm^{-1} for BKr.

We have searched, unsuccessfully, for BKr bands to the red of the spectral region displayed in Figure 7. Thus, we believe that the BKr($C^2\Delta$) state decays nonradiatively, in analogy with BAr($C^2\Delta$). The bands observed in the BKr FE spectrum are hence assigned to the $D - X$ transition, specifically the (v′,0) progression. The wavenumber difference between the B(^2D) + Kr threshold and the (0,0) band provides an estimate of the dissociation energy $D_0' = 121 \pm 2$ cm^{-1} for the BKr($D^2\Pi$) state. It is possible that some of the weak bands near the dissociation limit belong to the $E - X$ transition.

Similar to the BAr $D - X$ FE spectrum, an irregular variation of the intensities of the BKr bands is seen in Figure 7. We have measured decay lifetimes for the BKr($D^2\Pi$) v′ = 0 – 4 levels and find that these are all smaller than the lifetime of the free B(^2D) atomic level. This indicates that all the BKr(D, v′) vibrational levels are predissociated to some degree. The lifetimes of the v′ = 1 and 2 level [8.1 and 6.8 ns, respectively] are considerably shorter than the lifetimes of the other vibrational levels [11 – 13.5 ns]. This implies that the fluorescence quantum yields for the decay of BKr(D, v′ = 1, 2) are small and provides an explanation for the weakness of the (1,0) and (2,0) bands in Figure 7.

We have also searched the spectral region around the B $2s^23s\,^2S \leftarrow 2s^22p\,^2P$ atomic transition for the FE spectrum of the BKr $B^2\Sigma^+ - X^2\Pi$ electronic transition. Figure 8 shows such a scan. No bands assignable to diatomic BKr were found. Rather, several complicated features were observed. Moreover, these features displayed different dependences on the Kr mole fraction in the seed gas mixture. Feature A appeared first with increasing Kr concentration, while features B and C were observable at higher concentrations. Feature A displays a complicated structure, which cannot be explained as rotational structure of a transition to a bound diatomic excited state (as for BAr) or free $\leftarrow$ bound excitation to a repulsive excited state (as for BNe). We thus believe that these FE features arise from excitation of higher BKr$_n$ clusters.

On the basis of these experiments, we believe that the BKr($B^2\Sigma^+$) state decays nonradiatively. While it is usually difficult to make arguments on the basis of the lack of observation of a signal, in this case our detection of BKr through its $D - X$ transition shows that we can indeed produce BKr in these supersonic expansions. The most likely cause of predissociation in the $B^2\Sigma^+$ state is coupling to the repulsive $^4\Pi$ state, as illustrated in Figure 6 for BAr. As discussed above, the spin-orbit coupling of the $B^2\Sigma^+$ and $^4\Pi$ states will be weak if we ignore the changes in the electronic structure of the atomic moieties as these species are brought together. However, the spin-orbit interaction in Kr is large, and a small polarization of its electronic wave function may allow for a significant $B^2\Sigma^+ \sim ^4\Pi$ coupling and provide a mechanism for predissociation of the BKr($B^2\Sigma^+$) state.

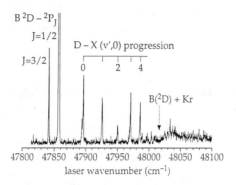

Figure 7. Low-resolution FE spectrum of BKr. The B atomic $^2D - {}^2P_{1/2,3/2}$ transitions are identified. BKr bands assigned to the $D - X$ (v',0) progressions and the threshold to the $B(^2D) + Kr$ continuum are identified.

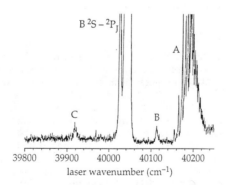

Figure 8. FE scan in the region of the B atomic $^2S - {}^2P$ transition for a B/Kr/He beam. Feature A, B, and C are assigned to BKr$_n$ clusters.

Discussion

In this paper, we have reviewed the considerable information now available on the interaction of the boron atom in both its ground and two excited electronic states with the rare gases. Laser spectroscopic studies of the BRg electronic transitions, in combination with electronic structure calculations, have provided considerable information on these interactions. Table III presents the BRg binding energies derived from both the experimental and theoretical investigations. For comparison, the binding energies of the rare gas dimers are included in Table III. We see that the dissociation energies of the $BRg(X^2\Pi)$ states are slightly larger than those of the corresponding rare gas dimer and that the difference in binding energy increases for the heavier rare gases. In similar fashion, the dissociation energies of the excited BRg states increases with increasing rare gas atomic number.

Table III. Dissociation energies D_0 (in cm^{-1}) for ^{11}BRg electronic states

Electronic state	BNe	BAr	BKr
$X^2\Pi$	21	102	159
$B^2\Sigma^+$	not bound	999	
$C^2\Delta$	111	3539	
$D^2\Pi$	8	63	121
$E^2\Sigma^+$		19	
Rg_2 [a]	18	83	129

[a] Potential energy curves for the rare gas dimers taken from: Ne–Ne, Ref. (19); Ar–Ar, Ref. (20); Kr–Kr, Ref. (21).

The most dramatic difference in the BRg binding energies is the large increase in the attractive nature of the potential energy curves of the $B^2\Sigma^+$ and $C^2\Delta$ states of BAr vs. BNe. Presumably, the corresponding BKr states are even more strongly attractive, but we have not been able to detect these states. To our knowledge, the binding energy of the $BAr(C^2\Delta)$ state is greater than that of any previously observed electronic state of a neutral metal atom – argon complex (1, 22). Recently, Massick and Breckenridge (22) observed a strongly bound $[D_e' \approx 2900$ cm$^{-1}]$ $^3\Sigma^-$ state of MgAr which has a doubly excited electron occupancy $3p\pi^2$. This is similar to the $2s\sigma2p\pi^2$ occupancy of $BAr(C^2\Delta)$. One reason for the strong binding in both of these states is that the lobes of $np\pi$ orbital are perpendicular to the internuclear axis, allowing a reduction in the electron repulsion. However, a similar reduced repulsion is operating for the corresponding $BNe(C^2\Delta)$ state, for which a strong binding is not found. This implies that the strong binding of these BAr states is specific to the more polarizable heavier rare gas atoms. It is unlikely that the enhanced binding is due to charge transfer in the BAr excited state because of the modest electron affinity of the B atom [EA = 0.28 eV (23)], despite the $BAr(C^2\Delta)$ significant excitation energy. A purely Coulombic ion-pair potential energy curve at an internuclear separation of 2 Å is > 2 eV higher than the energy of the $B(^2D)$ + Ar asymptote.

We have shown that the $BAr(C^2\Delta)$ state decays nonradiatively, because of predissociation induced by spin-orbit mixing with a repulsive $^4\Pi$ state. Several other examples of spin-orbit induced predissociation of excited states of metal – rare gas complexes, involving group 12 atoms, are known (1). These involve the $^1\Pi$ states of CdXe and ZnXe built upon the $nsnp$ ^1P atomic states. In both cases, predissociation occurs through spin-orbit mixing with a repulsive $^3\Sigma_1^+$ state correlating with the lower $nsnp$ $^3P_2$ atomic state (24). It is perhaps surprising that such interactions occur in van der Waals complexes since we might expect that molecular states correlating with different atomic asymptotes would not cross if the interactions were weak. Such crossings do occur because the non-bonding interactions can be strong compared to the energy differences between atomic states and can influence the decay of the excited state.

Our experiments demonstrate a heavy atom effect in the BRg excited state decay dynamics. There is no evidence for predissociation of the BNe($D^2\Pi$) state (5), while lifetimes of BAr($D^2\Pi$) vibrational levels are somewhat lower than that for the corresponding B($2s2p^2\ ^2D$) atomic level. The lifetimes of BKr($D^2\Pi$, v') levels are even smaller than in BAr. Similarly, the BKr($B^2\Sigma^+$) states appears to decay nonradiatively, while there is no evidence for predissociation of the bound vibrational levels of BAr($B^2\Sigma^+$) (2, 6).

Acknowledgments

This research was supported by the US Air Force Office of Scientific Research under Grant No. F49620-95-1-0055. We are indebted to Millard Alexander for encouragement and numerous conversations about boron – rare gas interactions. We are also grateful to Karl Sohlberg and David Yarkony for discussions about the electronic structure and decay dynamics of the BAr($C^2\Delta$) state and for carrying out calculations on the predissociation of this state. We thank Bill Breckenridge for helpful correspondence.

References

1.   Breckenridge, W. H.; Jouvet, C.; Soep, B. In *Advances in Metal and Semiconductor Clusters*; Duncan, M. A., Ed.; JAI Press: Greenwich, CT, 1995, Vol. 3; pp. 1-93.
2.   Hwang, E.; Huang, Y.-L.; Dagdigian, P. J.; Alexander, M. H. *J. Chem. Phys.* **1993**, *98*, 8484.
3.   Yang, X.; Hwang, E.; Alexander, M. H.; Dagdigian, P. J. *J. Chem. Phys.* **1995**, *103*, 7966.
4.   Yang, X.; Hwang, E.; Dagdigian, P. J.; Yang, M.; Alexander, M. H. *J. Chem. Phys.* **1995**, *103*, 2779.
5.   Yang, X.; Hwang, E.; Dagdigian, P. J. *J. Chem. Phys.* **1996**, *104*, 599.
6.   Hwang, E.; Dagdigian, P. J.; Alexander, M. H. *Can. J. Chem.* **1994**, *72*, 821.
7.   Hwang, E.; Dagdigian, P. J. *J. Chem. Phys. Lett.* **1995**, *233*, 483.
8.   Yang, X.; Dagdigian, P. J. *J. Chem. Phys.* **1997**, *106*, xxxx.
9.   Odintzova, G. A.; Striganov, A. R. *J. Phys. Chem. Ref. Data* **1979**, *8*, 63.
10. Yang, X.; Hwang, E.; Dagdigian, P. J. *J. Chem. Phys.* **1996**, *104*, 8165.
11. Basinger, W. H.; Lawrence, W. G.; Heaven, M. C. *J. Chem. Phys.* **1995**, *103*, 7218.
12. Loomis, R. A.; Schwartz, R. L.; Lester, M. I. *J. Chem. Phys.* **1996**, *104*, 6984.
13. O'Brian, T. R.; Lawler, J. E. *Astron. Astrophys.* **1992**, *255*, 420.
14. Parson, J. M.; Siska, P. E.; Lee, Y. T. *J. Chem. Phys.* **1972**, *56*, 1511.
15. Pouilly, B.; Alexander, M. H. *J. Chem. Phys.* **1987**, *86*, 4790.
16. Lefebvre-Brion, H.; Field, R. W. *Perturbations in the Spectra of Diatomic Molecules*; Academic: New York, 1986.
17. Sohlberg, K.; Yarkony, D. R. *J. Phys. Chem.* (submitted).
18. Sohlberg, K.; Yarkony, D. R. *J. Phys. Chem.* **1997**, *106*, xxxx.
19. Siska, P. E.; Parson, J. M.; Schafer, T. P.; Lee, Y. T. *J. Chem. Phys.* **1971**, *55*, 5762.
20. Parson, J. M.; Siska, P. E.; Lee, Y. T. *J. Chem. Phys.* **1972**, *56*, 1511.
21. Barker, J. A.; Watts, R. O.; Lee, J. K.; Shafer, T. P.; Lee, Y. T. *J. Chem. Phys.* **1974**, *61*, 3081.
22. Massick, S.; Breckenridge, W. H. *J. Chem. Phys.* **1996**, *104*, 784.
23. Hotop, H.; Lineberger, W. C. *J. Phys. Chem. Ref. Data* **1975**, *4*, 539.
24. Bililign, S.; Kaup, J. G.; Breckenridge, W. H. *J. Phys. Chem.* **1995**, *99*, 7878.

Chapter 10

Infrared Laser Snapshots

Vibrational, Rotational, and Translational Energy Probes of High-Energy Collision Dynamics

George W. Flynn, Chris A. Michaels, H. Charles Tapalian, Zhen Lin, Eric T. Sevy, and Mark A. Muyskens[1]

Department of Chemistry and Columbia Radiation Laboratory, Columbia University, New York, NY 10027

Extremely high resolution infrared diode lasers have been employed to study fundamental collision dynamics. High energy molecules, produced by excimer laser pumping, are used as reagents to investigate collisional excitation into individual rotational and vibrational states of molecules. Translational recoil energy of the bath molecules is probed by measuring the time dependent Doppler profile of the molecular infrared transitions. Quenching of highly vibrationally excited molecules such as C_6F_6 having chemically significant amounts of energy (50-100 kcal/mole) has been studied by observing the uptake of energy by a small bath molecule such as CO_2 on a single collision time scale. Two processes are observed in which the energy transferred in the collision is large ($\Delta E \gg kT$). Vibrationally excited $CO_2(00^01)$ bath molecules produced in these collisions ($\Delta E = 11\ kT$) exhibit a rotational population distribution and a velocity recoil profile that are both close to the initial ambient distributions, indicating that excitation is dominated by "soft" collisions mediated by long range attractive forces. Such "soft" collisions are ubiquitous for high energy donors and stiff vibrational modes in the bath. Translational/rotational excitation of the ground state $CO_2(00^00)$ is much more efficient than vibrational excitation of $CO_2(00^01)$, and a large increase in both translational and rotational energy ($\Delta E = 10\text{-}20\ kT$) is observed for scattering into the high J tail of the vibrationless ground state, strongly suggestive of impulsive collisions. Different distributions are observed for the probability of transferring a given amount of energy ΔE to recoiling $CO_2(00^00)$ from other hot donors of the same total energy, suggesting sensitivity to the nature of the donor. Finally, the mapping of quantum state resolved bath energy gain data into an energy transfer distribution function, $P(E,E')$, is discussed. Particular emphasis is placed on both the respective kinetic signatures of the dominant modes of energy transfer in $P(E,E')$ and the connection between these energy transfer experiments and unimolecular reaction kinetics.

[1]Permanent address: Department of Chemistry and Biochemistry, Calvin College, Grand Rapids, MI 49546

Unimolecular reactions of gas phase molecules are a class of transformations which have attracted the attention of physical chemists for over 70 years (*1*). Included in this wide class of reactions are decompositions, isomerizations, and eliminations, all of which play a role in atmospheric and combustion chemistry. The qualitatively correct, basic mechanism for the thermal unimolecular reaction of molecule A was worked out by Lindemann in 1922 (*1*),

$$A + M \xrightarrow{k_a} A^* + M \qquad \text{(Collisional Activation)}$$

$$A^* + M \xrightarrow{k_d} A + M \qquad \text{(Collisional Deactivation)}$$

$$A^* \xrightarrow{k_E} B \qquad \text{(Reaction)}$$

where B is the product of the reaction and M is the collider gas. The standard form for the first order, thermal unimolecular reaction rate constant can be easily derived (*2*) yielding

$$k_{uni} = \frac{k_a \, k_E \, [M]}{k_d[M] + k_E} . \qquad (1)$$

Work in the field of unimolecular reaction kinetics continues to focus on extending the basic Lindemann mechanism into a quantitatively correct theory of unimolecular reactions. The key simplification inherent in the Lindemann model is the assumption of only two reactant states, A and A^*. Generally, a reactant molecule A with internal energy sufficient to allow reaction has a huge number of accessible states indexed by its energy E. The reaction rate constant, k_E , is quite sensitive to this energy, and there are an enormous number of activation and deactivation rate constants to consider involving transitions between different states (energies) of the reactant molecule. Proper consideration of all of these contributions to the kinetics leads to the so called master equation (*3*),

$$\frac{d[A(E,t)]}{dt} = [M] \int_0^\infty P(E,E') \, [A(E',t)] - P(E',E) \, [A(E,t)] \, dE' - k_E \, [A(E,t)]. \qquad (2)$$

Here $P(E,E')$, the energy transfer distribution function, gives the probability that a reactant molecule at energy E' will, upon collision with a bath molecule, end up with an energy E. Thus, given the function $P(E,E')$ and the energy dependent reaction rate constants, it is possible to solve the master equation and specify the reaction kinetics completely. (Note that solving the master equation can be quite difficult; the interested reader is referred to reference *3* for a full discussion of this problem). It is clear that the energy transfer distribution function, $P(E,E')$, is a key ingredient of a complete, quantitative formulation of the kinetics of unimolecular reactions.

Several direct experimental methods have been devised to study the energy transfer kinetics of molecules with large internal energies (i.e. near the barrier for unimolecular reaction). Generally, these involve spectroscopic probes of the donor (reactant) molecule as it loses internal energy through collisions with bath molecules. These methods, including Ultraviolet Absorption (*4*), Infrared Fluorescence (*5*), and Kinetically Controlled Selective Ionization (*6, 7*), have been used to measure <ΔE>, the average energy transfer per donor/bath collision and the first moment of $P(E,E')$, for a wide variety of systems. Additionally, it has recently become feasible to calculate $P(E,E')$ functions by analyzing a large number of quasiclassical donor/bath collision trajectories using a suitable potential energy surface (*8-10*).

Recent experiments in our laboratory have employed quantum state resolved spectroscopic probes of the bath molecules (11-18) rather than of the highly vibrationally excited donor molecules. This alternative strategy for studying energy transfer kinetics simply involves monitoring the uptake of energy by the bath molecules as opposed to the energy loss of the donor. Recently, a prescription for obtaining energy transfer distribution functions from such state resolved data on energy uptake by the bath was presented (19). This bath probe method is a quite new approach to measuring $P(E,E')$ functions for the collisional relaxation of highly vibrationally excited molecules. At least in principle, this inversion process is capable of yielding the actual shape of $P(E,E')$ for certain systems and does not require the assumption of an empirical form for the distribution function.

Collisions between highly vibrationally excited molecules and thermal bath molecules are also of great interest from a dynamical perspective. Note that typical donor molecules in these experiments can possess as much as 5 or 6 eV of internal energy. Collisions involving molecules with such a "chemically significant" vibrational energy might reasonably be expected to display interesting dynamics. It is of considerable interest to test the energy transfer models developed for collisions between two relatively low energy molecules (20-22) in the case where one molecule has an enormous amount of internal energy. The bath probe experiments developed in our laboratory, due to their state resolved nature, reveal detailed information about the dynamical mechanism of donor/bath collisions. The technique allows a separate determination of the bath molecules' vibrational, rotational, and translational energy following a collision but prior to any subsequent relaxation. That is, it is possible to measure the total energy transferred in a collision as well as the partitioning of this energy into the bath degrees of freedom. The bath probe technique thus provides information about $P(E,E')$ as well as a clear picture of the energy transfer mechanism. This in turn yields unique insight into the connection between the intermolecular potential energy surface dictating the donor/bath interaction and the energy transfer kinetics.

The bath probe technique can be summarized in the following three equations for the specific case of a CO_2 bath molecule:

$$\text{Donor} + h\nu\,(\text{UV}) \longrightarrow \text{Donor}^E \qquad\qquad\qquad \text{(Hot Donor Production)}$$

$$\text{Donor}^E + CO_2\,(00^00,J',V') \longrightarrow$$
$$\text{Donor}^E + CO_2\,(00^01,J,V) \qquad \text{(Collisional Energy Transfer)}$$

$$CO_2\,(00^01,J,V) + h\nu\,(\text{IR}) \longrightarrow CO_2\,(00^02,J\pm1,V). \qquad \text{(Diode Laser Probe)}$$

In the general notation for CO_2 (nm^lp,J,V), n, m, and p, respectively, refer to the number of quanta in the symmetric stretching (ν_1), bending (ν_2), and antisymmetric streching (ν_3) vibrational modes; l, and J are the "vibrational" and rotational angular momentum; and V refers to the average velocity of the CO_2 molecule as projected onto the diode laser axis. The first step involves production of a highly vibrationally excited donor molecule (e.g. NO_2 (11-13), pyrazine (11-13, 15, 16, 18), C_6F_6 (14, 17), C_6H_6 (14)) via excitation to an electronically excited state followed by rapid internal conversion back to high vibrational levels of the ground electronic state. Subsequently, the donor molecule at energy E' collides with a bath molecule resulting in an energy transfer of magnitude $\Delta E = E-E'$. The final step involves a transient absorption measurement using an infrared diode laser to probe the bath molecule (N_2O, CO_2) following a single donor/bath collision. The ultra high resolution of the diode laser ($\Delta\nu = 0.0003$ cm^{-1}) makes it possible not only to probe individual bath rovibrational states but also to measure the Doppler broadened absorption lineshape yielding the bath molecules' post-collision velocity distributions. This paper will focus on several general trends, which are evident in the product state distributions for all donor/bath systems studied thus far, and the implications of these results regarding the dominant energy transfer mechanisms. The energy transfer distribution functions describing the donor/bath collisions will also be discussed with an emphasis on understanding the effects of the operative modes of energy transfer on various regions of $P(E,E')$.

Experimental

These experiments employ a UV pump/IR diode laser probe, transient absorption technique which has been described in detail elsewhere (23). A 1:1 mixture of donor and bath (acceptor) molecules flows through a 3.0 m long jacketed Pyrex collision cell. Collision systems which have been studied include (donor/bath) NO_2/N_2O, pyrazine/CO_2, and C_6F_6/CO_2. The collision cell temperature is controlled by circulating either water or methanol through the outer jacket of the cell.

Donor molecules are excited by pulses from either a 248 nm KrF excimer laser (Lambda Physik EMG 201 MSC) or an excimer-pumped dye laser. Pyrazine and C_6F_6 are pumped directly by the 248 nm pulses, resulting in electronic excitation followed by rapid intersystem conversion to highly vibrationally excited states of the ground electronic state. Highly vibrationally excited NO_2 molecules are produced by pumping with 450.4 nm pulses from an excimer-pumped dye laser. Pulsed laser intensities of ≤ 2 MW/cm^2 were used to insure only single photon excitation of the donor molecules. A cw lead salt diode laser beam is propagated co-linearly with the UV laser beam through the collision cell. The frequency of the diode is tuned such that absorption occurs on a specific rovibrational transition of the bath molecules. The IR light intensity transmitted through the cell is measured using a liquid nitrogen cooled InSb photodetector which, together with its preamplifier, has a rise time of approximately 0.5 μsec. In order to insure that the detected IR light is from a single laser mode, the transmitted diode laser beam is passed through a monochromator before being focused onto the photodetector. The total gas pressure in the collision cell is selected such that IR measurements can be performed following the pump pulse at approximately 1/4 the average time between gas-kinetic collisions (usually ~1 μsec) without being detector response time limited.

The diode laser frequency is actively locked to the desired bath molecule IR absorption line using a 1 m long reference gas cell. Approximately ten percent of the diode laser beam is directed by a beamsplitter into the reference cell, through a monochromator, and then focused onto an InSb or HgCdTe photodetector with a high-gain preamplifier. As the diode laser current is rapidly modulated over the peak of the absorption line (~1 KHz), the IR absorption signal from the reference detector is input to a lock-in amplifier. An error signal is then generated and sent to the diode laser controller, thus keeping the diode laser frequency locked to the desired absorption transition. Fluctuations in the diode laser intensity over the range of the modulation frequency are corrected by the use of a dual-channel signal acquisition technique (16).

Transient absorption measurements are acquired by pulsing the excimer pump laser at the point of peak absorption during the diode laser modulation cycle. In order to provide ample time for collision products to be pumped out of the cell between laser pulses, the maximum excimer laser repetition rate used was 1 Hz. Dual-channel signal acquisition provides a normalized measure of the IR absorption following each excimer laser pulse. The detected absorption is digitized on a LeCroy 9400A digital oscilloscope and then transferred to a personal computer for further data analysis.

Translational energies, sensed through the velocity recoil profiles of scattered bath molecules, are determined by measuring the Doppler-broadened absorption line-shapes of the rovibrational states. These lineshapes are measured by acquiring transient absorption signals at a series of diode laser frequencies distributed evenly across the absorption line. Since the diode laser linewidth is approximately 0.0003 cm^{-1}, very high resolution measurements of the Doppler lineshapes are possible (see Figure 1). The diode laser frequency is locked to the peak of a fringe from a scanning Fabry-Perot

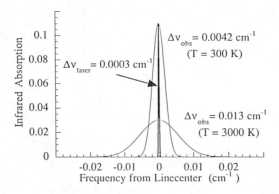

Figure 1. Shown are the laser linewidths for a typical infrared diode laser operating at 4.3 μm (0.0003 cm^{-1}), a Doppler broadened spectroscopic line for a transition at 4.3 μm for a molecule of mass 44 au at 300 K (width 0.0042 cm^{-1}), and a Doppler broadened line for a transition at 4.3 μm for a molecule of mass 44 at 3000 K (width 0.013 cm^{-1}). $\Delta\nu_{laser} = 1/10\Delta\nu_{Doppler}$ at 300 K. $\Delta\nu_{Doppler} = (8kT \ln2/mc^2)^{1/2}\nu_0$.

etalon (FSR = 289 MHz);   the fringe is then scanned across the lineshape in approximately 30 steps.   Roughly 100 transient absorption signals are collected and averaged at every step.

Recently, the question of the effect of donor photofragmentation on these energy transfer experiments has been raised.   Very high product quantum yields were measured from the 248 nm photodissociation of pyrazine using a molecular beam/time-of-flight apparatus (*24*), thus indicating the possibility of bath molecule excitation by dissociative products.   This question was addressed through the performance of a series of important UV pump/IR probe experiments designed to examine the time-dependent formation of photofragments from the donor molecules used in the energy transfer experiments (*25, 26*).   Results of these experiments indicate that the single collision energy transfer data presented here for pyrazine remains unaffected by this fragmentation process.

Research grade CO_2, N_2O, and NO_2 were used in these experiments.   Samples of pyrazine (Aldrich, 99+%) and C_6F_6 (Aldrich, 99+%) were purified by several freeze/pump/thaw cycles before use.

Results

The experiments presented here have been designed to provide a quantum-state resolved picture of the quenching of molecules with chemically significant amounts of energy. Donor molecules prepared with between 20,000 and 40,000 cm^{-1} of vibrational energy undergo collisions with bath molecules resulting in a transfer of energy from one molecular species to another.   Two types of quenching collisions between the donor and bath molecules are presented: those resulting in vibrationally-excited bath molecules and those resulting in ground vibrational state bath molecules.   The collisional transfer of vibrational, rotational, and translational energies are clearly delineated in this way because the narrowband diode lasers provide quantum-state and Doppler resolution.

Collisions Resulting in Vibrationally-Excited Bath Molecules.   Quenching collisions which result in the scattering of bath molecules into vibrationally-excited states are studied by probing the excited rotational-vibrational states of the bath molecules. These experiments explore the V-V (vibration-vibration) energy transfer process in which vibrational energy of the donor molecules is transferred to vibrational energy of the bath molecules.   Both the NO_2/N_2O and pyrazine/CO_2 collision systems were tested for V-V energy transfer (*13, 18*).   For the NO_2/N_2O system, scattering into the v_1 (10^00) and the v_3 (00^01) states of N_2O was probed by a 4.5 μm diode laser tuned to the v_3 antisymmetric stretch vibrational transitions,

$$N_2O\ (10^00,J,V) + hv\ (4.5\ \mu m) \longrightarrow N_2O\ (10^01,J\pm1,V)$$
$$N_2O\ (00^01,J,V) + hv\ (4.5\ \mu m) \longrightarrow N_2O\ (00^02,J\pm1,V)$$

while for the pyrazine/CO_2 system, scattering into the v_3 (00^01) state of CO_2 was probed by a 4.3 μm diode laser tuned to the v_3 antisymmetric stretch vibrational transitions,

$$CO_2\ (00^01,J,V) + hv\ (4.3\ \mu m) \longrightarrow CO_2\ (00^02,J\pm1,V).$$

In general, very little translational energy is transferred during these types of collisions, as is indicated by the near room temperature lineshapes observed see (Figure 2 (b) and (c)).   The amount of rotational energy transferred is also very small ($\Delta J \approx \pm1$). Experiments which explore the cell temperature dependence of the V-V collision rate constants were performed on the pyrazine/CO_2 system (*18*).   For the single-collision regime, rate constants were determined from the measured quantities using the equation

$$k_2^J = \frac{[CO_2\,(00^01,\,J,\,V)]}{[CO_2][\text{pyrazine}^E]_o\Delta t} \tag{3}$$

where $[CO_2]$ is the bulk CO_2 concentration, $[\text{pyrazine}^E]_o$ is the concentration of pyrazine molecules initially excited by the excimer laser pulse, and $[CO_2\,(00^01),J,V]$ is the quantum state population density determined from the infrared absorption measurement at $\Delta t = 1\ \mu sec$ following the excimer pulse. The single-collision rate constants for final rotational states ranging from $J = 3$ to $J = 55$ are presented in Table I. Although the total amount of energy transferred during such collisions is considerable, the probability of occurrence of these V-V collisions is only approximately 1% per gas kinetic collision. The average energy transferred per collision is therefore roughly equal to the bath excited-state vibrational energy divided by 100.

Collisions Resulting in Ground Vibrational State Molecules. This type of quenching collision is used to study V-R/T (vibration-rotation/translation) energy transfer mechanisms in which vibrational energy of the donor molecules is transferred to a combination of rotational and translational energies of the bath molecules. The V-R/T energy transfer mechanism was examined in the NO_2/N_2O, pyrazine/CO_2, and C_6F_6/CO_2 collision systems (13, 16, 17). Since these collisions leave the vibrational state of the bath molecules unchanged, diode probing is performed on the antisymmetric stretch vibrational transition from the ground vibrational state of the bath molecules. Figure 2 (a) shows the lineshape for the $00^00 \rightarrow 00^01$ R62 transition for N_2O molecules scattered into the 00^00, $J = 62$ state. The broad Doppler line profile, with a translational temperature of 881 K, represents the velocity recoil of the scattered ground-vibrational state bath molecules. Doppler linewidths measured as a function of final rotational state are shown in Figure 3 for the pyrazine/CO_2 and C_6F_6/CO_2 collision systems. Significant amounts of rotational energy are also transferred to the bath molecules through the V-R/T collision mechanism. Figure 4 illustrates the change in the rotational distribution of the ground vibrational state of CO_2 following collisions with vibrationally excited C_6F_6. The rotational temperature obtained from the data is approximately 800 K. Note that the total amount of energy transferred per collision is larger for the V-R/T mechanism than for the V-V mechanism because the probability is approximately 1 event per gas kinetic collision.

Discussion

Many energy transfer studies have focused on the importance of these events in the quenching of unimolecular reactions. Most of these studies have primarily probed the loss of energy from the donor, rather than the energy gain in the bath, as is the case in these studies. There are a number of advantages of probing the bath rather than the donor. First, since the excitation of the acceptor molecules is completely resolved with respect to the amount of energy going into the various degrees of freedom, a great deal of insight into the mechanism for energy loss from the donor can be obtained. Second, because only data from the first few microseconds after excitation of the donor are examined, single collision information is obtained for donor states of well defined energy

V–V Energy Transfer. Data from the present studies can be broken into essentially two groups. One describes collision processes where the energy transferred to the bath shows up in vibrational, rotational, and translational excitation of the acceptor.

$$\text{Pyr}^E + CO_2\,(00^00,J',V') \longrightarrow \text{Pyr}^{E-\Delta E} + CO_2\,(00^01,J,V)$$

Table I. Experimentally determined rate constants, k_2^J, for the collsional excitation of rotational states in the vibrationally excited anti-symmetric stretch state, (00^01) of CO_2 by vibrationally hot pyrazine: $Pyr^E + CO_2 (00^00) \rightarrow Pyr^{E-\Delta E} + CO_2 (00^01,J,V)$

Final CO_2 rotational state (J)	k_2^J (cm^3/ molecule s)[a]		
	T_{cell} = 243 K	T_{cell} = 298 K	T_{cell} = 364 K
3		$(8.7\pm2.9) \times 10^{-14}$	$(5.5\pm1.9) \times 10^{-14}$
5	$(2.5\pm0.9) \times 10^{-13}$	$(1.8\pm0.6) \times 10^{-13}$	$(9.6\pm3.2) \times 10^{-14}$
13	$(4.3\pm1.4) \times 10^{-13}$		
15	$(4.6\pm1.5) \times 10^{-13}$	$(2.9\pm0.9) \times 10^{-13}$	$(2.1\pm0.7) \times 10^{-13}$
17	$(4.4\pm1.5) \times 10^{-13}$	$(3.3\pm1.1) \times 10^{-13}$	$(2.4\pm0.8) \times 10^{-13}$
19	$(3.6\pm1.2) \times 10^{-13}$	$(2.8\pm0.9) \times 10^{-13}$	$(2.0\pm0.7) \times 10^{-13}$
21			$(1.5\pm0.5) \times 10^{-13}$
27	$(2.7\pm0.9) \times 10^{-13}$	$(2.7\pm0.9) \times 10^{-13}$	$(1.9\pm0.6) \times 10^{-13}$
29	$(2.4\pm0.8) \times 10^{-13}$	$(1.7\pm0.6) \times 10^{-13}$	$(1.2\pm0.4) \times 10^{-13}$
33	$(1.7\pm0.6) \times 10^{-13}$		$(1.4\pm0.5) \times 10^{-13}$
35	$(1.0\pm0.3) \times 10^{-13}$	$(1.4\pm0.5) \times 10^{-13}$	$(1.1\pm0.4) \times 10^{-13}$
37		$(9.9\pm3.3) \times 10^{-14}$	$(9.2\pm3.1) \times 10^{-14}$
39	$(7.0\pm2.3) \times 10^{-14}$	$(8.4\pm2.8) \times 10^{-14}$	$(8.0\pm2.6) \times 10^{-14}$
41	$(8.5\pm2.8) \times 10^{-14}$	$(7.7\pm2.6) \times 10^{-14}$	$(7.9\pm2.6) \times 10^{-14}$
43	$(4.6\pm1.6) \times 10^{-14}$	$(6.0\pm2.0) \times 10^{-14}$	$(5.1\pm1.7) \times 10^{-14}$
45	$(4.9\pm1.7) \times 10^{-14}$	$(5.2\pm1.7) \times 10^{-14}$	$(5.2\pm1.8) \times 10^{-14}$
47	$(2.6\pm0.9) \times 10^{-14}$	$(3.9\pm1.3) \times 10^{-14}$	$(3.5\pm1.2) \times 10^{-14}$
49	$(2.6\pm0.9) \times 10^{-14}$	$(3.4\pm1.2) \times 10^{-14}$	$(3.5\pm1.2) \times 10^{-14}$
51	$(2.3\pm0.8) \times 10^{-14}$	$(2.9\pm1.0) \times 10^{-14}$	$(3.3\pm1.1) \times 10^{-14}$
53	$(1.8\pm0.6) \times 10^{-14}$	$(1.8\pm0.6) \times 10^{-14}$	
55		$(2.0\pm0.7) \times 10^{-14}$	$(2.9\pm1.0) \times 10^{-14}$

[a] Rate constant calculated from experimentally determined quantities using equation 3.

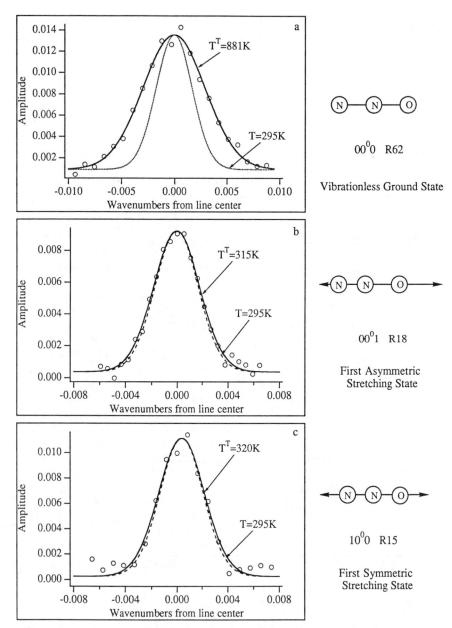

Figure 2. Nascent Doppler profiles of (a) the N_2O vibrationless ground state $00^00 \to 00^01$ R(62), (b) the asymmetric stretching state $00^01 \to 00^02$ R(18), and (c) the symmetric stretching state $10^00 \to 10^01$ R(15) spectral lines after collisions of N_2O with highly excited $NO_2^{(E)}$ at $E = 22{,}200$ cm⁻¹. The circles are data points. The solid curves are the best fits to a Gaussian function. The dotted lines are room temperature Doppler profiles. The measured nascent line width of the $00^00 \to 00^01$ R(62) transition is 0.0070 cm⁻¹, for $00^01 \to 00^02$ R(18) it is 0.0041 cm⁻¹, and for $10^00 \to 10^01$ R(15) it is 0.0041 cm⁻¹. For comparison, the room temperature line width is 0.004077 cm⁻¹.

Aromatic^E + CO$_2$ ----> Aromatic$^{E-\Delta E}$ + CO$_2$ (00^{0}0,J,V)

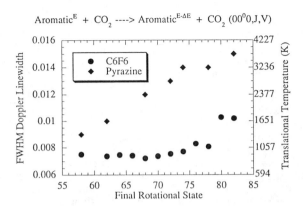

Figure 3. The FWHM Doppler linewidth of absorption transitions probing high rotational states of CO$_2$ following excitation by collisions with vibrationally hot perfluorobenzene (E_{int} = 41,822 cm^{-1}) and vibrationally hot pyrazine (E_{int} = 41,822 cm^{-1}) represented by the circles and diamonds, respectively, is plotted against the final rotational state J. The lab frame translational temperatures which are determined from the measured linewidths are given on the right y-axis. The linewidths are measured 1 μsec following 248 nm excimer laser excitation of the C$_6$F$_6$ or pyrazine, insuring that the measured velocity distributions are nascent (t_{coll} ≈ 4 μsec). The FWHM appropriate for a 298 K velocity distribution is 0.0042 cm^{-1}.

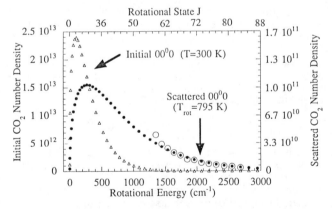

Figure 4. CO$_2$ ground vibrationless state number densities before (left y-axis) and after (right y-axis) collision with highly vibrationally excited perfluorobenzene (E_{int} = 41,822 cm^{-1}) plotted against rotational energy (bottom x-axis) and the rotational state J (top x-axis). The open triangles show a 300 K equilibrium rotational distribution describing the pre-collision CO$_2$ molecules. The open circles show the experimental points measured at 1 μsec following 248 nm excimer laser excitation of the C$_6$F$_6$ along with a fitted 795 K distribution (shaded circles). Number densities are given in units of molecules/cm^3. The number densities on the right y-axis are approximately two orders of magnitude smaller than those on the left y-axis due to the inefficiency of energy transfer into the high J states along with the fact that the number densities are measured at 1/4 of the mean gas kinetic collision time. Reproduced with permission from reference 17.

Here the thermal ($T = 300$ K) bath molecules, CO_2 (00^00), gain at least one quantum in the anti-symmetric stretch vibrational mode, but the final rotational state population distribution can be described by rotational temperatures (T_{rot}) of approximately 320 K, very near the initial, thermal distribution temperature, and a translational temperature (T_{trans}) also very near the initial precollision temperature. The fact that only low translational recoil and small changes in rotational angular momentum are observed strongly suggests that this vibration to vibration energy transfer proceeds via a long-range force mediated interaction employing the attractive tail of the pyrazine-CO_2 intermolecular potential (15).

Temperature dependent studies have shed a great deal of light on this energy transfer process. Of particular usefulness in determining the nature of the interaction inducing the V-V energy transfer is the temperature dependence of the rate constants for scattering into rotational states of the excited 00^01 level. Typically cross sections for V-V energy transfer processes mediated by long-range attractive forces show an inverse temperature dependence; this is considered one of the signatures of an attractive force mechanism (27-38). Conversely, vibrational energy transfer which proceeds via the repulsive wall of the intermolecular potential typically displays a small positive temperature dependence (log P $\propto T^{1/3}$), where P is the energy transfer probability (20, 39, 40).

The rate constants for scattering into the low J regions of the rotational distribution for vibrationally excited states of the bath show a strong inverse temperature dependence, while rate constants for scattering into the high J states of the distribution show essentially no temperature dependence (18). This is another indicator that scattering into low J levels ($J \leq 39$) of the vibrationally excited (00^01) rotational distribution is mediated by this long-range force mechanism. However, there is also strong evidence that scattering into the high J end of the vibrationally excited rotational distribution is significantly affected by a short-range repulsive mechanism, allowing for large changes in rotational angular momentum ($\Delta J \gg \pm 1$). Additionally, these encounters (where the bath molecule is vibrationally excited after a collision with the hot donor molecule) occur rather infrequently (0.1 to 0.001 per gas kinetic collision) and account for very small energy loss from the donor (11, 41-43).

V–RT Energy Transfer. The second category of energy transfer events are ones where only rotational and translational energy is transferred to the bath molecules, hence the bath molecules remain in the ground (vibrationless) state and are scattered to some other rotational and translational state.

$$Pyr^E + CO_2 (00^00, J', V') \longrightarrow Pyr^{E-\Delta E} + CO_2 (00^00, J, V).$$

In these studies, where various rotational states of the ground (vibrationless) level are probed, large amounts of rotational and translational energy are transferred to the bath molecules by vibrationally excited pyrazine ($E_{vib} = 40,640$ cm^{-1}). The initial (00^00) distribution is described by $T_{rot} = 298$ K, and has a mean rotational quantum number of $J = 22$. The scattered CO_2 (00^00) rotational states $J = 58$ to $J = 82$ are described by a much higher rotational temperature, $T_{rot} = 1200 \pm 180$ K. This substantial excitation is only weakly affected by temperature in the range studied ($T = 243$-364 K), indicating that the initial state of the scattered molecules is indeed near the center of the initial room temperature distribution ($J \approx 22$), rather than originating from the high energy wings of the initial thermal distribution.

As mentioned above, vibrational energy transfer which proceeds by a short-range repulsive mechanism gives rise to a slight positive temperature dependence in the rotational rate constants. A plot of the integrated rate constants of the various post collision rotational states ($J = 58$ to $J = 82$) as a function of cell temperature, shows a positive temperature dependence indicative of the short-range mechanism. The fitting of the data results in a precollision rotational state with $<J> \approx 28 \pm 3$ indicating first that the scattering process does indeed originate from the center of the distribution, and second that substantial angular momentum changes ($\Delta J \approx$ 30-54) and large translational recoil accompany these short-range impulsive V–RT collisions between CO_2 and vibrationally excited donors (16). All indicators suggest that these V–RT events are induced by a short-range force mechanism involving the repulsive part of the intermolecular potential. Due to the large energy transferred in these collisions they have been characterized as "supercollisions."

Energy Transfer Probabilities. Obtaining information about the energy transfer probability distribution, P(E,E'), for a donor/bath system from quantum state resolved scattering probabilities involves indexing the energy transfer probabilities, not by the final bath quantum state, but rather by the amount of accompanying energy transfer, ΔE. First we consider determining P(E,E') only for collisions which leave the bath molecule vibrationally unexcited (V–RT energy transfer); vibrational excitation of the bath (V-V energy transfer) will be considered separately. The experimental data is a set of probabilities of the type P(J_i, $g_i \rightarrow J_f$, g_f) where J_i and J_f are the initial and final bath rotational angular momentum quantum numbers; likewise g_i and g_f are the pre- and post-collision donor/bath relative speeds. For any collision described by a probability P(J_i, $g_i \rightarrow J_f$, g_f) there is a corresponding energy transfer of magnitude

$$\Delta E_{J_i\, g_i}^{J_f\, g_f} = B\, J_f\, (J_f + 1) - B\, J_i\, (J_i + 1) + \frac{1}{2}\, \mu(g_f^2 - g_i^2). \qquad (4)$$

The process of constructing a P(E,E') from a complete set of state resolved bath data involves evaluating an expression of the type

$$P(\Delta E)\, d\Delta E = \sum_{J_i\, g_i\, J_f\, g_f} P(J_i, g_i \rightarrow J_f, g_f)\, d(\Delta E - \Delta E_{J_i\, g_i}^{J_f\, g_f})\, d\Delta E \qquad (5)$$

for all $\Delta E = E\text{-}E'$. Since the experiments described here are done on the single collision time scale, E' is defined precisely by the initial optical excitation and thus P(E,E') is simply related to P(ΔE). The delta function in equation 5 insures that the probability associated with a collision involving a given set of quantum numbers, $\{J_i, g_i, J_f, g_f\}$, is distributed into the correct energy bin $d\Delta E$ of the energy transfer distribution function.

It is worth noting that contributions to the quantity P(ΔE) $d\Delta E$ may come from many terms in equation 5. Qualitatively it is easy to imagine two collisions, one involving a large change in bath rotational energy and a small change in the bath translational energy, and the other with just the reverse (small change in bath rotational energy, etc.), which yield the same total ΔE and thus contribute to the same energy bin in P(E,E') despite the different set of bath quantum numbers. To further illustrate this concept, consider the plots shown in Figure 5 for the C_6F_6/CO_2 system. These are P(ΔE) functions constructed counting only the probability for collisions populating a single final bath rotational state (i.e. $J = 64$ or 80). The shape of these single J_f state P(ΔE) distributions essentially reflects the distribution of recoil velocities for molecules scattering into each final rotational state and the corresponding distribution of translational energy transfers. Obviously, contributions to a given energy bin in a P(E,E') can come from collisions populating many final J states (i.e. terms in equation 5). Given plots of the type shown in Figure 5 for each final bath rotational state, a straightforward calculation gives the full P(E,E'); the contributions from each final bath quantum state are simply summed for each energy bin.

It is important to point out several characteristics of the available data, (which are also discussed in some detail in reference 19), that make the transformation discussed above somewhat complicated. First, both the initial rotational states, J_i, and speeds, g_i, which appear in equations 4 and 5 are not well defined in these experiments as there is a thermal distribution of these initial state quantities. Temperature dependent experiments have proven useful in identifying the average initial rotational states and speeds for collisions yielding bath molecules with particular final rotational quantum numbers and speeds, thus allowing a calculation of $P(E,E')$ for two systems, pyrazine/CO_2 and C_6F_6/CO_2 (16, 17, 19). Further efforts to specify the initial state distributions using state selective bath excitation prior to the donor/bath collision are also currently underway (44).

The other difficulty in deriving $P(E,E')$ from the experiments lies in the incomplete set of available data. Due to technical problems, probing bath molecule states which have significant thermal populations ($J_f < 56$ for CO_2) is difficult. Since collisions populating these low J_f states typically involve small energy transfers (low ΔE), the noted lack of data restricts the energy space over which the current experiments provide an accurate representation of $P(E,E')$ (19). The available data for the pyrazine/CO_2 and the C_6F_6/CO_2 systems allow energy transfer distribution functions to be calculated over the range -1500 cm^{-1} > E-E' > -7000 cm^{-1}. Figure 6 shows $P(E,E')$ for the pyrazine/CO_2 system where $E' = 40,640$ cm^{-1}; the broad "background" is, as indicated, due to V-RT transfer in which donor vibrational energy is transferred into the bath rotational and translational degrees of freedom. This part of the distribution function was calculated using the methodology discussed above and reveals that large ΔE collisions (E-E' < -1500 cm^{-1}) occur predominantly via the short-range V-RT mechanism with a probability on the order of 5 x 10^{-5} /cm^{-1} or smaller for this system.

The sharp resonance like feature in $P(E,E')$ at E-E' = 2350 cm^{-1} is due to the long range resonant excitation of the ν_3 (anti-symmetric stretch) vibrational mode of CO_2 (19). Pyrazine/CO_2 collisions which result in excitation of the ν_3 mode are relatively inefficient, occurring at the rate of roughly one per hundred Lennard-Jones collisions (18), yet this probability (~1%) is spread over a very narrow range of ΔE thus yielding the sharp feature in $P(E,E')$. In collisions which involve bath vibrational excitation, there are three contributions to the total energy transfer, one each arising from the changes in the bath molecules' vibrational, rotational, and translational energy. Clearly, the vibrational contribution to ΔE is just the energy of the vibrational state excited, while, as discussed earlier, the product state distributions reveal that V-V collisions are accompanied by very little rotational and translational excitation of the bath molecule. Thus, the total energy transfer is given largely by the change in vibrational energy, and the probability for this process is distributed into a narrow energy bin centered at $\Delta E = 2350$ cm^{-1} when constructing $P(E,E')$. (A discussion of the width of this feature can be found in reference 19). V-V resonance features like that shown in Figure 6 are likely to be found in $P(E,E')$ functions for all donor/polyatomic bath systems in which excitation of high frequency bath vibrational modes takes place to an appreciable extent.

Several additional characteristics of the $P(E,E')$ functions discussed here warrant comment. Attempts to fit the pyrazine/CO_2 $P(E,E')$ function shown in Figure 6 to a commonly used exponential model yields an average energy transfer over all collisions involving energy transfer from donor to bath, $<\Delta E>_d = 536$ cm^{-1} (19), which is in reasonably good agreement with the value obtained from Infrared Fluorescence experiments, $<\Delta E>_d = 380$ cm^{-1} (45), especially considering that only the high ΔE tail of the distribution function is available for fitting. This agreement is encouraging although it is clear that further development of methods to probe low ΔE collisions is necessary to provide a more detailed comparison between bath and donor probe experiments. Upon examination of $P(E,E')$ for both the pyrazine/CO_2 and the C_6F_6/CO_2 systems, the V-RT channel is clearly found to be a more efficient donor deactivation mechanism than the V-V channel. While this is expected to be a general conclusion for systems with small bath molecules, generalization to systems involving bath molecules with many more

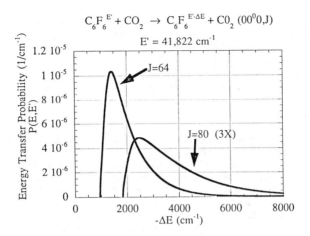

$$C_6F_6{}^{E'} + CO_2 \rightarrow C_6F_6{}^{E'-\Delta E} + CO_2 \ (00^00,J)$$

$E' = 41,822 \ cm^{-1}$

Figure 5. Shown are plots of the probability that the first Lennard-Jones collision between a highly vibrationally excited C_6F_6 molecule with $E' = 41,822 \ cm^{-1}$ and a carbon dioxide bath molecule results in excitation of the CO_2 rotational state $J = 64$ or $J = 80$ in the ground vibrationless level with an accompanying energy transfer of magnitude ΔE. The mean initial rotational state, J_i, for collisions populating high J_f states ($J_f \approx 58-82$) of CO_2 is estimated to be $J_i = 31$ based on temperature dependent experiments.

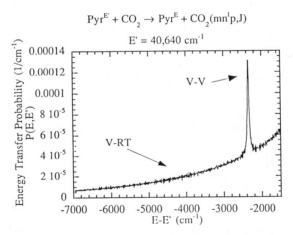

$$Pyr^{E'} + CO_2 \rightarrow Pyr^E + CO_2(mn^lp,J)$$

$E' = 40,640 \ cm^{-1}$

Figure 6. Shown is a plot of $P(E,E')$, the energy transfer distribution function for the pyrazine/CO_2 system with $E' = 40,640 \ cm^{-1}$, including the probability of vibrational excitation of the CO_2 anti-symmetric stretch, $v_3 = 2349 \ cm^{-1}$, vibrational mode. E is the energy of the hot pyrazine molecule following a single pyrazine/CO_2 collision. The long range, resonant V-V energy transfer process resulting in excitation of the v_3 CO_2 vibration is accompanied by little rotational or translational excitation of the bath, giving rise to the narrow resonance like feature in $P(E,E')$ centered at $E-E' = 2350 \ cm^{-1}$. The V-RT contribution is calculated based on an estimated value of $J_i = 28$ for the mean initial rotational state, J_i, for collisions populating high J_f states ($J_f \approx 58-82$) of CO_2 as determined from temperature dependent experiments. Reproduced with permission from reference 19.

acceptor vibrational modes may not be correct as V-V energy transfer may play a larger role in such cases (*46*). Finally, examination of Figure 6 reveals that the bath probe technique is capable, not only of identifying the major mechanisms of energy transfer in collisions between highly vibrationally excited donor molecules and bath molecules, but also of resolving the dynamic signatures of these mechanisms in P(*E,E*'), including broad "background" V-RT and sharp resonance feature V-V energy transfer processes.

Acknowledgments

The authors thank Drs. Klaus Luther and Thomas Lenzer for several stimulating discussions. Suggestions and encouragement from Professors J. Barker, H. Hippler, and I. Oref are greatly appreciated. The authors also thank Dr. James Chou, Dr. Liedong Zheng, and Professor Amy Mullin for contributions to earlier studies described here. This work was performed at Columbia University and supported by the Department of Energy (DE-FG02-88-ER13937). Equipment support was provided by the National Science Foundation (CHE-94-19465) and the Joint Services Electronics Program (U. S. Army, U. S. Navy and U. S. Air Force; DAAH04-94-40057).

Literature Cited

1   Lindemann, F. A. *Trans. Faraday Soc.* **1922**, *17*, 598.
2   Weston, R. E., Jr.; Schwarz, H. A. *Chemical Kinetics*; Prentice-Hall: Englewood Cliffs, 1972.
3   Gilbert, R. G.; Smith, S. C. *Theory of Unimolecular and Recombination Reactions*; Blackwell Scientific Publications: Oxford, 1990.
4   Hippler, H.; Troe, J. . In *Bimolecular Collisions*; Ashfold, M. N. R., Baggott, J. E., Eds.; Royal Society of Chemistry: London, 1989.
5   Barker, J. R.; Toselli, B. M. *Int. Rev. Phys. Chem.* **1993**, *12*, 305.
6   Löhmannsröben, H. G.; Luther, K. *Chem. Phys. Lett.* **1988**, *144*, 473.
7   Luther, K.; Reihs, K. *Ber. Bunsenges. Phys. Chem.* **1988**, *92*, 442.
8   Lendvay, G.; Schatz, G. C. *J. Phys. Chem.* **1990**, *94*, 8864.
9   Lenzer, T.; Luther, K.; Troe, J.; Gilbert, R. G.; Lim, K. F. *J. Chem. Phys.* **1995**, *103*, 626.
10 Clary, D. C.; Gilbert, R. G.; Bernshtein, V.; Oref, I. *Faraday Discuss.* **1996**, *102*, In Press.
11 Chou, J. Z.; Flynn, G. W. *J. Chem. Phys.* **1990**, *93*, 6099.
12 Chou, J. Z.; Hewitt, S. A.; Hershberger, J. F.; Flynn, G. W. *J. Chem. Phys.* **1990**, *93*, 8474.
13 Zheng, L.; Chou, J. Z.; Flynn, G. W. *J. Phys. Chem.* **1991**, *95*, 6759.
14 Sedlacek, A. J.; Weston, R. E., Jr.; Flynn, G. W. *J. Chem. Phys.* **1991**, *94*, 6483.
15 Mullin, A. S.; Park, J.; Chou, J. Z.; Flynn, G. W.; Weston, R. E., Jr. *Chem. Phys.* **1993**, *175*, 53.
16 Mullin, A. S.; Michaels, C. A.; Flynn, G. W. *J. Chem. Phys.* **1995**, *102*, 6032.
17 Michaels, C. A.; Lin, Z.; Mullin, A. S.; Tapalian, H. C.; Flynn, G. W. *J. Chem. Phys.* **1996**, submitted for publication.
18 Michaels, C. A.; Mullin, A. S.; Flynn, G. W. *J. Chem. Phys.* **1995**, *102*, 6682.
19 Michaels, C. A.; Flynn, G. W. *J. Chem. Phys.* **1997**, submitted for publication.
20 Yardley, J. T. *Introduction to Molecular Energy Transfer*; Academic Press: New York, 1980.
21 Moore, C. B. *Adv. Chem. Phys.* **1973**, *23*, 41.
22 Weitz, E.; Flynn, G. W. *Ann. Rev. Phys. Chem.* **1974**, *25*, 275.
23 Flynn, G. W.; Weston, R. E., Jr. *J. Phys. Chem.* **1993**, *97*, 8116.

24 Chesko, J. D.; Stranges, D.; Suits, A. G.; Lee, Y. T. *J. Chem. Phys.* **1995**, *103*, 6290.
25 Michaels, C. A.; Tapalian, H. C.; Lin, Z.; Sevy, E. T.; Flynn, G. W. *Faraday Discuss.* **1995**, *102*, In press.
26 Sevy, E. T.; Muyskens, M. A.; Lin, Z.; Michaels, C. A.; Tapalian, H. C.; Flynn, G. W. Manuscript in preparation.
27 Mahan, B. H. *J. Chem. Phys.* **1967**, *46*, 98.
28 Sharma, R. D.; Brau, C. A. *Phys. Rev. Lett.* **1967**, *19*, 1273.
29 Stephenson, J. C.; Moore, C. B. *J. Chem. Phys.* **1972**, *56*, 1295.
30 Zittel, P. F.; Moore, C. B. *J. Chem. Phys.* **1973**, *59*, 6636.
31 Lucht, R. A.; Cool, T. A. *J. Chem. Phys.* **1975**, *63*, 3963.
32 McGarvey, J. A. J.; Friedman, N. E.; Cool, T. A. *J. Chem. Phys.* **1977**, *66*, 3189.
33 Foster, T. J.; Crim, F. F. *J. Chem. Phys.* **1981**, *75*, 3871.
34 Zittel, P. F.; Masturzo, D. E. *J. Chem. Phys.* **1989**, *90*, 977.
35 Dillon, T. A.; Stephenson, J. C. *J. Chem. Phys.* **1974**, *60*, 4286.
36 Shin, H. K. *Chem. Phys. Lett.* **1975**, *32*, 218.
37 Shin, H. K. *J. Chem. Phys* **1993**, *98*, 1964.
38 Poulsen, L. L.; Billing, G. D.; Steinfeld, J. I. *J. Chem. Phys.* **1978**, *68*, 5121.
39 Lambert, J. D. *Vibrational and Rotational Relaxation In Gases*; Clarendon Press: Oxford, 1977.
40 Schwartz, R. N.; Slawsky, Z. I.; Herzfeld, K. F. *J. Chem. Phys.* **1952**, *20*, 1591.
41 Tang, K. Y.; Parmenter, C. S. *J. Chem. Phys.* **1983**, *78*, 3922.
42 Tardy, D. C.; Rabinovitch, B. S. *Chem. Rev.* **1977**, *77*, 369.
43 Jalenak, W.; Weston, R. E., Jr.; Sears, T.; Flynn, G. W. *J. Chem. Phys.* **1988**, *89*, 2015.
44 Tapalian, H. C.; Michaels, C. A.; Sevy, E. T.; Flynn, G. W. Work in progress.
45 Miller, L. A.; Barker, J. R. *J. Chem. Phys.* **1996**, *105*, 1383.
46 Lenzer, T.; Luther, K. *J. Chem. Phys.* **1996**, *104*, 3391.

Chapter 11

Ab initio Calculations of the Interaction Potentials of Ar–HCN and Ar–HCO

Jianxin Qi, Max Dyksterhouse, and Joel M. Bowman[1]

Department of Chemistry and Cherry L. Emerson Center for Scientific
Computation, Emory University, Atlanta, GA 30322

First results of *ab initio* calculations of potential surfaces of Ar-HCN
and Ar-HCO are reported. Single reference, configuration interaction
(CISD) calculations, using double zeta and triple zeta basis sets, and
counterpoise correction are done using the code MOLPRO. Limited
test calculations are also done using the CEPA method, and also
limited CISD calculations are done with triple zeta basis sets for all
atoms. The electronic energies are fit using a novel sum-of-pairs
potential. In this fit the pair potential depends on the Ar-atom
internuclear position vector. This representation is shown to be
excellent for fitting the interaction potential surface for fixed
molecular geometries of HCN and HCO. Potential surfaces are given
for HCN in the vicinity of its equilibrium geometry, HNC, and an
approximate HCN/HNC saddle point geometry. The potentials for Ar-
HCO are reported for three geometries, equilibrium HCO, the H-CO
saddle point geometry, and one for a nearly dissociated H+CO. A
comparison of the fits to the *ab initio* energies with previous
conventional Lennard-Jones, and new *ab initio* two-body, sum-of-pairs
potentials is given for Ar-HCN.

The rare gas-molecule interaction is the simplest one that can induce molecular
energy transfer, association, dissociation, isomerization, etc. Thus, rare gas-molecule
systems have been used in many theoretical and experimental studies of these
important chemical processes. In recent years, there have been several quasiclassical

[1]Corresponding author

150

trajectory calculations on the energy transfer process of highly excited SF_6 [1] and CS_2 [2] in collision with He, Ar, Xe, and the energy transfer and dissociation from highly excited OCS in collision with Ar. More recently, there have been quantum state specific scattering studies on collision induced isomerization of Ar-HCN system [3], (with HCN treated as a semi-rigid bender), collision induced dissociation and energy transfer of highly vibrationally excited HCO in collisions with Ar [4], similar studies for Ar-HCN (with the HCN vibrations treated exactly) [5], and quantum/classical comparisons of energy transfer in CS_2 [6]. In all cases, because *ab initio* interaction potentials were not available for these systems, these dynamical studies used simple Lennard-Jones sum-of-pairs potentials.

There have been a number of *ab initio* calculations of rare gas-polyatomic van der Waals systems. Specifically for Ar-HCN, Clary *et al.* [7] reported an *ab initio* interaction potential surface with a fixed equilibrium HCN geometry, based on the CEPA method. This potential cannot be used to calculate the collision processes mentioned above that involve geometry changes of HCN. It also cannot describe vibrational predissociation of van der Waals complexes, nor the interesting dependence of van der Waals spectra on internal excitation. More recently, Tao *et al.* [8] reported *ab initio* SCF, MP2, and MP4 calculations of the Ar-HCN van der Waals complex. These authors investigated the effect on the potential surface of bending HCN by 15 and 30 degrees from the linear geometry. No *ab initio* calculations have been reported for the Ar-HCO potential.

In this chapter, *ab initio* calculations of the Ar-HCN and Ar-HCO interaction potentials are reported for three HCN and three HCO geometries. The calculations were done using MOLPRO [9]; details of the calculations will be given below. The ultimate goal of these calculations are realistic six degree-of-freedom potentials for these systems that will be used in dynamics calculations. Thus, an efficient and accurate method to fit the *ab initio* electronic energies is crucial. We present a novel fitting approach to the *ab initio* calculations, and also compare the resulting potential surfaces to two sum-of-pairs potentials, one based on the standard Lennard-Jones form, and the other based on new *ab initio* two-body potentials. Although the emphasis in this work is not on the van der Waals complexes, we do make some comparisons with previous calculations on the Ar-HCN system, which did focus on the van der Waals system [7, 8].

Calculations

The coordinate system, details of the *ab initio* method, basis sets, and tests are described in this section.

Coordinate System. Before going into the details of the calculation, we describe the coordinates used to represent our results. The coordinates are body fixed spherical coordinates with the origin at the center of mass of CN(O), with the z axis along CN(O), and H in xz plane. As shown in Fig.1 the CN(O) distance is denoted by $r_{CN(O)}$, the position of H is $(R_H, \gamma, 0)$, and the position of Ar is (R, θ, φ). In order to represent the relative position of Ar to a given atom we also introduce a local, atom-centered coordinate system. The position of Ar relative to this coordinate system is denoted $(R_{ArX}, \theta_X, \varphi_X)$, where X denotes the atom H, C, N (or O).

Ab initio **Calculations.** The *ab initio* calculations were performed using the MOLPRO program package [9]. The calculations were done at the CISD level with the Davidson correction. The interaction energies were calculated using the supermolecule model, in which the whole system is treated as a molecule. The reference energies were calculated for a separation between Ar and the molecule of 20 angstroms, i.e., R = 20 angstroms.

Basis Sets. The choice of basis sets was based on the consideration of polarization of atoms. The polarization of Ar and H atoms are harder to describe using small basis sets, so the relatively large AVTZ basis sets [10] were used for these atoms. For other atoms AVDZ basis sets were used. The basis set superposition error (BSSE) was corrected using the standard counterpoise method [11].

Test of the CISD Method. We performed several tests of the CISD method and the basis sets. The CISD interaction energies, with the Davidson correction, were checked against CEPA calculations for Ar-HCN and Ar-HCO, using the same basis sets. Fig. 2 compares interaction energies for three sets of calculations (without the counterpoise corrections) for the collinear Ar-HCN configuration, and for the geometries indicated for Ar-HCO, with HCO at its equilibrium and transition state geometries. As seen the agreement using these two methods is excellent.

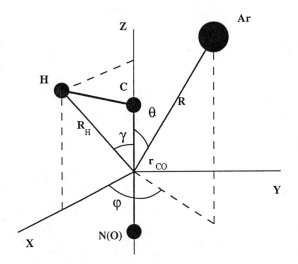

Fig. 1: Coordinate system for Ar-HCN(O)

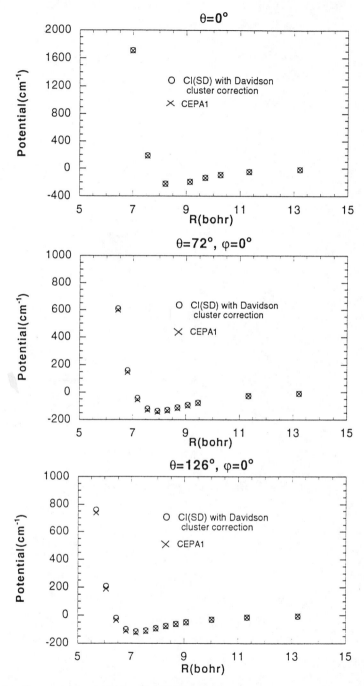

Fig. 2: Comparison of CISD and CEPA1 interaction energies. Upper panel: Ar with linear HCN; middle panel: Ar with minimum HCO geometry; lower panel: Ar with transition state HCO geometry.

It should be noted that CISD is a single reference method, which for HCN should be perfectly adequate. For HCO there are two conical intersections with the ground state X ^{2}A'. Fortunately the intersection regions are at the linear HCO and HOC geometries that are far removed in energy from the HCO minimum and the dissociation reaction path, which are the regions of the HCO potential of interest to us. We did perform CISD calculations on isolated HCO to determine the equilibrium HCO geometry and dissociation energy, D_e. The results are shown in Table 1, along with previous results from Keller *et al.* who use a CASSCF+MRCI method with quite large basis set [12], and also from earlier CISD calculations of Harding [13]. As seen, there is good agreement with these previous calculations, especially with the recent ones of Keller *et al.*

Next we examine the adequacy of the basis sets chosen for Ar-HCN and Ar-HCO.

Table 1. Stable HCO geometry and dissociation energy

	this work	WKS[a]	BBH[b]
r_{CO}(bohr)	2.229	2.233	2.259
r_{CH}(bohr)	2.094	2.110	2.124
< HCO(deg)	125.4	124.5	124.2
D_e(eV)	0.852	0.834	0.785

[a] Ref. 12

[b] Ref. 13

Test of the Basis Sets. As noted above, we used AVTZ basis sets for Ar and H and AVDZ basis sets for C , N and O. The latter might be considered relatively small, and perhaps also lead to some basis set imbalance. Therefore, we performed some comparisons of interaction energies using the AVTZ/AVDZ basis with energies using AVTZ basis sets for all atoms. These calculations were done with the counterpoise correction. The results of these comparisons are shown in Fig. 3 for Ar-HCN and Ar-HCO, for the angles indicated. As seen, there is very good agreement, except at the highest energies shown, where the energies for the smaller basis are a few percent higher than for the bigger basis. (Note, that the AVTZ calculations take about three times the computer time as ones with the smaller basis.)

Thus, based on these tests and the one shown in Fig. 2, we conclude that for both Ar-HCN and Ar-HCO, the CISD method, with the Davidson and counterpoise corrections, and the smaller basis should provide realistic interaction potential surfaces.

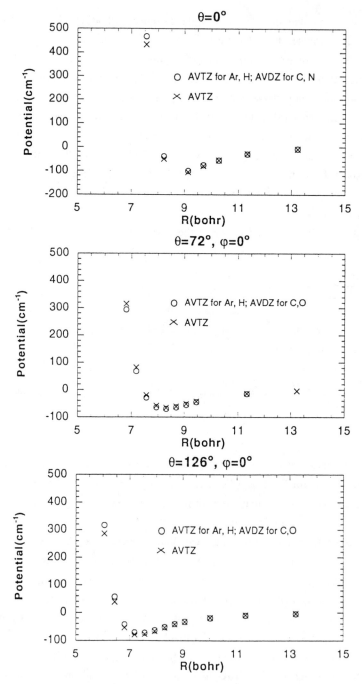

Fig. 3: Comparison of CISD interactions energies for basis sets indicated. Upper panel: Ar with linear HCN geometry; middle panel: Ar with minimum HCO geometry; lower panel: Ar with transition state HCO geometry.

Results and Surface Fitting

The difficulties in constructing a potential surface from *ab initio* energies are well known, and our ultimate goal of having six degree-of-freedom potentials for both Ar-HCN and Ar-HCO is still a distant one. Among the various strategies we are considering is one based on a novel sum-of-pairs representation. Recall that in the usual sum-of-pairs approach the potential is written as

$$V_{int} = \sum_X V_{ArX}(R_{ArX}),$$

(1)

where X represents atoms H,C,N or H,C,O, and $V_{ArX}(R_{ArX})$ is the (isotropic) two-body interaction between isolated atom X and Ar, e.g., a Lennard-Jones potential. The appeal of the sum-of-pairs potential is that the potential is full dimensional, and can be used in studies in which the molecular geometry changes. However, the potential is not expected to be very realistic, and for van der Waals systems it has long been abandoned even as a zero-order model. It has not been examined yet in any detail for rare-gas polyatomic molecule interactions where the molecular geometry changes. We will do that in a limited way below.

Table 2. Molecular geometries used in the calculations

	HCN		
	HCN	HNC	transition
r_{CN}(bohr)	2.196	2.196	2.196
R_H(bohr)	3.282	2.918	2.258
γ(deg)	0.0	180.0	76.9

	HCO		
	HCO	Saddle point	H+ CO
r_{CO}(bohr)	2.288	2.153	2.288
R_H(bohr)	3.043	4.213	5.669
γ(deg)	34.77	47.26	34.77

Clearly an isotropic two-body potential cannot in general represent the interaction of the atom in the molecule with a rare-gas atom, since the true interaction depends on the direction as well as the magnitude of the interatomic position vector. Thus, a reasonable generalization of the isotropic two-body potential is a two-body interaction that depends on the interatomic distance position vector. We use this approach to fit the Ar-HCN and Ar-HCO interaction potentials for molecular geometries given in Table 2. For Ar-HCN the three HCN geometries are linear HCN, linear HNC, and a bent structure that is close to HCN/HNC saddle point [14].

Details of the Fitting. The functional form of the new two-body representation of the interaction potential is

$$V_{int}(R,\theta,\varphi,R_H,r,\gamma) = \sum_{X=C,H,O(N)} V_{ArX}(R_{ArX},\theta_X,\varphi_X),\tag{2}$$

where

$$\begin{aligned}V_{ArX}(R_{ArX},\theta_X,\varphi_X) &= V^0_{ArX}(R_{ArX}) + V^1_{ArX}(R_{ArX})\cos(\theta_X) + V^2_{ArX}(R_{ArX})\cos^2\theta_X\\ &+ V^3_{ArX}(R_{ArX})\sin\theta_X\cos\varphi_X + V^4_{ArX}(R_{ArX})\sin^2\theta_X\cos^2\varphi_X\\ &+ V^5_{ArX}(R_{ArX})\cos\theta_X\sin\theta_X\cos\varphi_X + V^6_{ArX}(R_{ArX})\sin^2\theta_X\sin^2\varphi_X\end{aligned}$$

$$\tag{3}$$

and where each $V^l_{ArX}(R_{ArX})$ is expressed in the form adopted from Clary *et al.* [7], i.e.,

$$V^l(R_{ArX}) = \sum_{i=1}^{n-1} C^l_i R^{i-1}_{ArX} e^{-\alpha R_{ArX}} + C^l_n[1-\exp(-(\frac{R_{ArX}}{\beta})^{n_\beta})]/R^6_{ArX}.\tag{4}$$

The C^l_i's are the undetermined coefficients, α, β, and n_β are pre-determined parameters. The n is the number of the expansion terms; we choose n=5 for all of our fitting. For linear molecule geometries, in order to have cylindrical symmetry, only the first three terms in (3) are non-zero.

It should be noted that the above non-isotropic pairwise form for the potential bears some resemblance to a form suggested by Naumkin and Knowles[15] to represent the interaction of a rare gas with a halogen molecule. They suggested a simple geometric-weighted sum of two body ground and excited state electronic potentials.

Ar-HCN

The Fit. For the linear HCN and HNC geometries the calculations were done for 110 Ar geometries with θ varying from 0 to $180°$, in steps of $18°$ and 10 values of R per θ.

For the bent HCN configuration φ varied from $0°$ to $180°$ in steps of $30°$ and so 770 Ar geometries were done for this HCN configuration.

The calculated *ab initio* energies for Ar-HCN were fit using the functional form of eqs. (2) - (4). After some experimentation fitting radial cuts, we used for the non-linear parameters the following values: $\alpha = 2.0$ angstrom^{-1} $\beta = 4$ angstrom, and $n_\beta = 8$. The linear parameters were all determined by a standard least squares method.

The fits are all quite good, with average absolute error of 10 cm^{-1} or less. Fig. 4 shows two fits of the interaction energies versus θ at fixed R = 7.56 bohr for Ar-HCN(bent). We have also examined radial cuts for the fitted potential for the 11 values of θ where calculations were done, and also at the values of φ where calculations were done, and found very good agreement with the *ab initio* points.

Polar contour plots of the fitted surfaces are given in Figs. 5 -7. As seen, for Ar-HCN there is a single well at the linear Ar-H-C-N geometry. For HNC, there is a single well at the Ar-H-N-C geometry, and for bent HCN there is also a single well in the vicinity of the H atom. Thus, it appears that the minimum van der Waals structure is largely determined by the position of the H atom.

Comparison with Isotropic Sum-of-Pairs Potential. In order to examine the realism of the conventional isotropic sum-of-pairs potential we calculated potentials for the three HCN geometries considered above using simple Lennard-Jones pair potentials, with parameters taken from a previous quantum scattering study of energy transfer in Ar collisions with HCN [5]. Comparisons of the contour plots are given in Figs. 8, 9, 10 and 11. From these figures, we see that the conventional sum-of-pair potentials are qualitatively reasonable, but not at all quantitative in the attractive and repulsive region.

We also calculated the sum-of-pairs potential using *ab initio* diatomic potentials (using the CISD method with the same basis sets used in the full calculations). We do not find a significant improvement over the Lennard-Jones sum-of-pairs potential.

Van der Waals Minima. Approximate values of the Ar-HCN minima for the three HCN geometries have been determined based on the fits. For Ar-HCN the minimum is for the linear geometry, i.e., $\theta = 0$, with R equal to 8.9 bohr, corresponding to an ArH distance of 5.6 bohr, and a well depth of 104 cm^{-1}. These values can be compared to previous results; however, it must be keep in mind that the HCN geometry we have considered is somewhat different from the equilibrium geometry (the difference is in the value of the CH bond length). Clary *et al.* [7] reported the

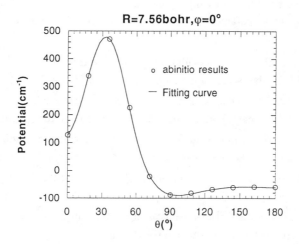

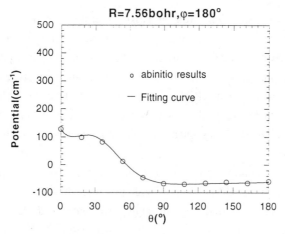

Fig. 4: The interaction energies vs q for Ar-HCN with HCN at transition state geometry, circles: *ab initio* , solid line: fit.

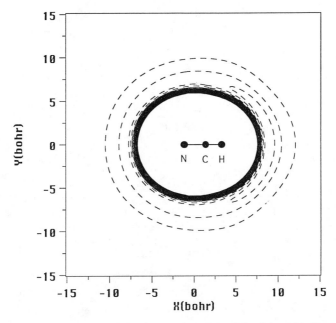

Fig. 5: Polar contour plot of the interaction potential of Ar with linear HCN. Contour values are in cm^{-1}.

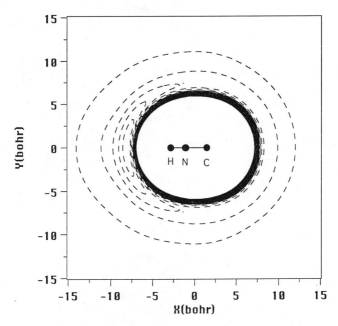

Fig. 6: Same as Fig. 5, but of Ar with linear HNC.

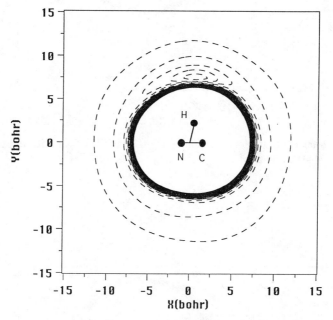

Fig. 7: Same as Fig. 5, but of Ar with HCN at transition state geometry

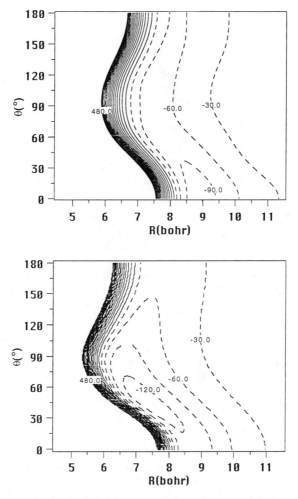

Fig. 8: Comparison of *ab initio* potentials and Lennard-Jones sum-of-pairs potentials for Ar with linear HCN. Upper: *ab initio*. Lower: sum-of-pairs. Contour values are in cm^{-1}.

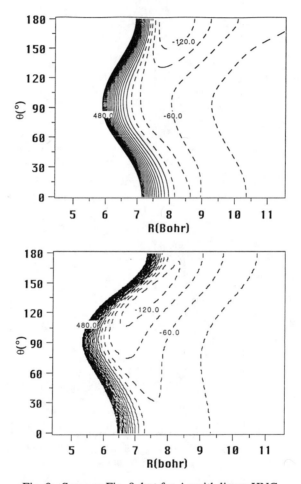

Fig. 9: Same as Fig. 8, but for Ar with linear HNC.

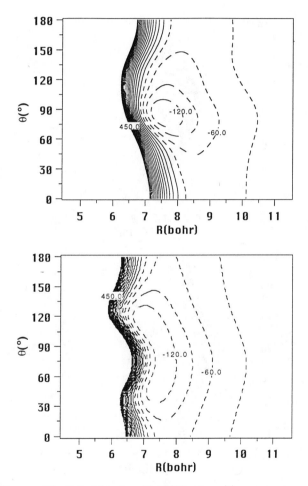

Fig. 10: Same as Fig. 8, but for Ar with HCN at transition geometry, and $\varphi = 0^{\circ}$.

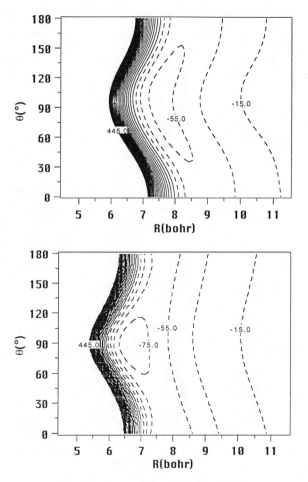

Fig. 11: Same as Fig.10 But φ=180°.

minimum for the linear geometry with an ArH distance of 5.68 bohr and well depth of 84.95 cm^{-1}. The more recent calculations of Tao *et al.* [8] also find the minimum for the linear Ar-HCN geometry; with a well depth of 135 cm^{-1} and an ArH distance of 5.56 bohr. We are in agreement with Tao *et al.*, and in disagreement with Clary *et al.*, in not finding a minimum for $\theta = 180°$. [We performed a more systematic determination of the van der Waals minimum using the coupled-cluster method, CCSD(T), with AVTZ basis sets for all atoms. For the calculated HCN minimum geometry, we find a well depth of 130 cm^{-1}, and an ArH distance of 5.50 bohr, in very good agreement with the results of Tao *et al* .]

For Ar-HNC the structure of the van der Waals minimum is also a linear one with the Ar at the H side, and we estimate that R is approximately 8 bohr and the well depth is approximately 155 cm^{-1}. The minimum point of interaction energy of Ar and the bent HCN geometry is a planar structure with θ approximately 90°, R approximately 7.5 bohr. and a well depth of approximately 155 cm^{-1}.

Ar-HCO. We calculated three potential surfaces corresponding to equilibrium HCO, a nearly dissociated H+CO geometry, and the H-CO saddle point geometry. These HCO geometries were taken from the modified BBH surface[13] and are given in Table 2. To date we have applied the counterpoise correction to the planar Ar-HCO geometries only, and so results are shown only for that configuration, i.e., $\varphi = 0°$ and 180°. The calculations were done using the same values of θ described above for Ar-HCN. The fit was done exactly as described for Ar-HCN; however, the values of the non-linear parameters are $\alpha = 3.0$ angstrom^{-1}, $\beta = 3.0$ angstrom, and $n_\beta = 8$. The fits are all quite good, with average absolute error again of 10 cm^{-1} or less.

Two plots of the fit and the calculated interaction energies vs θ are shown in Fig.12. We have also examined all radial cuts where the *ab initio* calculations were and agreement of the fits and *ab initio* energies is very good.

Polar contour plots of the fitted potentials are given in Figs. 13-15. for $\phi = 0$, i.e., for the Ar on the same side of the CO axis as the H atom. As seen, the van der Waals minima are not localized near the H atom, as they are in Ar-HCN, nor are they as deep and generally as localized as they are in Ar-HCN. Based on the fitted potential we estimate for the HCO equilibrium geometry that at the minimum θ i s approximately 108°, the well depth is approximately 91 cm^{-1}, and R is approximately 7.1 bohr. For the HCO saddle point geometry we estimate the same value of the θ, a well depth of 76 cm^{-1}, and R equal to 7.6 bohr. In order to determine the geometry of the van der Waals complex and well depth more precisely, CCSD(T) level calculations with the same basis set have been done to manually search the minimum

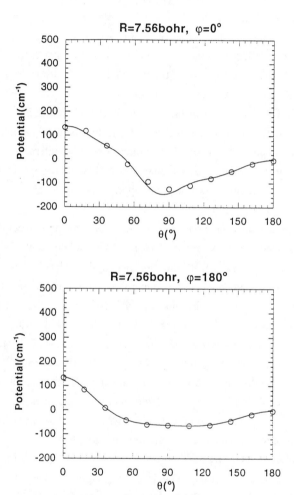

Fig. 12: The interaction energies vs θ for Ar- equilibrium HCO, circle: *ab initio* , solid line: fit.

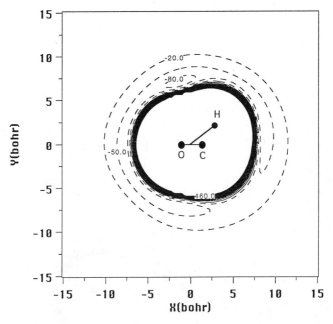

Fig. 13: Polar Contour plot of interaction potential of Ar with equilibrium HCO. Contour values are in cm^{-1}.

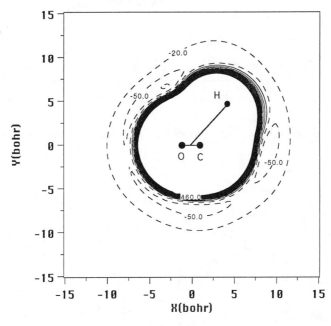

Fig. 14: Same as Fig.13, but of Ar with HCO at transition state geometry.

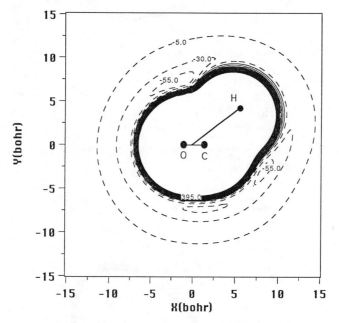

Fig. 15: Same as Fig.13, but of Ar with HCO at a nearly dissociated geometry.

around the estimated minimum. The minimum θ is $90°$, the well depth is approximately 110 cm^{-1}, and R is 7.37 bohr.

It is interesting to note that the characteristics of the Ar-HCO minimum are quite similar to those calculated very recently for Ar-CO [16], i.e., R = 7.073 bohr, θ = 98°, and a well depth = 96.3 cm^{-1}. This similarity is probably not surprising given that the CH bond in HCO is quite weak, unlike the CH bond in HCN.

Summary and Conclusions

We presented results of *ab initio* calculations of the interactions potentials for Ar-HCN and Ar-HCO considering three molecular geometries in each case. For Ar-HCN the three geometries were linear HCN, linear HNC, and a bent geometry in the vicinity of the HCN/HNC transition state. For Ar-HCO the three geometries were the equilibrium HCO configuration, the H-CO transition state, and a nearly dissociated H+CO. The electronic energies were fit using a novel anisotropic sum-of-pairs form. Contour plots were presented, and compared to potentials based on Lennard-Jones sum-of-pairs potentials and *ab initio* sum-of-pairs potentials. Neither of these approximate representations gave satisfactory agreement with the *ab initio* potentials. Several tests of the *ab initio* method and basis sets give us confidence that the energies obtained are quite realistic for use in dynamics calculations, which is the ultimate use we intend to make of these surfaces. Nevertheless, we did make several comparisons of the van der Waals complexes for the Ar-HCN system with previous calculations that focused on those complexes. Generally good agreement was found with the previous results. Characteristics of the van der Waals minimum for Ar-HCO were reported for the first time, and found to be quite similar to those of Ar-CO.

Acknowledgments. JMB thanks the Department of Energy (DEFG05-86ER13568) for support of work on Ar-HCO and the National Science Foundation (CHE-940815030) for support of the work on Ar-HCN. We thank Professor Keiji Morokuma, Dr. Jamal Musaev, and Professor H-J. Werner for helpful discussions. We also thank Professor Werner for pointing out reference 15, and Susan Culik for proofreading the manuscript.

Literature Cited

1. Lendvay, G.; Schatz, G. C. *Chem. Phys.*, **1993**, *98*, 1034
2. (a) Gibson, L. L.; Schatz, G. C. *J. Chem. Phys.*, **1985**, *83*, 3433; (b) Bruehl, M., Schatz, G. C. *J. Chem. Phys.*, **1988**, *89*, 770; (c) Bruehl, M., Schatz, G. C. *J.*

Phys. Chem., **1988**, *92*, 7223; (d) Lendvay, G. Schatz, G. C. *J. Phys. Chem.*, **1991**, *95*, 8748.

3. Lan, B. L.; Bowman, J. M. *J. Chem. Phys.*, **1994,** *101*, 8564.

4. Pan, B; Bowman, J. M. *J. Chem. Phys.*, **1995**, *103*, 9661.

5. Bowman, J. M.; Padmavathi, D. A. *Mol. Phys.,* **1966**, *88*, 21.

6. (a) Schatz, G. C.; Lendvay, G. *J. Chem. Phys.*, submitted; (b) Lendvay, G.; Schatz, G. C.; Takayanagi, T. Chap xx in this book.

7. Clary, D. C.; Dateo, C. E.; Stoecklin, T. *J. Chem. Phys.*, **1990**, *93*, 7666.

8. Tao, F.-M.; Drucker, S.; Klemperer, W. *J. Chem. Phys.,* **1995**, *102*, 7289.

9. MOLPRO is a package of ab initio programs written by H.-J. Werner and P. J. Knowles, with contributions from J. Almlof, R. D. Amos, M. J. O. Deegan, S. T. Elbert, C. Hampel, W. Meyer, K. Peterson, R. Pitzer, A. J. Stone, and P. R. Talor.

10. (a) Woon, D. E.; Dunning, T. H. Jr. *J. Chem Phys.,* **1993,** *98*, 1358; (b) Kendall, R. A.; Dunning, T. H. Jr.; Harrison, R. J. *J. Chem Phys.,* **1992,** *96*, 6796.

11. Boys, S. F.; Bernadi, F. *Mol. Phys.*, **1970**, *19*, 553.

12. Werner, H.-J.; Bauer, C.; Rosmus, P.; Keller, H.-M.; Stumpf, M.; Schinke, R. *J. Chem. Phys.,* **1995**, *102*, 3593.

13. Bowman, J. M.; Bittman, J. S.; Harding, L. B. *J. Chem. Phys.*, **1986**, *85*, 911.

14. Bowman, J. M.; Gazdy, B.; Bentley, J. A.; Lee, T.J.; Dateo; C.E., *J. Chem. Phys.,* **1993**, *99*, 308.

15. Naumkin, T.; Yu,; Knowles, P. J. *J. Chem. Phys.,* **1995**, *103*, 3392.

16. Shin, S.; Shin, S. K.; Tao, F.-M, J. *Chem. Phys.,* **1996**, *104*, 183.

Chapter 12

Infrared and Collisional Relaxation of Highly Vibrationally Excited NO

Applications of Stimulated Emission Pumping in Crossed and "Uncrossed" Molecular Beams

Marcel Drabbels[1] and Alec M. Wodtke

Department of Chemistry, University of California, Santa Barbara, CA 93106

The first results from a newly constructed crossed-molecular-beam apparatus specifically designed for use with stimulated emission pumping are presented. We report the direct determination of the infrared Einstein A-coefficients of highly vibrationally excited NO molecules. In addition, results on the rotationally inelastic scattering of NO $X\,^2\Pi$(v=20, J=0.5, e/f) with He and the near resonant V-V energy transfer of highly vibrationally excited NO $X\,^2\Pi$(v=20-22) with N_2O are presented.

The techniques of stimulated emission pumping (SEP) (*1*) and crossed-molecular-beams (*2*) have appeared to be on a collision course for years. The prospect of single-quantum-state preparation of highly excited molecules in practically arbitrary vibrational state, together with collision energy control and the single collision conditions of crossed beams would seem to provide high quality data that can be compared most directly to *ab initio* theory. Surprisingly such experiments are, thus far, extremely rare (*3*). This can be largely understood if one appreciates the technical complexity of constructing such an experimental apparatus. SEP requires two tunable laser systems to prepare the reactant, and presumably a third laser for probing the consequences of collisions. Crossed-molecular-beams experiments represent another set of challenges. A so-called "universal crossed-molecular-beam machine" (*4, 5*) requires a minimum of five diffusion pumps and a relatively complex set of differentially pumped chambers "just to get the beams crossed". The two techniques combined would appear to represent a nearly insurmountable technical feat.

In this paper we present initial results from a new crossed-molecular-beam machine specifically designed for use with stimulated emission pumping. This apparatus relies on pulsed molecular beams and is vastly simplified in comparison to the implications of the above paragraph.

[1]Current address: FOM Institute for Molecular and Atomic Physics, Kruislaan 407, 1098 SJ Amsterdam, Netherlands

Single collision conditions as well as control of collision energy is obtained within a single vacuum chamber pumped by a six-inch diffusion pump. SEP provides rovibrational and parity state selection of NO in vibrational states as high as v=22 in a molecular beam with a narrow velocity distribution.

We present several examples of the kind of measurements that can be performed with this arrangement. First, by monitoring the population of the SEP prepared NO under collision free conditions in a single "uncrossed" beam geometry, infrared radiation of highly vibrationally excited molecules can be monitored in the time domain, providing a direct route to infrared Einstein A-coefficients (6). Second, rotational energy transfer of highly vibrationally excited NO has been studied in crossed beams with He (7) and N_2O (8). The results for He can be compared to new *ab initio* calculations which rely on the rigid rotor approximation (7). In these calculations, the highly vibrationally excited NO molecule is frozen at the average internuclear distance of the excited vibrational state. This approximation gives quite good agreement with experiment for rotational energy transfer. Third, near resonant V-V energy transfer has been studied in crossed beams between NO(v=20-22) with N_2O (8). The rotationally resolved state-to-state cross sections for this process reveal the role of NO rotation in accommodating the energy mismatch of the near resonant energy transfer.

Experimental

A schematic overview of the experimental setup is given in Figure 1. A primary molecular beam was formed by expanding 3 atm. neat NO into a vacuum chamber through the 0.8 mm diameter orifice of a pulsed valve (General Valve) and passing it through a conical skimmer with a diameter of 2.5 mm placed 2 cm downstream of the nozzle. The pulsed valve was operated at 10 Hz and had a pulse duration of 250 μs full width. The vacuum chamber was equipped with a 6-inch diffusion pump with a pumping speed of 2000 ℓ/s. The base pressure in the chamber was $5 \cdot 10^{-7}$ torr and rose to $1 \cdot 10^{-5}$ torr with the primary molecular beam running.

Highly vibrationally excited NO molecules were produced by means of SEP. Six centimeters downstream from the nozzle and 4 centimeters after the opening of the skimmer, the NO molecules were excited from the $X\,^2\Pi$(v=0) ground state to the $B\,^2\Pi$(v=5) valence state with the PUMP laser. They were subsequently de-excited by the DUMP laser to produce molecules in single quantum states of $X\,^2\Pi$(v≈20). As PUMP laser, a pulse amplified single-mode ring dye laser was employed. The output of an Argon ion (Spectra Physics 171) pumped single mode ring dye laser (Coherent CR-699) operated with Rhodamine 6G was pulse amplified in a home built 3-stage amplifier chain, filled with a mixture of Rhodamine 610 and Rhodamine 640 dyes. This system was pumped by a frequency doubled, Q-switched, injection seeded Nd:YAG laser (Continuum Powerlite 7010-IS) with a pulse energy of 400 mJ at 532 nm and a pulse duration of approximately 5 ns. The amplified visible radiation was frequency doubled in a KDP crystal to yield 15 mJ of UV light around 300 nm. After passing a λ/2-plate, the doubled and residual fundamental radiation were frequency summed in a BBO crystal to yield light near 198 nm with a pulse energy of typically 2 mJ and a Fourier transform limited bandwidth of less than 200 MHz.

An excimer (Lambda Physik EMG 200) pumped dye laser (Lambda Physik FL3002) operating on C-153, was used to stimulate (DUMP) molecules from the electronically excited B state down to high vibrational levels of the ground electronic state. The DUMP laser, which had an output energy of 10-20 mJ per pulse and a bandwidth of 0.15 cm^{-1}, counterpropagated the PUMP laser and was fired 5 ns after the PUMP laser. The fluorescence from the B state was collected with an F/1.6 lens system and imaged onto a photomultiplier tube (Hamamatsu R212UH) placed at right angles to both the molecular and preparation laser beams. In order to reduce the scattered light from both the PUMP and DUMP laser, a UG 5 colored glass filter was placed in front of this phototube. Fluorescence dip measurements could thus be carried out with this arrangement, to ensure periodically that the SEP procedure was effective.

The secondary beam was formed by expanding the colliding gas under study through a similar pulsed valve and conical skimmer. The secondary beam crosses the primary beam at an angle of 30° at a distance of approximately 12.5 cm and 6.5 cm from the primary and secondary valves, respectively. At this crossing region a third laser beam intersects both molecular beams to PROBE the rotational and vibrational distribution of the NO molecules. To this end a similar excimer pumped dye laser system also operating on C-153 was employed which excites the molecules from vibrational levels in the electronic ground state to either the $A\ ^2\Sigma^+$ or $B\ ^2\Pi$ state.

Laser induced fluorescence (LIF) from the electronically excited states was collected

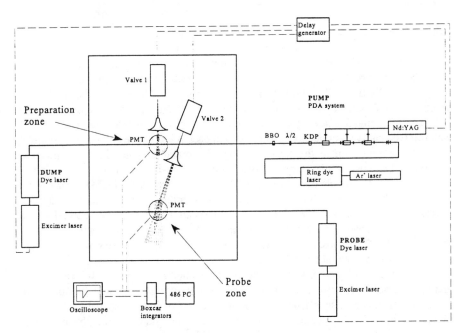

Figure 1: Schematic diagram of the crossed-molecular-beam arrangement. Two pulsed molecular beams are crossed in a single chamber. SEP is used to prepare highly vibrationally excited NO in the primary (downward moving) beam at the preparation zone of the machine. LIF is used at the probe zone to determine the final state distribution of the molecules.

by an F/0.7 quartz lens system placed at right angles to both the primary molecular beam and the PROBE laser beam and imaged onto a solar blind photomultiplier tube (Hamamatsu 166UH). The signals from the photomultiplier tubes at the preparation region and the collision region are both processed by a digital oscilloscope (LeCroy 9430) and boxcar integrator (SRS 250) interfaced with a 486 PC.

In the experiments for the determination of the Einstein A-coefficients for highly excited vibrational levels of NO the DUMP laser was triggered at 5 Hz, while all other components of the experiment were triggered at 10 Hz. Operating the boxcar integrator in "toggle mode", where the difference of odd and even shots was integrated and averaged, enabled the detection of only those processes that depend on the DUMP laser. This then allows for direct determination of the Einstein A-coefficients.

In the collisional energy transfer experiments, both molecular beams and all three lasers were operated at 10 Hz. However, the secondary pulsed molecular beam was delayed by 1 ms at 5 Hz. Under the delayed conditions, collisions between the two molecular beams are not possible. For these shots, any collisional redistribution of the prepared state of NO is due to collisions of NO with the background gas. Using the boxcar integrator in "toggle mode," which averages the difference of odd and even shots, the effect of background gas can be completely removed from the data and only the energy transfer due to collisions between the molecules in the two molecular beams is observed.

Initial State Preparation

With SEP we are able to prepare NO in a single vibrational, rotational spin-orbit and parity state. Figure 2 shows a representative LIF spectrum of the band head of the B $^2\Pi_{1/2}(v=5) \leftarrow X\ ^2\Pi_{1/2}(v=0)$ vibronic band used for the PUMP step of the SEP. This spectrum was recorded at the preparation zone using the frequency tripled output of the pulse amplified ring dye laser system. The doublet structure in the spectrum is due to the well resolved parity components (Λ-doublets) of each rotational level. These Λ-doublet states are present both in the lower and upper vibrational states involved in these transitions. Only two of the four conceivable transitions between these sets of doublets are allowed electric dipole transitions (9). The output power of the PUMP laser system is high enough to saturate this transition and line broadening becomes observable at pulse energies of more than 1 mJ. We interpret this to mean that 50% of the population in a given parity component of a specific rotational level in the ground state is transferred to the B state. From the observed $B\ ^2\Pi_{1/2}(v=5) \leftarrow X\ ^2\Pi_{1/2}(v=0)$ LIF spectrum a rotational temperature of 15 K could be deduced.

In the absence of the DUMP laser beam, the excited B state will radiate to various vibrational levels in the ground electronic state. The resulting vibrational distribution in the ground electronic state is determined by the Franck-Condon factors and the transition dipole moment function connecting the two electronic states, while the rotational distribution in these vibrational states is governed by the Hönl-London factors and the parity selection rule for electric dipole transitions $\pm \rightarrow \mp$. The states prepared by this "Franck-Condon pumping", travel in the molecular beam to the PROBE zone, where the rovibrational distribution is probed by means of LIF. Although the Franck-Condon factors of high vibrational states of the ground electronic state and the low vibrational levels of

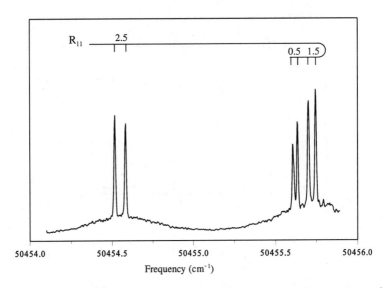

Figure 2: Representative LIF spectrum in the region of the NO B $^2\Pi_{1/2}$ (v=5) ← X $^2\Pi_{1/2}$ (v=0) band head recorded with the frequency tripled output of the pulse amplified ring dye laser. The doublet structure is due to the Λ-doublets (e/f levels) associated with each rovibronic state.

the A $^2\Sigma^+$ state are very small (~10^{-4}), these transition could be saturated by the unfocussed output of the PROBE laser. When probing via the A-X system, we can make use of the fact that the main P- (or R-) branch lines and Q-branch lines are well separated and probe different e/f-symmetry (parity components) levels of a given rotational state forthis $^2\Sigma^+$ ← $^2\Pi$ system (10). This allows us to obtain information about the population in both parity levels which results from energy transfer despite the fact that the PROBE laser has a bandwidth of 5 GHz. This should be contrasted with LIF detection of highly vibrationally excited NO using the B-X system. For this $^2\Pi$ ← $^2\Pi$ transition, the main P, Q, and R branches probe both parity components of a given rotational state, and thus the lines appear as closely-spaced doublets (10). Although the B ← X bands generally have more favorable Franck-Condon factors (~10^{-1}), the resolution of our probe laser is too low to allow us to resolve the Λ-doublets in either P, Q, or R branch. Therefore, we were unable to obtain information about the populations in individual parity components when probing through the B-X system.

The upper panel of Figure 3 shows an LIF spectrum of the A $^2\Sigma^+$(v=3) ← X $^2\Pi_{1/2}$(v=20) vibronic band after Franck-Condon pumping through the B $^2\Pi_{1/2}$(v=5, J=1.5, e) level. The observed population in the v=20 level of the electronic ground state clearly reflects the Hönl-London factors for the X $^2\Pi$ ← B $^2\Pi$ transitions, where both states belong to Hund's case (a): strong ΔJ=±1 and weak ΔJ=0 transitions (10). The lower panel of Figure 3

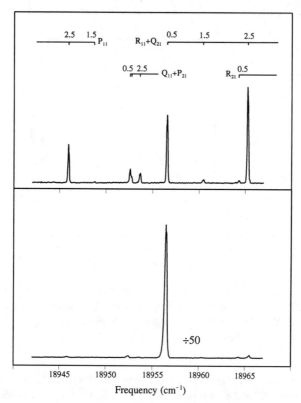

Figure 3: Preparation of single quantum states of highly vibrationally excited NO X $^2\Pi(v=20)$ in a molecular beam. The upper panel illustrates the result of "Franck-Condon pumping". The lower panel shows how SEP can be used to produce a single quantum state. We estimate that greater than 3% of the molecular beam ($3 \cdot 10^{12}$ cm^{-3}) is transferred to the desired state X $^2\Pi(v=20, J=0.5, e)$. The population in other nearby rotational states is almost two orders of magnitude lower.

shows the population in the $X(v=20)$ level when the DUMP laser is tuned to the B $^2\Pi_{1/2}(v=5, J=1.5, e) \rightarrow X\ ^2\Pi_{1/2}(v=20, J=0.5, e)$ transition. This transition can easily be saturated which allows the enhancement of the number of molecules in the $J=0.5$, e-symmetry level by almost two orders of magnitude. With SEP we find that 95% of all the molecules in $v=20$ are in the $J=0.5$, e-symmetry state. The maximum population in any of the other $v=20$ rotational states is ~1%. Using the measured rotational temperature in the molecular beam (15 K) and the facts that both the PUMP and DUMP transitions are saturated, we calculate that 3% of *all* the molecules present in the beam at the preparation zone are optically pumped to the desired state. 94% remain in the vibrational ground state, and 3% are in other, mainly low, vibrational states. With these numbers and the known geometry, we estimate the number density of prepared molecules in the collision zone to be $3 \cdot 10^{12}$ molecules/cm^3.

Determination of IR Einstein A-Coefficients

For the determination of the Einstein A-coefficients an experiment was performed using only the primary molecular beam. The principle of this technique is best explained by inspection of Figure 4 which shows some of the results of this work. The upper panel of Figure 4 is similar to the upper panel of Figure 3. Here the population of NO in v=19 produced by Franck-Condon pumping through the $B\ ^2\Pi$(v=5, J=2.5) level is probed 80.9 μs after preparation at the PROBE zone of the machine via the $B\ ^2\Pi$(v=4) ← $X\ ^2\Pi$(v=19) transition. The rotational levels of $X\ ^2\Pi$(v=19) populated in this way are J=1.5, 2.5 and 3.5. The middle panel shows the effect of introducing the DUMP laser at the preparation zone of the machine and tuning it to the $B\ ^2\Pi$(v=5, J=2.5) → $X\ ^2\Pi$(v=21, J=3.5) transition. The enhancement of the population in J=2.5 and 4.5 rotational levels of v=19 is very clear. Less obvious is the decrease of the population in the J=1.5 and 3.5 levels due to the large fraction of B-state population that is now unavailable for spontaneous visible emission to v=19. Figure 4 presents experimental evidence of a process where Nitric Oxide loses two vibrational quanta while only exchanging a single quantum of angular momentum during the 80.9 μs collision free flight time. Such a process can only be due to infrared emission.

The bottom panel of Figure 4 shows the difference-signal of the upper and middle panel. If we now define the difference signal for the J=3.5 population as $\Delta_{19,3.5}$ and that of J=4.5 as $\Delta_{19,4.5}$, it can be shown that:

$$\frac{\Delta_{19,4.5}}{\Delta_{19,3.5}} = -\ \frac{Fraction\ X(v=21, J=3.5)\ \rightarrow\ X(v=19, J=4.5)}{Fraction\ B(v=5, J=2.5)\ \rightarrow\ X(v=19, J=3.5)} \tag{1}$$

where *Fraction* X(v=21, J=3.5) → X(v=19, J=4.5) is the fraction of molecules that have undergone the indicated infrared transition during the flight time and *Fraction* B(v=5, J=2.5) → X(v=19, J=3.5) is the fraction of molecules that have undergone the indicated electronic transition in the absence of the DUMP laser.

In principle, this expression is approximate since a part of the signal originating from J=3.5 also results from infrared emission. In practice, the infrared Q-branch is ~20 times weaker than the observed P and R branches and thus, this contribution to the J=3.5 signal is smaller than the noise of the measurement. Likewise, $\Delta\Omega$=+1 infrared emissions could not be observed as expected. Therefore the observed signal loss induced by the DUMP laser for J=3.5 is essentially only due to the competition between stimulated emission to v=21 and spontaneous visible emission to v=19. In contrast the signal increase for J=4.5 is due entirely to the collaboration of stimulated visible emission to v=21 and spontaneous infrared emission from v=21 to 19. Similar data were obtained for IR radiation from v=21 to v=20,18 and 17.

The quantity, *Fraction* B(v=5, J=2.5) → X(v=19, J=3.5), the fraction produced in J=3.5 by Franck-Condon pumping, can be calculated from known spectroscopic data on the B-X emission system. To accomplish this a standard RKR analysis (11-13) was carried out to obtain the vibrational wave-functions for the $B\ ^2\Pi$(v=5) and $X\ ^2\Pi$(v) vibrational states. The semi-empirical transition dipole moment function reported recently

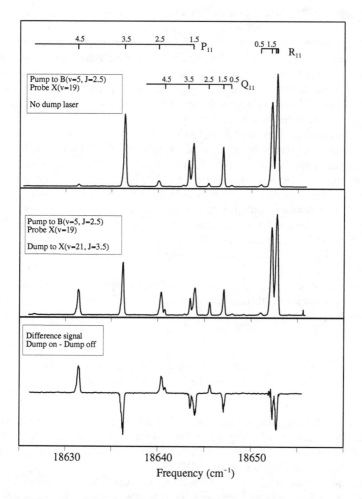

Figure 4: Infrared emission's influence on the observed LIF spectra. Upper Panel: The resulting $B\,^2\Pi(v=5) \leftarrow X\,^2\Pi(v=19)$ probe spectrum after the PUMP laser is tuned to the $R_{11}(1.5)$ line of the $B\,^2\Pi(v=5) \leftarrow X\,^2\Pi(v=0)$ band in the absence of the DUMP laser. Levels are populated by spontaneous emission from the $B\,^2\Pi$ state. Due to the selection rules of this rovibronic transition only $J=1.5$, 2.5 and 3.5 are observed. Middle panel: The resulting PROBE spectrum of $v=19$ with the DUMP laser populating $v=21$, $J=3.5$. Population observed in $J=4.5$ and 2.5 is seen to increase due to infrared emission between $v=21$ and $v=19$, which is governed by $\Delta J=\pm 1$ infrared selection rules. Lower Panel: The difference signal from the upper and middle panel. The increase of the $J=4.5$ and decrease of the $J=3.5$ level is clearly visible. The ratio between the increase and decrease for these rotational levels can be used to determine the IR Einstein A-coefficients, see text.

by Langhoff *et al.* (*14*) was fit to a fourth order polynomial damped by an exponential function to ensure smoothness and proper behavior in the asymptotic region. There is excellent agreement between this semi-empirical transition dipole moment and that determined experimentally by Piper *et al.* (*15*). This fit function was used together with the wave functions to calculate the final state distribution produced by Franck-Condon pumping.

Knowing the fraction that has undergone infrared radiation in a predetermined flight time allows the direct reduction of the data to infrared Einstein A-coefficients. These results are shown in Table I for all of the bands measured in this work. These may be summed over final vibrational states to give the infrared radiative lifetime of the X $^2\Pi$(v=21) level, which is found to be 5.9± 0.4 ms. Also shown in this table are the results of *ab initio* calculations by Langhoff *et al.* (*16*). The agreement between the results of this work and these most recent *ab initio* calculations is nothing short of spectacular. This serves as clear confirmation that both the ab initio calculations and this new technique of SEP molecular beam time-of-flight can be used to determine Einstein A-coefficients with good accuracy and precision. The Einstein A-coefficients derived from this work also clearly show that the ratio of 1st overtone-to-fundamental emission continues to rise above v=11, in agreement with *ab initio* results (*16*) and in contrast to chemiluminescence experiments (*17*).

Table I: Einstein A-coefficients (s^{-1}) for highly vibrationally excited NO

v → v′	This Work [a]	Langhoff *et al.* (*16*)
21 → 20	70.0 ± 3.6	63.9
21 → 19	66.7 ± 3.4	59.7
21 → 18	27.2 ± 2.4	24.5
21 → 17	5.9 ± 1.7 [b]	-
7 → 6	55.7 ± 6.8	60.0

[a] 95% confidence intervals, based on statistical error analysis.
[b] Derived radiative lifetime for v=21, J=3.5, Ω=0.5 is 5.89±0.38 ms.

Observation of Inelastic Scattering of NO X $^2\Pi_{1/2}$(v=20, J=0.5, *e/f*) with Helium

Relative state-to-state cross-sections have been determined for rotationally inelastic collision of NO X $^2\Pi_{1/2}$(v=20, J=0.5, *e/f*) with He at a collision energy of 195 cm^{-1}. For energy transfer within the initially populated Ω=1/2 spin-orbit manifold, state-to-state cross-sections were extracted from analysis of the A $^2\Sigma^+$(v=3) ← X $^2\Pi_{1/2}$(v=20) PROBE LIF spectrum. State-to-state cross-sections for spin-orbit changing collisions, Ω=1/2 → Ω=3/2, were found to be more than an order of magnitude weaker than the spin-orbit conserving collisions, Ω=1/2 → Ω=1/2. This necessitated use of B $^2\Pi$(v=5) ← X $^2\Pi$(v=20) band for the PROBE step, which was found to be a more sensitive detection scheme.

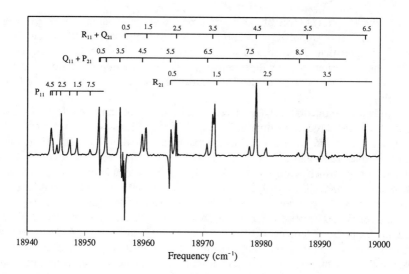

Figure 5: NO A $^2\Sigma^+(v=3) \leftarrow X\,^2\Pi(v=20)$ LIF spectrum after collisional redistribution of NO X $^2\Pi_{1/2}(v=20, J=0.5, e)$ by He. The negative signals (off-scale) represents population removed from the prepared state(s) while the positive signals represent collisionally populated states.

However as explained above, this meant that no information on the parity changing energy transfer propensities could be obtained in these experiments.

Figure 5 shows a typical A–X spectrum obtained from the crossed beam scattering of NO X $^2\Pi(v=20, J=0.5, e)$ with He. The "negative signals" in the spectrum indicate a decrease of population (of both the prepared state as well as states populated by Franck-Condon pumping), whereas positive signals indicate an increase in population. It should be noted that the negative signals in this figure are not a reliable measure of the collision induced depletion of the prepared state since the detector is saturated when the laser is tuned to these transitions. In a separate set of experiments, we attempted to measure the collision induced depletion of the prepared state. This depletion was too small to be observed but is certainly smaller than 10%, which ensures that the experiments are performed within the single collision regime.

From the intensities of the rotational transitions in the PROBE spectrum, we can obtain relative state-to-state energy transfer cross-sections. In transforming from intensities in the LIF spectrum to relative cross-sections, one must take the following factors into account: the variation in line strengths in the A–X band system, the spectral sensitivity to alignment of the scattered molecules and the density to flux transformation. Since the A–X transition could be saturated by the output of the PROBE laser, it is not necessary to correct for a variation in line strengths. Thus the measured intensities are a direct measure of the densities of the states probed, once the degeneracies of the levels are taken into account. Furthermore, saturating the PROBE transitions greatly reduces the sensitivity of the LIF signal to possible alignment of the scattered molecules. Measurements of the LIF signals on various rotational transitions with different polarizations of the probe laser, at laser powers low enough to avoid saturation, showed that no alignment effects could be detected within the experimental uncertainty. This validates the direct conversion from LIF intensities to densities. The density to flux transformation is also negligible, as the He atom is much lighter than the NO. The error introduced by neglecting this transformation is estimated to be smaller than the uncertainty in the observed LIF signals.

To increase the precision of the relative cross-sections, the results of 10 to 15 scans were averaged. Additionally, various choices of PUMP and DUMP transitions were used to prepare the initial state via different spectral pathways. This helped to eliminate any influence of Franck-Condon pumping on the derived relative cross-sections.

The experimental state-to-state cross-sections for the spin-orbit conserving transitions are presented graphically in Figure 6, and compared with calculated cross-sections obtained from a close-coupled (CC) quantum scattering formalism (*7, 18, 19*). The CC calculations were performed on two different potential energy surfaces (PES's) calculated with the CEPA method (*20-22*). For one PES the NO internuclear distance was fixed at the equilibrium distance of $r_e = 2.1746$ bohr (*23*), whereas for the second PES, the average internuclear distance in v=20, $< r >_{20} = 2.6037$ bohr was chosen to fix the NO bond length. Both CC calculations represent an effective rigid rotor approximation; however, the two reduced dimensionality PES's are markedly different (*7*).

State-to-state cross-sections were obtained for scattering out of both e and f Λ-doublet levels of J=0.5. Since the calculations show that the cross-sections for scattering out of the e and f levels are nearly equal for the NO-He system, the experimental cross-sections, averaged over both initial states, are compared in Figure 6 with the average of the theoretical cross-sections for e/f conserving and changing transitions. To normalize the experimental cross-sections for spin-orbit conserving transitions, we set the sum of the measured inelastic cross-sections, out of the J=0.5 state into all accessible rotational states in the Ω=0.5 spin-orbit manifold, equal to the calculated value for the $< r >_{20}$ PES.

We see in Figure 6 that the agreement between theory and experiment is excellent, pronounced oscillatory structure. This oscillation is a consequence of the near-homonuclear character of the NO molecule and has been discussed in detail in several publications.(*19, 24, 25*). We also see that the cross-sections determined with the $< r >_{20}$ PES's agree better with the experimental results in Figure 6, particular for transitions with large ΔJ. This is a direct consequence of the increased angular anisotropy in the $< r >_{20}$ PES as compared with the r_e PES (*7*). There are small differences between experiment and theory. The experimentally observed rotational excitation is somewhat larger than that of the theory.

There are several possible explanations for this discrepancy: First, the angular anisotropy may still be underestimated in the $< r >_{20}$ PES's. A second explanation may be due to sampling of collision energies substantially higher than the mean value of 195 cm^{-1}. One would expect an increased degree of rotational excitation with increasing collision energy. Finally the rigid rotor approximation used in the calculations might not be valid. Given the substantial improvement in the agreement between theory and experiment by fixing the NO bond length at 2.60 bohr, $< r >_{20}$, in comparison to 2.17 bohr, r_e; it is clear that the effect of extending the NO bond length is important to the

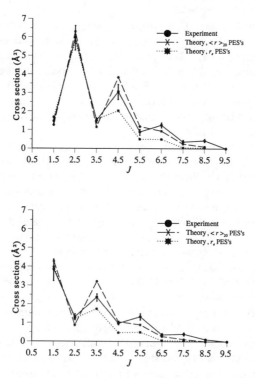

Figure 6: Comparison of the experimental and theoretical cross-sections for spin-orbit conserving (Ω=0.5 → 0.5) collisions between NO X $^2\Pi_{1/2}$(v=20, J=0.5, e or f) and He. To guide the eye the points have been connected by straight lines. The upper panel shows the cross-sections for e/f conserving transitions; both the experimental and theoretical values plotted are the average of the e → e and f → f cross-sections. The lower panel shows the cross-sections for e/f changing transitions; both the experimental and theoretical values plotted are the average of the e → f and f → e cross-sections.

theoretical calculations of the rotational scattering dynamics. A substantial portion of the NO v=20 wave function extends out much further than 2.6 bohr. Indeed, the last maximum in the v=20 vibrational wavefunction occurs at $r \approx 3.2$ bohr. Thus a significant fraction of the collisions will sample NO molecules stretched substantially beyond the equilibrium bond length. Therefore, it seems most likely that the small remaining differences between experiment and theory are due to the treatment of NO(v=20) as a rigid rotor.

As mentioned before, the cross-sections for spin-orbit changing collisional energy transfer are more than an order of magnitude smaller than those for spin-orbit conserving.

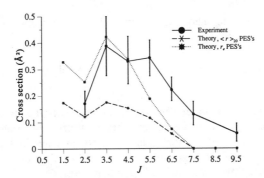

Figure 7: Comparison of the cross-sections for spin-orbit changing ($\Omega=0.5 \rightarrow 1.5$) collisions between NO X $^2\Pi_{1/2}(v=20, J=0.5, e$ or $f)$ and He. To guide the eye the points have been connected by straight lines. The plotted theoretical values for the cross-sections are averaged over both the initial and final Λ-doublet components.

Consequently, the former were monitored using the more sensitive LIF through the *B-X* band. Because the resolution of individual Λ-doublet levels of the final state was not possible using this detection scheme, no information could be obtained on the *e*/*f* dependence of the $\Omega=0.5 \rightarrow 1.5$ cross-sections. As a result, we compare the experimentally measured cross-sections to the averaged sum of the theoretical cross-sections into both Λ-doublet levels of the final state. To normalize the cross-sections for the spin-orbit changing relative to those for spin-orbit conserving energy transfer, we use the B $^2\Pi(v=5)$ state to probe the collision-induced population in the $J=4.5$ level of both spin-orbit states (again summed over both Λ-doublets).

Figure 7 compares the experimental cross-sections for $\Omega=0.5 \rightarrow 1.5$ energy transfer with the theoretical predictions. Examining now the cross-sections for spin-orbit changing transitions we observe that the theoretical cross-sections determined for the $< r >_{20}$ PES are smaller than those for the r_e PES. The agreement with experiment is less satisfactory than in the case of the spin-orbit conserving transitions. Although the variation of the magnitude of the cross-sections with ΔJ is well reproduced in the theoretical simulation, the $< r >_{20}$ cross-sections themselves are too small, by roughly a factor of 2. This discrepancy is most likely caused by errors in the calculated PES governing the spin-orbit changing transitions. It is known that this PES which is the difference between the A' and A" PES's, is much more sensitive to errors than the PES governing the spin-orbit conserving collisions, given by the average of the A' and A" PES's (7). Alternatively, the effects mentioned above might contribute to the observed difference between theory and experiment.

Resonant V-V Energy Transfer in Collisions of NO(v>>0) with N₂O

Relative state-to-state cross-sections for near resonant vibrational energy transfer between highly vibrationally excited NO and ground state N_2O was also studied in the crossed beam setup at a collision energy of 125 cm^{-1}. Resonant vibrational energy transfer is expected to take place via:

$$NO \ X \ ^2\Pi(v=22,21,20) + N_2O(0,0,0) \rightarrow NO \ X \ ^2\Pi(v-1) + N_2O(0,0,1) \qquad (2)$$

Here (0,0,0) indicates N_2O in its vibrationless ground state while (0,0,1) indicates N_2O with one quantum excited in the antisymmetric stretch. V-V transfer between NO molecules and ground state N_2O was observed for NO prepared in the vibrational levels v=20, 21 and 22, where the energy mismatch is -49, -18 and +14 cm^{-1}, respectively (26, 27). Here a positive energy mismatch indicates that the process is endothermic, whereas a negative mismatch indicates that it is exothermic. For the next lower and higher vibrational states, which have an energy mismatch of -80 and +47 cm^{-1} respectively, no evidence of vibrational energy transfer could be observed.

The efficiency of the V-V transfer depends strongly on the energy mismatch and therefore on the prepared vibrational level of the NO molecule. Since the intensity of the recorded spectra depend on factors that are difficult to control in the present set-up, we could only approximately determine the relative vibrational cross-sections for V-V energy transfer. It was found that the V-V energy transfer was most efficient for NO molecules in v=21 and approximately five times less efficient for molecules in v=20 and 22. As mentioned before, no energy transfer was detectable for v=19 and 23, suggesting that these processes are at least another factor of five less probable. This observed resonance for energy transfer for v=21 is clear evidence that we are indeed observing the V-V energy transfer to N_2O, despite the fact that none of our measurements directly probe the N_2O molecule.

The fact that the cross-section for v=22, which is *endo*thermic by 14 cm^{-1} is not as large as for v=21, which is *exo*thermic by 18 cm^{-1}, indicates that initial translational energy does not effectively compensate for the energy mismatch. This suggests that the rotational degrees of freedom may be more important for eliminating the energy mismatch than is translation.

This becomes clearer when looking at the rotational energy distribution of the vibrationally relaxed NO molecules. The upper panel of Figure 8 shows the relative cross-sections for spin-orbit conserving rotational energy transfer of NO $X \ ^2\Pi_{1/2} \ J=0.5$ with N_2O for the v=20 and 21 vibrational levels as a function of rotational energy. Although the cross-sections cannot be put on the same relative scale, it is clear that the rotational energy transfer is quite similar for both vibrational levels. In light of the results in the section on rotational energy transfer between NO and He, these results are not surprising. We have seen that the vibrational influence on the rotational energy transfer can be largely accounted for by the variation of the average bond length with vibrational quantum number. Since this variation is small between v=20 and 21, one expects the rotational energy transfer to be similar, as is observed.

This should be contrasted with the lower panel of Figure 8 which shows the relative cross-sections for the V-V energy transfer as function of final rotational energy. The cross sections are given for the spin-orbit conserving, $\Omega=0.5 \rightarrow 0.5$, transitions. No spin-orbit changing collisions could be observed, although they are energetically accessible for the v=20 and 21 vibrational levels (26). For ease of comparison the cross-sections are put on

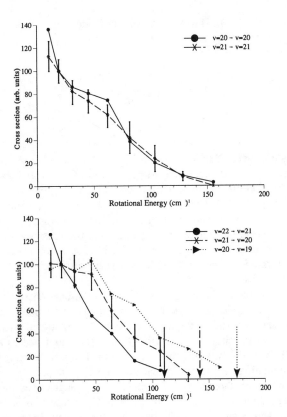

Figure 8: Upper panel: Cross-sections as function of the NO rotational energy for spin-orbit conserving rotational inelastic collisions of NO X $^2\Pi_{1/2}(v, J=0.5)$ with N_2O at a collision energy of 125 cm^{-1}. Lower panel: Cross-sections as function of the NO rotational energy for vibrational inelastic collisions of NO X $^2\Pi_{1/2}(v, J=0.5)$ with N_2O at the same collision energy. To guide the eye the points have been connected by straight lines.

the same scale but it should be remembered that the absolute values for these cross-sections are quite different for the different vibrational levels. It is estimated that for NO in v=21, the vibrational level which is the most efficient, the cross-section for V-V energy transfer is a factor of 5-10 smaller than for rotational energy transfer. It is seen from this figure that the shape of the curves for the three vibrational levels is quite similar, except for the fact that they seem to be shifted by about 30 cm^{-1} between each successive vibrational level. This difference corresponds closely with the change in energy mismatch for the V-V energy transfer process for the successive vibrational levels. Furthermore all three curves persist to the energetic limit, indicated by the arrows in the figure. Both of these observations indicate that NO rotation plays an important role in compensating the energy mismatch.

A more quantitative analysis can be performed by looking at the average amount of rotational energy transferred:

$$\Delta E_{rot} = \frac{\sum_{J'} \sigma_{J\text{-}J'} (\epsilon_{J'} - \epsilon_J)}{\sum_{J'} \sigma_{J\text{-}J'}} \tag{3}$$

Here the summation over J' should start at $J'=1.5$; but since no information on this rotational level could be obtained due to population of this rotational level via IR radiation, see the section on the determination of IR Einstein A-coefficients, the summation is started at $J'=2.5$. The error introduced due to this truncation is expected to be small, since the energy associated with this rotational level is small compared to other rotational levels. Furthermore it can be expected that the cross-section for $J'=1.5$ is comparable for all three vibrational levels as is also seen for $J'=2.5$. The average rotational energy transfer calculated in this way is given in Table II for the different vibrational levels.

Inspection of this table shows that the average rotational energy transfer for the rotationally inelastic collisions is independent of the vibrational level. For the *vibrationally* inelastic collisions, the average amount of rotational energy transferred exceeds the energy mismatch for all three vibrational levels and increases with increasing exothermicity. This again indicates that the rotational degrees of freedom are more important for eliminating the energy mismatch than is translation.

V-V energy transfer is thought to be governed by long range forces, more specifically the dipole-dipole interaction in the case of $NO + N_2O$. In case of near resonances, the theory of Sharma and Brau (28) usually describes the observed vibrational transition probabilities quite well. However, since this theory is only developed up to first order of time dependent perturbation theory it does not allow for the large rotational quantum number changes we have observed in this work. The more general theory of Dillon and

Table II: Average rotational energy transfer (cm^{-1}) between collisions of NO $X\,^2\Pi(v) + N_2O$ at a collision energy of 125 cm^{-1}

$v \rightarrow v'$	ΔE_{rot}
$20 \rightarrow 20$	39.2 ± 1.7
$21 \rightarrow 21$	39.8 ± 2.1
$20 \rightarrow 19$	48.2 ± 2.3
$21 \rightarrow 20$	39.1 ± 2.4
$22 \rightarrow 21$	30.7 ± 2.1

Stephenson (*29, 30*) may yield better agreement with observation. This theory was originally developed for a dipole-dipole vibrational interaction and a dipole-quadrupole rotational interaction. Recently, this theory has been adapted to treat more general types of multipole interactions (*31*). At least this theory allows for large rotational quantum number changes. A comparison of this theory for the NO + N_2O system with the experimental results will be the subject of a forthcoming publication (*8*).

Conclusion

We have used SEP in a crossed-molecular-beam setup to prepare highly vibrationally excited NO molecules in single quantum states. The efficiency of this process is high enough that one may study several different kinds of energy transfer processes. In this way the IR Einstein A-coefficients of high vibrational levels of the NO molecule could be determined. Furthermore rotationally and vibrationally inelastic scattering could be studied with He and N_2O in a crossed-molecular-beam setup. This allowed detailed information to be obtained which can be compared to high level theoretical predictions. The possibility of using this approach to study the single collision chemistry of highly vibrationally excited molecules appears quite promising and is being pursued as work for the future.

Acknowledgments

The authors wish to thank the Air force Office of Scientific Research for their support under grant #F49620-95-1-0234. as well as the UCSB Department of Chemistry Laser Pool under NSF grant CHE-9411302.

References

1. Hamilton, C. E., Kinsey, J. L., and Field, R. W., *Ann. Rev. Phys. Chem.* **1986**, 37, 493
2. Lee, Y. T. in *Atomic and Molecular Beam Methods I;* Scoles, G., Ed.; Oxford University Press: Oxford, 1988
3. Ma, Z., Jons, S. D., Giese, C. F., and Gentry, W. R., *J. Chem. Phys.* **1991**, 94, 8608
4. Lee, Y. T., McDonald, J. D., LeBreton, P. R., and Herschbach, D. R., *Rev. Sci. Instr.* **1969**, 40, 1402
5. Yang, X., Blank, D., Lin, J., Suits, A. G., Lee, Y. T., and Wodtke, A. M., *Rev. Sci. Instr.*, In press
6. Drabbels, M. and Wodtke, A. M., *J. Chem. Phys.*, In press
7. Drabbels, M., Wodtke, A. M., Yang, M., and Alexander, M. H., To be published
8. Drabbels, M. and Wodtke, A. M., To be published
9. Drabbels, M. and Wodtke, A. M., *Chem. Phys. Lett.* **1996**, 256, 8
10. Herzberg, G., *Spectra of Diatomic Molecules,* Van Nostrand, Princeton, 1968
11. Rydberg, R., *Ann. Phys.* **1931**, 73, 376
12. Klein, O., *Z. Phys.* **1932**, 76, 226
13. Rees, A. L. G., *Proc. Phys. Soc. London Ser. A* **1947**, 59, 998
14. Langhoff, S. R., Partridge, H., Bauschlicher, C. W., and Komornicki, A., *J. Chem. Phys.* **1991**, 94, 66638

15. Piper, L. G., Tucker, T. R., and Cummings, W. P., *J. Chem. Phys.* **1991**, 94, 7667
16. Langhoff, S. R., Bauschlicher, C. W., and Partidge, H., *Chem. Phys. Lett.* **1994**, 223, 416
17. Rawlins, W. T., Foutter, R. R., and Partker, T. E., *J. Quant. Spectrosc. Radiat. Transf.* **1993**, 49, 423
18. Arthurs, A. and Dalgarno, A., *Proc. Roy. Soc. London Ser. A* **1960**, 256, 540
19. Alexander, M. H., *J. Chem. Phys.* **1982**, 76, 5974
20. Meyer, W., *Int. J. Quant. Chem. Symp.* **1971**, 5, 341
21. Meyer, W., *J. Chem. Phys.* **1973**, 58, 1017
22. Meyer, W., *Theoret. Chim. Acta* **1974**, 35, 277
23. Yang, M. and Alexander, M. H., *J. Chem. Phys.* **1995**, 103, 6973
24. Alexander, M. H., *Chem. Phys.* **1985**, 92, 337
25. Degli Esposti, A., Berning, A., and Werner, H.-J., *J. Chem. Phys.* **1995**, 103, 2067
26. Amiot, C., *J. Mol. Spectrosc.* **1982**, 94, 150
27. Herzberg, G., *Infrared and Raman Spectra;* Van Nostrand: Princeton, 1945
28. Sharma, R. D. and Brau, C. A., *J. Chem. Phys.* **1969**, 50, 924
29. Dillon, T. A. and Stephenson, J. C., *J. Chem. Phys.* **1972**, 58, 2056
30. Dillon, T. A. and Stephenson, J. C., *Phys. Rev. A* **1972**, 6, 1460
31 Seoudi, B., Doyennette, L., and Margottin-Maclou, M., *J. Chem. Phys.* **1984**, 81, 5649

Chapter 13

A New Method for Measuring Vibrational Energy Distributions of Polyatomics

Derek R. McDowell[1], Fei Wu, and R. Bruce Weisman[2]

Department of Chemistry and Rice Quantum Institute, Rice University, Houston, TX 77005

A new kinetics-based method is described for determining the distribution of vibrational energy contents in a highly excited, precollisional polyatomic sample. The method is suitable for excited species whose population decay constant depends on vibrational energy. In such cases a sample's decay kinetics will show a distribution of rate constants that can reveal the underlying initial distribution of vibrational energies. This method is applied to low pressure pyrazine samples whose total T_1 population is measured by triplet-triplet transient absorption. The resulting highly defined kinetic data are analyzed to obtain a decay constant distribution, using a sum of Gaussian peaks model. The result is then transformed into a vibrational energy distribution using an independently determined calibration curve. The sample's narrow nascent energy distribution appears to evolve into a bimodal form through collisions with helium atoms. It is estimated that approximately 0.7% (on average) of the gas kinetic encounters between an excited triplet pyrazine molecule and a helium atom lead to vibrational deactivation of *ca.* 2000 cm^{-1}.

For many decades, physical chemists have recognized that vibrational excitation can be a crucial factor in controlling unimolecular and bimolecular processes of polyatomic molecules. An issue of central concern in systems activated either by thermal or optical excitation is the collisional flow of vibrational energy to and fromthe reactants. In recent years, "direct" methods have been developed that allow the average vibrational energy of relaxing samples to be monitored through molecular

[1]Current address: Department of Chemistry, Wittenberg University, P.O. Box 720, Springfield, OH 45501

[2]Corresponding author

infrared emission, ultraviolet-visible absorption, or kinetic characteristics. Although these methods have provided unprecedented insight into vibrational energy transfer from highly excited polyatomics, they measure only the first moment of the sample's full vibrational energy distribution. This distribution, which describes the inhomogeneous variation in energy contents among the sample molecules, is of deeper interest than its first moment. New experimental information about such distributions has recently become available from extensions of the infrared emission method to deduce the second, as well as first, moment (1,2) and from the KCSI energy-selected photoionization technique of Luther and co-workers (3), which can reveal full energy distributions during the approach to thermalization.

We describe here a new approach for deducing the distribution of vibrational energy among molecules in an excited, pre-collisional sample. This method, which we term "vibrational energy distributions from kinetic analysis," or "VEDKA," is an outgrowth of our previous studies of average collisional energy loss from triplet state pyrazine. As illustrated schematically in Figure 1, optical excitation of selected vibrational levels of pyrazine's S_1 state is followed by efficient, subnanosecond intersystem crossing that produces the T_1 state with a well-defined amount of vibrational excitation equal to the difference between the photon energy and the T_1 origin. These activated triplet state molecules subsequently undergo a second, much slower intersystem crossing process that returns them to the ground electronic state.

Key to our methods is the strong dependence of the rate constant for this second process on the vibrational energy content of the triplet pyrazine molecule (4,5). Having constructed a "calibration curve" representing this dependence, we are able to assess average energy loss by analyzing the kinetics of total triplet population (measured through triplet-triplet transient absorption). The research reported here focuses on the population decay kinetics of low-pressure triplet pyrazine samples, with the goal of deducing not the average energy content, but instead the full vibrational energy distribution in the sample molecules. Our VEDKA method attempts to analyze kinetic data as superpositions of first-order rate constants that reflect the inhomogeneous variation of energy contents in the sample. Once the decay constant distribution has been experimentally deduced, it is mapped into the corresponding distribution of molecular vibrational energies. Our data confirm that the nascent energy distribution in the triplet pyrazine sample is much narrower than a thermal distribution of equal average energy. Although the VEDKA approach is rigorously valid only in the pre-collisional limit, we have found intriguing changes in the decay constant distribution caused by early collisions with helium atoms. These results suggest that the pyrazine energy distribution is quickly converted into an unexpected bimodal form through a surprisingly efficient collisional relaxation channel.

Experimental Method

Our measurements were made on low pressure static samples of pyrazine vapor contained in a 68 cm optical cell. Sample molecules were excited to specific vibrational levels of the S_1 state using tunable ultraviolet pulses generated by frequency-doubling the output of a Lumonics Q-switched Nd:YAG / dye laser system. Following excitation, sub-nanosecond intersystem crossing leaves the excited pyrazine molecules in the T_1 electronic state with a vibrational energy content of 4056 or

Pyrazine Triplet Relaxation

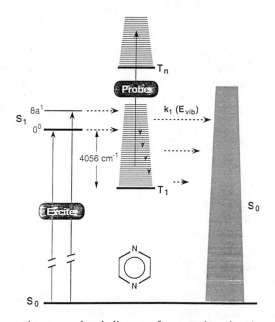

Figure 1. Schematic energy level diagram for pyrazine showing excitation and probe transitions, rapid intersystem crossing from S_1 to T_1, slower, energy-dependent intersystem crossing from T_1 to S_0, and collisional relaxation within T_1.

5433 cm^{-1} (for excitation of the 0_0^0 or $8a_0^1$ transition). The total triplet pyrazine population is measured as a function of delay after excitation through its weak absorption of a continuous 676 nm diode laser probing beam, which co-propagates at a slight angle to the excitation beam in order to avoid transient absorption signals from the suprasil cell windows. A baffled and spectrally filtered silicon photodiode (EG&G model FND-100Q) detects the probe beam after it emerges from the sample cell. This photodiode's output waveform is recorded at 1 ns intervals and averaged over several thousand excitation pulses with a Tektronix TDS-744A digitizing oscilloscope. The resulting averaged traces are transmitted to a laboratory computer for conversion into induced absorbance and subsequent kinetic analysis. The instrumental response function is accurately determined by measuring the excitation pulse's shape with the detection system normally used for the probe beam.

Random noise in the induced absorbance traces is reduced by the use of an amplitude-stable probe laser, a detection bandwidth limit of 20 MHz, and repetitive signal averaging. Avoidance of systematic errors is more difficult, requiring rigorous suppression of electronic and optical cross-talk between the excitation and probe systems. Our optical setup is also designed to reduce artifacts from thermal lensing in the sample as well as from cell window transients. The instrument's performance is checked with empty-cell scans, which typically show systematic and random induced absorbance errors below *ca.* 2 x 10^{-6}. For most sample conditions, peak signal to r.m.s.-noise ratios reach several hundred.

Data Analysis

The central target of the data analysis is $f_E(E,t)$, a function whose energy-dependence at a given time describes the inhomogeneous distribution of vibrational energies in the triplet population and whose integral over energy gives the total T_1 population at time t. We note that the $T_n \leftarrow T_1$ spectrum probed in this experiment is broad and diffuse, so triplet sample molecules in all stages of vibrational relaxation contribute to the induced absorption signal. The relatively mild variation of their absorption strengths is expressed by an energy-dependent absorption cross-section function, $\sigma(E)$. Each differential energy range therefore contributes an absorption component $\sigma(E)f_E(E,t)dE$, giving a total measured signal

$$S(t) \propto \int_0^\infty \sigma(E)f_E(E,t)dE. \qquad [1]$$

The monotonic increase of the triplet molecules' decay constant with vibrational energy is represented by the independently measured "calibration" function, $k(E)$. The population distribution function can change through two types of processes: (1) collisional transfer of population among the energy components; and (2) first-order decay of each isolated energy component.

At pressures low enough that collisional redistributions of type (1) may be neglected relative to unperturbed decays of type (2), each energy component will undergo independent exponential decay:

$$f_E(E,t) = e^{-k(E)t} f_E(E,0) \cdot \quad [2]$$

The measured time-dependent signal therefore assumes the form

$$S_t \propto \int_0^\infty e^{-k(E)t} f_E'(E,0) \, dE \, . \quad [3]$$

Here, $f_E'(E,t)$ denotes the sensitivity-weighted energy distribution function $f_E(E,t)\sigma(E)$. In a low pressure sample in which all molecules have the same energy, the observed kinetics will be simple exponential decay with a rate constant corresponding to that energy. By contrast, a sample with a broadened energy distribution will show kinetics composed of a superposition of exponential decays.

Given low-pressure experimental kinetic data, one can analyze to find the initial rate constant distribution function, $f_k(k,0)$, which is defined through the relation

$$S_t \propto \int_0^\infty e^{-kt} f_k(k,0) \, dk \, . \quad [4]$$

Once found, $f_k(k)$ may be transformed into the corresponding energy distribution function, $f_E'(E)$, through the following relation:

$$f_E(E(k)) = f_k(k) \frac{dk}{dE} \cdot \quad [5]$$

Here $k(E)$ is the "calibration" function and $E(k)$ is its inverse.

It is possible to extract rate constant distributions from kinetic data through inverse Laplace transform approaches such as CONTIN (*6-8*) or through the maximum entropy method (*9*). However, we have instead chosen to model the decay constant distributions as sums of Gaussian peaks because Gaussian distributions are mathematically tractable as well as commonly observed in trajectory and master equation simulations of vibrational relaxation (*1,10,11*). In our Gaussian model, each peak has an independent center position k_i, width parameter σ_i, and amplitude A_i:

$$f(k) = \sum_{peaks \; i} \frac{A_i}{\sqrt{2\pi} \, \sigma_i} e^{-\frac{1}{2}\left(\frac{k-k_i}{\sigma_i}\right)^2} \quad [6]$$

The A_i values are scaled so that they sum to one. We note that this model reduces to a familiar discrete sum of exponential decays when the peak widths become small relative to their center positions. Fitting of our data to the model is accomplished with a Marquardt routine (*12*) that iteratively adjusts the Gaussian parameters to achieve the lowest least-squares deviation between an experimental kinetic trace and the convolution of the calculated S_t with our measured instrument response function. We find that three peaks are commonly needed to generate an adequate fit to the data.

Results and Discussion

Near-Nascent Conditions. We have tested the VEDKA method on pyrazine triplet states produced with 5433 cm^{-1} of vibrational energy through $8a_0^1$ excitation of a sample at 0.01 torr, the lowest pressure that allowed the high signal-to-noise ratio needed for a VEDKA analysis. Figure 2 displays the resulting induced absorbance data. In the upper frame, unsmoothed data points are shown as open circles and the optimized fit, based on equation (6) with $i = 3$, is drawn as a solid line. The lower frame displays on an expanded scale the difference between data and fit. These residuals have an r.m.s. value of 1.7×10^{-6}, significantly smaller than could be obtained by fitting to a sum of three simple exponentials. In Figure 3 we plot the $f(k)$ distribution that corresponds to the optimized fit of Figure 2. The main peak in this distribution, centered at 5.0×10^6 s^{-1}, corresponds to $T_1 \rightarrow S_0$ radiationless decay of the nearly nascent triplet pyrazine molecules with *ca.* 5433 cm^{-1} of vibrational energy. At the sample pressure used for this measurement the mean Lennard-Jones collision interval is 4.4 μs, or 22 times the nascent triplet lifetime. One would therefore expect a nearly monoenergetic initial distribution to have undergone little collisional perturbation, and as a result to show essentially exponential decay with a single, well-defined rate constant. However, careful analysis indicates that the main peak is somewhat broadened, with a width parameter of $0.7 \pm 0.2 \times 10^6$ s^{-1}. This apparent non-zero width may in part result from vibrationally inelastic collisions. We also find that the decay constant distribution must contain two other peaks to accurately fit the data. The first of these peaks is centered near 0.5×10^6 s^{-1} and contains only 0.5% of the total probability, a value too small to allow its width parameter to be determined. As discussed in the following section, we assign this minor feature to a small fraction of triplet pyrazine molecules that have undergone partial collisional relaxation. The final peak in the distribution lies at 17×10^6 s^{-1}, beyond the range graphed in Figure 3, and represents 8% of the total probability. We suggest that this small, rapidly decaying component can be understood through the radiationless decay model proposed by Amirav (*13*). In this view, optical excitation generates an electronically inhomogeneous population, as the $K \neq 0$ pyrazine molecules undergo rapid (Coriolis-assisted) statistical-limit dephasing into pure triplet states while those with $K = 0$ retain substantial S_1 character and show distinct decay properties. We assign the high frequency decay component in our data to the $K = 0$ subset and conclude that this component has no direct bearing on the pure-triplet relaxation processes of interest here.

Equation 5 allows us to convert the deduced decay rate distribution into the corresponding vibrational energy distribution. The result is shown in the top frame of Figure 4. For comparison, the bottom frame shows the energy distribution calculated for a thermalized triplet pyrazine sample (at 841 K) containing the same average vibrational energy as the nascent sample's. In performing this calculation we used the set of triplet vibrational frequencies compiled earlier (*5*), with corrections from the latest spectroscopic analysis of Fisher (*14*). The strikingly narrower distribution in the top frame compared to the bottom qualitatively confirms our expectation of a sharply

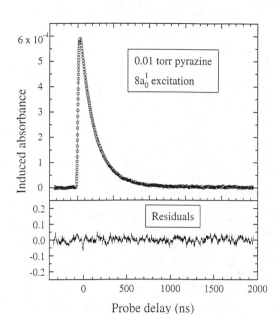

Figure 2. Induced absorbance measured at 676 nm after $8a_0^1$ excitation of a sample containing 0.01 torr pyrazine. Top frame: data (open circles) and fit (solid curve) computed as described in the text. Bottom frame: difference between data and fit.

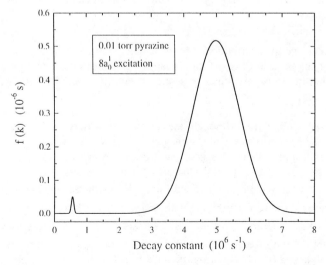

Figure 3. Decay constant distribution function corresponding to the fit of Fig. 1.

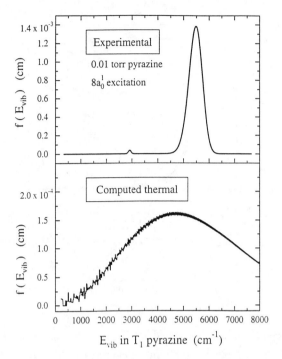

Figure 4. Top frame: vibrational energy distribution function deduced from the decay constant distribution function of Fig. 3. Bottom frame: calculated vibrational energy distribution function for a T_1 pyrazine sample thermally equilibrated at 841 K (having the same average energy as provided by $8a_0^1$ excitation).

defined nascent energy distribution. From a quantitative standpoint, we view the width of the main peak in the top frame as an upper limit to the actual nascent width and expect that future experimental refinements will give the VEDKA method finer energy resolution.

Effects of Early Collisions. Although the analysis described above is rigorously valid only in the pre-collisional limit, it may also be useful for revealing the effects of the earliest collisions. With this goal we have performed decay rate distribution analyses for samples containing 0.01 torr of pyrazine plus small pressures of added helium. Figure 5 shows the results from this series of measurements, in which 0_0^0 excitation prepared the triplet pyrazine with 4056 cm[-1] of vibrational energy. (Using 0_0^0 rather than $8a_0^1$ excitation gives increased signal-to-noise ratio and a lower nascent decay constant.) Two effects of the added helium are apparent. First, the nascent, dominant peak near 2.0×10^6 s[-1] is significantly broadened. We view this broadening as a kinetic consequence of vibrationally inelastic collisions. The second apparent effect is growth of the low frequency peak near 0.15×10^6 s[-1], which, according to our knowledge of $k(E)$, corresponds to triplet pyrazine molecules that have lost approximately 2000 cm^{-1} of vibrational energy. This partially relaxed population evidently arises from efficient V $\rightarrow$ T, R processes in collisions with helium. As for why population accumulates near 2000 cm^{-1}, we note that this energy matches the sharp threshold in relaxation susceptibility found in previous studies of triplet pyrazine's energy-dependent collisional energy loss (*4,5,15,16*). Excited triplet molecules can therefore undergo efficient relaxation until their energy content approaches 2000 cm^{-1}, after which they become relatively resistant to further relaxation. This results in a kinetic bottleneck and a resulting bimodal energy distribution function.

Figure 6 plots the growth of the slow-decaying peak's integral as a function of helium pressure. The sloped fit to the data implies a significant probability for removal of approximately half of the vibrational energy of T_1 pyrazine donor molecules by collisions with helium atoms. To estimate the rate constant for this surprising relaxation process, we have constructed a simplified kinetic cascade model containing only three species: the nascent triplet pyrazine with 4056 cm^{-1} of vibrational energy, partially relaxed triplet pyrazine with ca. 2000 cm^{-1} of vibrational energy, and ground electronic state pyrazine (invisible at our probing wavelength). In this model, both of the triplet species are assigned appropriate rate constants for unimolecular decay to the ground state, and a bimolecular rate constant describes the collisional conversion of the nascent triplet into the partially relaxed species. Adjustment of the model's parameters allows very good fits to the measured kinetic traces. From the optimized parameter values obtained with several helium pressures, we deduce the bimolecular rate constant for partial vibrational relaxation to be 1.3 x 10^5 s^{-1} torr^{-1} , or 0.7% of the gas kinetic encounter rate between helium and pyrazine. We believe that this result is semi-quantitatively valid despite the simplicity of the model used to derive it. The deduced efficiency must be viewed as very large for a 2000 cm^{-1} V $\rightarrow$ T, R "supercollision" process. Nevertheless, it lies at least two orders of magnitudes below the efficiency for fluorescence quenching in pyrazine (*17*), confirming the distinction between these two processes. In interpreting our result one should note that individual collision intervals can be considerably shorter than the mean value, triplet Lennard-Jones parameters may exceed the ground state values assumed in this estimate, and Lennard-Jones parameters may underestimate relaxation cross-sections in general.

Conclusions

In summary, we have introduced a method, called VEDKA, that allows near-nascent vibrational energy distributions to be deduced from kinetic data taken under low pressure conditions. The VEDKA method may in principle be applied to any excited species that undergoes irreversible decay (through chemical reaction or radiationless relaxation) with a rate constant that varies with vibrational energy content. In its first application, the method confirms that the energy distribution in a sample of T_1 pyrazine formed through intersystem crossing is much narrower than a thermal distribution carrying the same average energy. At pressures high enough to cause significant collisional energy level changes during the observation period, the method

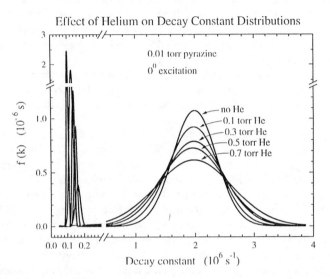

Figure 5. Decay constant distributions derived from pyrazine samples excited at 0_0^0 in the presence of various pressures of helium buffer gas.

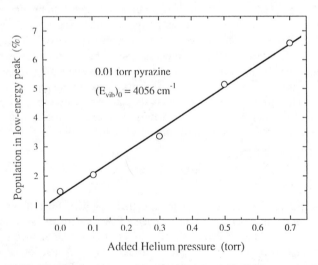

Figure 6. Population percentage represented by the low-frequency peak in Fig. 5 as a function of the added helium pressure.

loses validity. A complete interpretation of pressure-dependent decay rate distributions must therefore rely on master equation modeling. Even without such modeling, however, our results strongly suggest the rapid formation of a bimodal energy distribution through collisions of triplet pyrazine with helium, the mildest relaxer species. The rate of strong deactivation appears remarkably high: an average of 0.7% of gas kinetic collisions with helium atoms remove ca. 2000 cm^{-1} of vibrational energy from the triplet pyrazine. We also note that a bimodal structure of the type deduced here would likely remain hidden from experimental methods that deduce only two or three moments of the energy distribution. We believe that with future refinements, the method of high-definition kinetic analysis should prove a powerful tool for investigating the early stages of collisional vibrational relaxation in excited polyatomics.

Acknowledgments

We are grateful to Dr. Michael Barnes for introducing us to kinetic distribution analysis, and to the National Science Foundation and the Robert A. Welch Foundation for support of this research.

Literature Cited

1. Brenner, J. D.; Erinjeri, J. P.; Barker, J. R. *Chem. Phys.* **1993**, *175*, 99.
2. Hartland, G. V.; Quin, D.; Dai, H.-L. *J. Chem. Phys.* **1994**, *100*, 7832.
3. Lohmannsroben, H. G.; Luther, K. *Chem. Phys. Lett.* **1988**, *144*, 473.
4. Bevilacqua, T. J.; Andrews, B. K.; Stout, J. E.; Weisman, R. B. *J. Chem. Phys.* **1990**, *92*, 4627.
5. Bevilacqua, T. J.; Weisman, R. B. *J. Chem. Phys.* **1993**, *98*, 6316.
6. Provencher, S. W. *Comput. Phys. Commun.* **1982**, *27*, 213.
7. Provencher, S. W. *Comput. Phys. Commun.* **1982**, *27*, 229.
8. Gregory, R. B.; Zhu, Y. *Nucl. Instr. and Meth.* **1990**, *A 290*, 172.
9. Livesey, A. K.; Brochon, J. C. *Biophysical Journal* **1987**, *52*, 693.
10. Troe, J. *J. Chem. Phys.* **1982**, *77*, 3485.
11. Lendvay, G.; Schatz, G. C. *J. Phys. Chem.* **1990**, *94*, 8864.
12. Bevington, P. R.; Robinson, D. K. *Data Reduction and Error Analysis for the Physical Sciences*, 2nd ed.; McGraw-Hill: New York, 1992.
13. Amirav, A. *Chem. Phys.* **1986**, *108*, 403.
14. Fischer, G. *Can. J. Chem.* **1993**, *71*, 1537.
15. Weisman, R. B. Vibrational energy loss from triplet state polyatomics. In *Advances in Chemical Kinetics and Dynamics*; Barker, J. R., Ed.; JAI Press: Greenwich, Connecticut, 1995; Vol. 2B; pp 333.
16. McDowell, D. R.; Weisman, R. B. to be published.
17. McDonald, D. B.; Fleming, G. R.; Rice, S. A. *Chem. Phys.* **1981**, *60*, 335.

Chapter 14

Quantum Scattering Studies of Collisional Energy Transfer from Highly Excited Polyatomic Molecules: Collinear He + CS$_2$ at Energies up to 92 kcal/mol

György Lendvay[1], George C. Schatz[2], and Toshiyuki Takayanagi[2,3]

[1]Central Research Institute for Chemistry, Hungarian Academy of Sciences, H–1525 Budapest, P.O. Box 17, Hungary
[2]Department of Chemistry, Northwestern University, Evanston, IL 60208–3113

We present the results of an accurate quantum scattering study of collisional energy transfer in the collinear He + CS$_2$ system, considering energies up to 92 kcal/mol. These results are generated using a coupled channel calculation with a basis of 1000 vibrational states from a discrete variable representation calculation. Detailed comparisons with the results of classical trajectory calculations are performed so as to assess classical/quantum correspondence for energy transfer moments, and for the energy transfer probability distribution function. We find very good agreement of the energy averaged first moments over a wide range of molecular vibrational energies, provided that the translational energy is not too low (translational temperatures below 300K). The second moments, as well as $<\Delta E>_{up}$ and $<\Delta E>_{down}$ show poorer agreement, especially at low temperatures. The quantum energy transfer distribution functions show considerable mode-specific behavior, but the overall envelope is approximately exponential for $|\Delta E| > 2$ kcal/mol, with a faster than exponential drop for $\Delta E < -20$ kcal/mol and $\Delta E > +5$ kcal/mol, and with a broad spike for $|\Delta E| < 2$ kcal/mol. All of these results accurately match the corresponding classical trajectory distributions. The connection of the shape of these distributions with the individual state-to-state probabilities is studied in detail, and it is found that much of this behavior may be connected with the probability of symmetric stretch excitation and deexcitation.

Collisional energy transfer from highly excited polyatomic molecules has received considerable interest recently. This type of energy transfer process, which provides a mechanism for dissipation and accumulation of energy in excited molecules, is important

[3]Permanent address: Japan Atomic Energy Research Institute, Takai-mura, Naka-gun, Ibaraki-ken 319–11, Japan

in the kinetics of gas-phase chemical reactions and in the relaxation of photoexcited molecules. The reason for the increased interest in collisional energy transfer is twofold: first, experimental techniques for studying the evolution of ensembles of highly excited molecules using spectroscopic techniques have achieved considerable progress recently (*1-4*), second, theoretical methods have been developed which make it possible to simulate the experiments (*5-7*). The theoretical simulations are based on classical mechanics, in which trajectories are used to simulate collisions, with initial conditions sampled using Monte Carlo techniques. The comparison of the experimental and theoretical average energy transfers (in the few cases when both are available) show reasonable agreement (*5b,5c*), although not systematically better than a factor of two in comparisons of values of the average energy transfer per collision $<\Delta E>$. An open question thus remains concerning the accuracy of the classical mechanics for the description of collisional energy transfer. Comparison with experiment does not help to settle this question, because of inaccuracies in the potential surfaces which are normally used in the calculations. In addition, the interpretation of the experiments may also carry inaccuracies due to calibration problems. The most reasonable way to judge whether classical mechanics really describes correctly the dynamics of energy transfer is direct comparison of classical and quantum mechanical scattering results done with the same Hamiltonian.

The present work is aimed at the question of quantum effects and their importance in collisional energy transfer involving highly excited polyatomic molecules. We have performed classical and quantum scattering calculations for a model system that is designed to mimic a realistic energy transfer problem, and have compared the resulting values of $<\Delta E>$ and the energy transfer distribution P(E, E ′). The main challenge in making this comparison is that quantum calculations, by their nature, provide information about transitions at the state-resolved level, while classical calculations use microcanonical sampling of the internal phase space of the excited molecule so that state-to-state information is not calculated. In order to make comparison possible, we need a method to transform the quantum results to the microcanonical level. This way we can mimic "quantum microcanonical" distributions that are directly comparable to the classical results. This transformation involves a double average for either $<\Delta E>$ or P(E, E ′), one average over final states and the other average over a microcanonical distribution of initial vibrational states. Although there have been several theoretical studies concerning the quantum dynamics of collisional energy transfer in the past (*8-14*), these calculations have not considered information for microcanonical ensembles, and they have not examined P(E, E ′).

The specific application we consider here is to a collinear model of He + CS_2 at total energies up to 92 kcal/mol. The two stretch modes of CS_2 are explicitly described in this model, and both the intermolecular and intramolecular potentials are chosen to be realistic. This study extends an earlier study (*15*) of collinear He + CS_2 in which we considered energies up to 75 kcal/mol, and concentrated on the classical/quantum correspondence for energy transfer moments. That work demonstrated very good classical/quantum correspondence for the moments, but the results were less clear for P(E, E ′) due to the low state density at lower energies. The present study examines the same problem for energies where the state density is much higher, so the nature of the classical/quantum correspondence is more transparent. In addition, we will analyze the behavior of state-to-state transition probabilities so that the functional dependence of P(E, E′) can be understood on a microscopic level, and its relationship with the intramolecular dynamics can be established.

To extend the calculations from 75 to 92 kcal/mol necessitates increasing the number of basis functions in the scattering calculations from 500 to 1000. This requires significant computational resources, as scattering calculations must be done on a fine energy grid in order to convert transition probabilities as a function of total energy into translational energy accurately. Even energies of 92 kcal/mol are well below the CS_2 dissociation energy (129 kcal/mol on the surface we used (16)), so the energy regime we are studying is well away from that relevant to recombination measurements. However the vibrational eigenstates at 92 kcal/mol experience much stronger intramolecular couplings than at 75 kcal/mol, so the results we present are useful for demonstrating the influence of anharmonic effects and Fermi resonance effects on the energy transfer dynamics.

Theory

Quantum Calculations. The details of our quantum scattering calculations for collinear He + CS_2 are given elsewhere (15). Briefly, we used a standard coupled channel (CC) method for the Hamiltonian corresponding to the complete system:

$$H = -\frac{\hbar^2}{2\mu_R}\frac{\partial^2}{\partial R^2} + H_{vib}(CS_2) + V \qquad (1)$$

where R is the Jacobi coordinate between He and the center of mass of CS_2, μ_R is the translational reduced mass, H_{vib} is the CS_2 vibrational Hamiltonian and V is the intermolecular potential. The vibrational Hamiltonian of CS_2 is:

$$H_{vib} = -\frac{\hbar^2}{2\mu_1}\frac{\partial^2}{\partial r_1^2} - \frac{\hbar^2}{2\mu_2}\frac{\partial^2}{\partial r_2^2} + V_{CS_2} \qquad (2)$$

in terms of the Jacobi coordinates r_1 (between C and the center of mass of SS) and r_2 (between the two S's) where μ_1 and μ_2 are the reduced masses associated with r_1 and r_2, and V_{CS2} is the intramolecular potential.

To solve the Schrödinger equation associated with this Hamiltonian, we first solve for the eigenfunctions associated with H_{vib}, using a discrete variable representation (DVR) (17), with a sequential truncation-diagonalization solution of the vibrational eigenvalue problem. The CC equations are solved using a standard propagation method (18) with the boundary conditions that the wavefunction should go to zero in the R → 0 limit. The scattering matrix elements S_{nm} are then used to define transition probabilities using

$$P_{nm} = |S_{nm}|^2 \qquad (3)$$

The transition probabilities P_{nm} may be used to define initial state-selected energy transfer moments using

$$\langle \Delta E_n \rangle = \sum_m (E_m - E_n) P_{nm} \tag{4}$$

$$\langle \Delta E_n^2 \rangle = \sum_m (E_m - E_n)^2 P_{nm} \tag{5}$$

$$\langle \Delta E_n \rangle_{down} = \frac{\sum_{m(down)} (E_m - E_n) P_{nm}}{\sum_{m(down)} P_{nm}} \tag{6}$$

$$\langle \Delta E_n \rangle_{up} = \frac{\sum_{m(up)} (E_m - E_n) P_{nm}}{\sum_{m(up)} P_{nm}} \tag{7}$$

Note that in defining the up (down) moments, the sums are constrained to those transitions which lead to an increase(decrease) in the vibrational energy. One point of ambiguity in this definition concerns how or whether to include the elastic (m=n) transition in the denominator of Eqs. (6) and (7). This ambiguity does not occur in the classical definition of these energy transfer moments, but we have found that it can be quite important in determining the quantum moments, due to the fact that the quantum elastic transition probability is often much larger than any of the inelastic probabilities. In Ref. 15 we studied several possible choices for this calculation, and concluded that including ½ of P_{mm} in the denominator of Eqs. (6) and (7) gives close to the best possible agreement with the classical results. We will use the same choice in the present study.

In presenting the second moment, we will always convert it to the variance σ, which is defined by the usual expression:

$$\sigma = \sqrt{\langle \Delta E^2 \rangle - \langle \Delta E \rangle^2} \tag{8}$$

To compare with trajectory results for a microcanonical ensemble, it is necessary to define microcanonical moments. The microcanonical average means that the energy shell over which the averaging is done is narrow, ideally having the width of a δ-function. We have chosen to define the microcanonical moments by averaging the state-selected quantum moment over a small range of initial energies, using an exponential gap weight function. Thus for the first moment we use

$$\langle \Delta E(E) \rangle_\mu = \frac{\sum_n \langle \Delta E \rangle_n e^{-|E - E_n|/\gamma}}{\sum_n e^{-|E - E_n|/\gamma}} \tag{9}$$

where γ is a width parameter that we choose empirically to be as small as possible while yielding an energy averaged moment that is approximately a smooth function of energy. Generally we chose this width parameter to be 1.9 kcal/mol, but the results are not sensitive to this choice provided that it is larger than the mean energy level spacing while at the same time smaller than the energy range associated with large-scale changes in the moments. Expressions similar to (8) may be written for the other moments. The choice of an exponential gap weighting function is arbitrary, but we found that other choices give similar energy averaged moments.

To determine the distribution function $P(E, E')$ for a microcanonical ensemble, we apply an averaging procedure similar to that in Eq. (8) to the state-to-state probability P_{nm}. In Ref. 15 we tested several possible methods for performing this average, including: (A) averaging P_{nm} over a range of initial states to each final state, with equal weighting for each initial state, and then averaging over final states using the exponential gap weight; (B) using the exponential gap weight function to average both initial and final states; and (C) averaging over initial and final states using the weight function $\exp\{-[(E_n - E_c)^2 + \Delta E^2]^{1/2}/\delta\}$, where E_c is the energy of the central state in the range to be averaged over, ΔE is the energy gap between initial and final states, and δ is a parameter similar to γ. Note that in method C, the probability is represented as a function of the energy *gap* between the initial and final state so that state combinations with the same gap contribute to the same energy bin, while in methods A and B the average refers to the initial and final energies. Once the method C averaging has been performed, we convert the distribution back to an absolute energy scale by adding in the energy of the central state E_c to the energy gap.

In Ref. 15 we found that the $P(E, E')$ distributions from these three methods were all quite similar, so in the present application we only consider method C. To test convergence of the averaging procedure, we will present results in which different values of the width parameter δ have been used, and we will also consider different choices for the range of states about the central state to be used in doing the average.

Classical Calculations. The collinear trajectory calculations are based on the same Hamiltonian as the quantum calculations. The methods and calculations are the same as described in Ref. 15, so we omit details here.

Details of Applications. The calculations were based on the CS_2 potential surface due to Carter and Murrell (*16*) (hereafter denoted CM), which is derived from a spectroscopic force field at low energies, and which is also chosen to dissociate correctly. The two harmonic stretch frequencies associated with this potential are 674 cm^{-1} and 1532 cm^{-1} and the dissociation energy to CS + S is 129 kcal/mol. Equipotential contours of the CM potential for collinear SCS are plotted in Figure 1 as a function of the normal mode coordinates. The intermolecular potential is one of several potentials that were considered by Bruehl and Schatz (*5a*), namely the one labelled He-2.

The DVR calculations were based on $E_{max} = 113$ kcal/mol, and a 100x100 grid of DVR points. Coupled channel calculations were based on $N_{max} = 1000$ basis functions. Calculations based on smaller basis sets indicate that this should yield fully converged results up to the highest energy we considered (92 kcal/mol). Calculations with this basis set were done over the energy range 63-92 kcal/mol on an energy grid of 0.6 kcal/mol. The resulting transition probabilities were then spline interpolated to convert from total to translational energy, and to do Boltzmann averaging.

Results

Moments. Figure 2 presents the initial state-resolved first moment $<\Delta E>$ as well as its microcanonical counterpart $<\Delta E>_\mu$ versus initial vibrational energy E from the quantum calculations at a fixed translational energy E_{trans} of 0.63 kcal/mol. The curves drawn

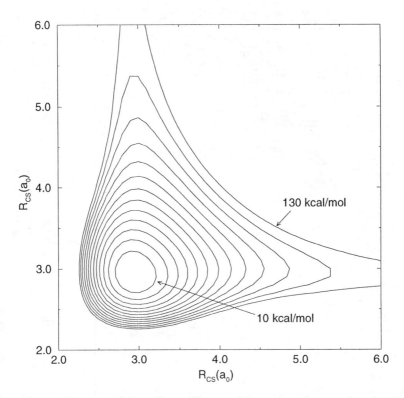

Figure 1.   Contours of the collinear CS_2 potential energy surface as a function of the
two CS distances (in a_0). Contours are in 10 kcal/mol increments, with
the bottom of the well taken to be zero energy.

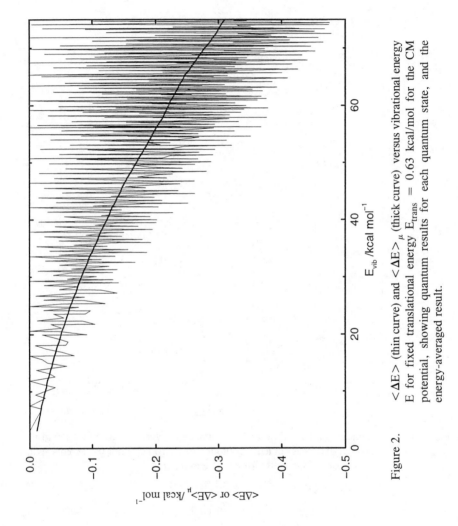

Figure 2. $\langle \Delta E \rangle$ (thin curve) and $\langle \Delta E \rangle_\mu$ (thick curve) versus vibrational energy
E for fixed translational energy E_{trans} = 0.63 kcal/mol for the CM
potential, showing quantum results for each quantum state, and the
energy-averaged result.

connecting the different values of $<\Delta E>$ have no physical significance; they are included to aid in visualizing the dependence of $<\Delta E>$ on initial state. This figure shows the expected strong state-specific variations in $<\Delta E>$, which for the most part arises because the symmetric stretch mode can be collisionally deexcited much more easily than the asymmetric stretch. $<\Delta E>_\mu$, by contrast, is a smoothly decreasing function of E, with only slight fluctuations around a function that could easily be fit by a quadratic or exponential function of E. This stronger-than-linear variation of $<\Delta E>_\mu$ with E is similar to what was found by Bruehl and Schatz (*19*) in their earlier study of collinear He + CS_2.

The results in Figure 2 are for E_{trans} = 0.63 kcal/mol. To show how the results vary with E_{trans}, in Figure 3 we present all four of our energy transfer moments, namely $<\Delta E>_\mu$, σ, $<\Delta E>_{\mu,up}$ and $<\Delta E>_{\mu,down}$, as a function of E_{trans} for a fixed vibrational energy of E = 63 kcal/mol. This value of E corresponds to the energy of the 269th state of collinear CS_2, but, of course, the energy averaged moments are not sensitive to specific quantum states. Figure 3 includes both the quantum scattering and classical trajectory results, and we note that the first moment $<\Delta E>_\mu$ shows very good agreement between quantum and classical mechanics over a wide range of translational energies. The agreement is better than 20%, which is almost within statistical uncertainty, except for E_{trans} less than about 0.7 kcal/mol. At low values of E_{trans} the disagreement between classical and quantum results becomes quite large on a relative basis, but the magnitude of $<\Delta E>_\mu$ is very small in this limit. The dependence of $<\Delta E>_\mu$ on E_{trans} is similar to what has been seen in earlier trajectory studies in three dimensions (*5b,5d*).

The moments other than $<\Delta E>_\mu$ in Figure 3 show less good agreement between classical and quantum results. The best relative agreement is at high energy, where differences are less than 20%. Below 1 kcal/mol the relative differences are substantial, especially for σ. For $<\Delta E>_\mu$, σ and $<\Delta E>_{\mu,down}$, the quantum moments are larger in magnitude than their classical counterparts at low E_{trans}. This result is the expected behavior if tunnelling were to enhance penetration of the He closer to the CS_2 where coupling is larger.

Figure 4 shows the dependence of all four moments $<\Delta E>_\mu$, σ, $<\Delta E>_{\mu,up}$ and $<\Delta E>_{\mu,down}$ on E for a translational temperature of T = 1000K. Here we see that the classical/quantum correspondence is best for $<\Delta E>_\mu$, still very good for σ, and poorer for $<\Delta E>_{\mu,up}$ and $<\Delta E>_{\mu,down}$. The quality of the comparisons does not change significantly with E, although there is a slight worsening of the agreement at the highest E's considered.

State-to-state probability distribution P_{nm}. Figure 5 shows the state-to-state probabilities P_{nm} for initial states n = 493, 496, 502 and 504 as a function of energy of the final state m. The translational temperature in each figure is 300K. These particular states were chosen in part because in the next section we will consider microcanonical probabilities for energies that include these states. In addition, the particular states chosen illustrate some limiting cases that are important for interpreting the dependence of the probabilities on specific vibrational modes.

Consider first the plot for n = 493. This "spectrum" is especially simple, as the initial state has only two quanta of symmetric stretch excitation, plus a large number of quanta of the less active asymmetric stretch. The most intense peak in this plot is the elastic peak at about 83 kcal/mol. On either side of the elastic peak is a series of intense peaks, spaced at 1.5 kcal/mol intervals, corresponding to symmetric stretch excitation or

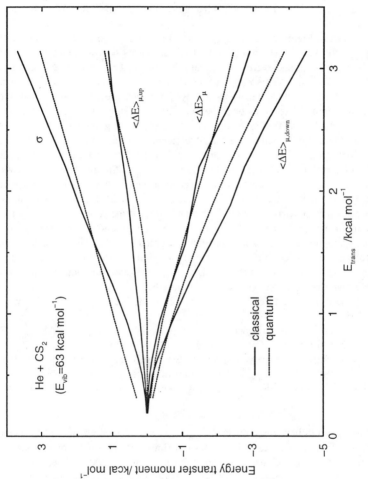

Figure 3. $\langle\Delta E\rangle_\mu$, σ, $\langle\Delta E\rangle_{\mu,up}$ and $\langle\Delta E\rangle_{\mu,down}$ versus E_{trans} for $E = 63$ kcal/mol, comparing quantum results (dashed curves) with classical results (solid curves).

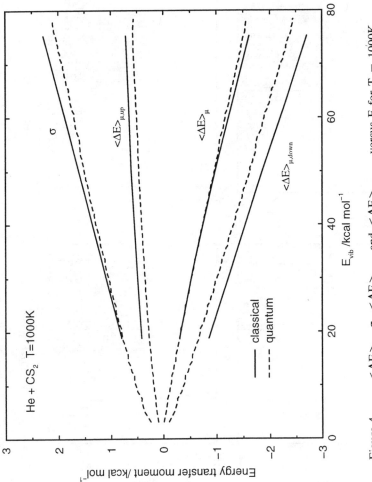

Figure 4. $\langle \Delta E \rangle_\mu$, σ, $\langle \Delta E \rangle_{\mu,up}$ and $\langle \Delta E \rangle_{\mu,down}$ versus E for T = 1000K. Solid curves are classical results, dashed are quantum results.

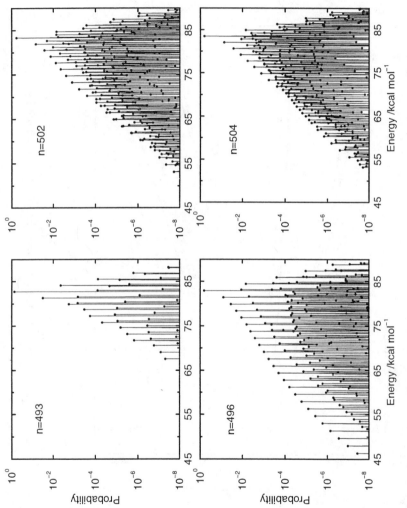

Figure 5.   State-to-state probability distributions P_{nm} for n = 493, 496, 502 and 504, for a translational temperature of 300K.

deexcitation relative to the initial state. This series of peaks cuts off at 80 kcal/mol in the down branch after two quanta have been lost, thereby establishing the initial state as having just two quanta of symmetric stretch excitation. In addition to this progression of intense peaks which involves excitation or deexcitation of the symmetric stretch, the 1000K results show three additional progressions with the 1.5 kcal/mol spacing, one which cuts off at 76 kcal/mol, one at 72 kcal/mol and one at 68 kcal/mol. Each of these progressions is associated with deexciting successive numbers of quanta (1, 2 and 3) of the asymmetric stretch (which has a 4 kcal/mol quantum), plus excitation or deexcitation of quanta of the symmetric stretch, and in each case the cutoff occurs when two quanta of symmetric stretch have been deexcited. Another progression occurs in the up branch of the spectrum that corresponds to exciting one quantum of the asymmetric stretch. These progressions which involve exciting or deexciting the asymmetric stretch are much less probable than the progression which involves only symmetric stretch excitation or deexcitation. However, there is not much dependence of the transition probability on the symmetric stretch quantum number once one or more quanta of asymmetric stretch have been transferred.

Now consider the n = 496 plot. The initial state here has a large number of quanta of the symmetric stretch, but none in the asymmetric stretch, so there is a long progression of peaks separated by 1.5 kcal/mol that correspond to deexciting as many as 23 quanta of the symmetric stretch relative to the elastic peak. A second series that corresponds to deexcitation of one quantum of asymmetric stretch, plus excitation or deexcitation of the symmetric stretch is also quite visible, and there are additional progressions below that. Note that some of the states have "erratic" intensities, such as the third most intense peak in the group of states near 74-75 kcal/mol. These indicate that the final states being produced are part of a Fermi-coupled pair.

The plots for states n = 502 and 504 are examples of energy transfer from a Fermi-coupled pair of initial states. Plots of the wavefunctions for these states are nearly identical, and they have the characteristic structure of a 5:2 periodic classical trajectory. The existence of 5:2 Fermi resonant states was suggested previously for collinear CS_2 by Bruehl and Schatz (*19*) based on an analysis of classical surfaces of section, and the present results confirm this. Since the wavefunctions of the Fermi-coupled pair have similar spatial characteristics, energy transfer out of each state should be similar, and this is what the n = 502 and 504 plots indicate. It is interesting to note that the spectrum for state 503 is sufficiently different from 502 or 504 that distinguishing it from the Fermi-coupled pair is straightforward. The spectrum of final states in each plot is quite complex. This makes sense since the initial states in each case are the superposition of two states, which means that the spectra are the superposition of two spectra. In spite of this complexity, the overall envelope of the spectra looks much like that for n = 496. One interesting point concerning the n = 502 and 504 plots is that the transition probability between these two states is two orders of magnitude below the elastic probability. This means that the Fermi-coupled states are not very strongly coupled to each other, even though the energy gap is quite small. Simple perturbation theory arguments indicate that this can happen if the elastic amplitudes associated with the underlying zeroth order states precisely interfere with each other.

The overall envelope of the plots in Figure 5 consists of an extended exponential decay for $|\Delta E| > 2$ kcal/mol with somewhat faster than exponential drop-off at very large $|\Delta E|$ ($\Delta E < -20$ kcal/mol in the down branch). At small $|\Delta E|$, the envelope shoots up above the exponential envelope; this is especially noticeable for the elastic peak.

We will refer to this behavior as a "broad spike" near $|\Delta E| = 0$. The spike is most noticeable in the intense progression of peaks that corresponds to excitation of just the symmetric stretch relative to the initial state. Transitions that require multiple quanta to be transferred between modes such as those for which $|\Delta E|$ is within 1.5 kcal/mol of zero are down by many orders of magnitude relative to the elastic peak, and do not contribute, at least visually, to the spike.

Energy Transfer Distribution Function P(E, E´). Figure 6 presents the microcanonical energy transfer probability distribution function $P(E, E´)$ at an initial energy of 83 kcal/mol (the same energy as the n = 500 state of CS_2) and a translational temperature of 300K. These results were obtained using the method C averaging procedure, with widths δ of 0.06, 0.3 and 0.6 kcal/mol. This figure shows how the state-to-state probability distribution evolves into the microcanonical average as more and more states are added together. We see that this evolution is very simple, with the primary effect being to smooth out the pronounced fluctuations that are seen in the state-resolved distributions. Indeed, the fluctuations are almost completely quenched for the two largest choices of width parameters. This result is to be expected, as 0.3 kcal/mol is larger than the average spacing of energy levels (roughly 0.06 kcal/mol) for states in the vicinity of 83 kcal/mol.

Another measure of convergence of the microcanonical average is given in Figure 7. Here we show $P(E, E´)$ for different choices of the size of the interval of states used in performing the average used in method C, all for the same value (0.3 kcal/mol) of the width parameter δ. With this δ, and a large number of states in the interval, we recover the same smooth distribution as in Figure 6. Narrowing the interval leads to fluctuations about this average. The results in Figures 6 and 7 indicate that $P(E, E´)$ is well defined by the averaging procedure. The $\Delta E = 0$ peak is more sensitive than the rest to details of the averaging, as makes sense given that only a few states contribute, and their probabilities vary widely. In general we expect that very close to the elastic peak, the transition probabilities should be quite small, as the only transitions that can give ΔE's near zero are ones that involve simultaneous transitions in both vibrational modes, and the probabilities of such multiquantum processes are often small as we have seen in Figure 5. However the small energy gap also serves to enhance the probabilities of these transitions as compared to the energies where the states are less densely packed, and in the end, this effect dominates in the microcanonical average, leading to a broadened spike in the distribution near $\Delta E = 0$. This spike is actually quite robust in the present results, being quite noticeable in essentially all the results in Figures 6 and 7, and similar to what is seen in Figure 5. This is in contrast to what we found for the same Hamiltonian, but at lower energies, in Ref. 15, where the lower state density caused the spike to be much more sensitive to the averaging procedure.

On the semilog scale used in Figures 6 and 7, the probabilities clearly show an exponential dependence on the difference in energy between the initial and final states for ΔE's between -2 and -20 kcal/mol, and between +2 and +5 kcal/mol. The slope of the "down branch" of the distribution is smaller than that of the "up branch", as is expected due to detailed balance. For ΔE's less than -20 kcal/mol and more than +5 kcal/mol, the quantum distribution drops more quickly than the exponential result. However the probabilities for such transitions are very small, so these do not contribute significantly to energy transfer moments. All of these results are in one-to-one correspondence with what we concluded in analyzing the state-to-state probabilities in Figure 5, and the

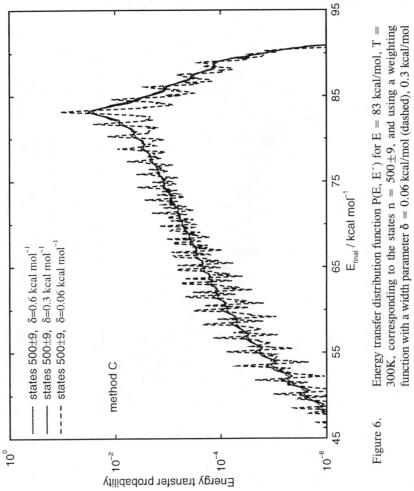

Figure 6.   Energy transfer distribution function P(E, E´) for E = 83 kcal/mol, T = 300K, corresponding to the states n = 500±9, and using a weighting function with a width parameter δ = 0.06 kcal/mol (dashed), 0.3 kcal/mol (solid) and 0.6 kcal/mol (dotted).

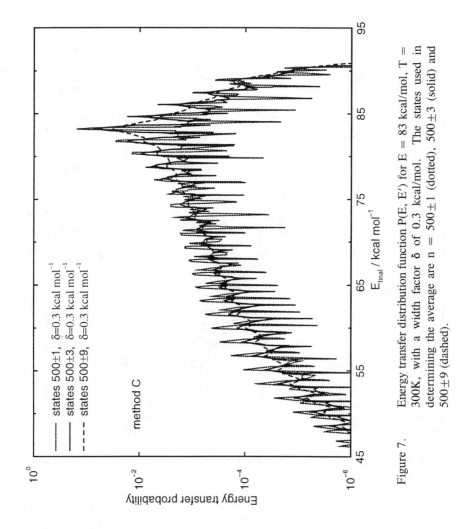

Figure 7.   Energy transfer distribution function P(E, E') for E = 83 kcal/mol, T = 300K, with a width factor δ of 0.3 kcal/mol. The states used in determining the average are n = 500±1 (dotted), 500±3 (solid) and 500±9 (dashed).

distributions we have obtained are also similar to biexponential distributions that have commonly been observed in 3D trajectory calculations (*20*).

Figure 8 presents the comparison of the δ = 0.3 kcal/mol results for state n = 500 ± 9 with the corresponding classical trajectory distribution at T = 300K and 1000K. The classical distribution was obtained by dividing a very large number of trajectories into bins that are 0.12 kcal/mol in width. The figure shows that the quantum distributions overlap very well with their classical counterparts at both temperatures, with both showing the broad spike at small ΔE and the large ΔE exponential tail. One very encouraging point concerning the comparisons in Figure 8 is that the spike at small ΔE has about the same width and amplitude in the quantum and classical results. This result was not apparent in the results we presented in Ref. 15, and apparently the problem there was the that state density was not sufficiently high to give a quantum distribution function that was independent of the averaging process on an energy scale of the order of the width of the spike. In the present application, this problem is less important because of the larger state density, so now we see good agreement.

Conclusions

This paper has presented new results concerning the collisional relaxation of highly excited CS_2, extending the calculations reported earlier in Ref. 15 to much higher energies. This extension has proven useful for several reasons, although many of the conclusions, such as the good agreement of classical and quantum moments, are identical to what was seen before. One important feature of the new results is that the much higher state density has enabled us to determine the appearance of the broad spike in P(E, E′) near $|\Delta E|$ = 0 much more clearly than before. The result shows that the good classical/quantum correspondence between the classical and quantum distributions seen for large $|\Delta E|$ also extends to small $|\Delta E|$. In addition, we see that this spike, as well as other features of the energy transfer distributions, are already properties of the state-to-state distributions, and particularly of the single mode transitions that involve excitation or deexcitation of the symmetric stretch mode relative to the elastic peak. This suggests that an uncoupled two mode model of CS_2 may provide a good approximation to the results we have obtained.

One other new feature of the present study arises from the presence of Fermi resonant states in the vibrational manifold. The state-to-state probability distributions associated with Fermi coupled states are found to be very similar to each other, but the probability of transfer between the Fermi-coupled states is surprisingly small. Fermi-coupled initial states give probability distributions that are much like the superposition of two uncoupled state distributions, but the overall shape of the distributions is otherwise not unusual. Fermi-coupled final states are associated with final state probabilities that are significantly perturbed compared to what would be expected by extrapolating from lower or higher energy. However this has essentially no effect on the microcanonical energy transfer distribution function.

An important question that we cannot answer completely at this point is what it is that controls the shape of the P(E, E′) distribution. Linearly coupled harmonic oscillators subject to exponential potentials typically give exponential distribution functions, so the results that we obtain are consistent with this behavior except for the small $|\Delta E|$ spike and the faster than exponential behavior at large $|\Delta E|$. The small $|\Delta E|$ spike is universally seen in 3D trajectory simulations of energy transfer in

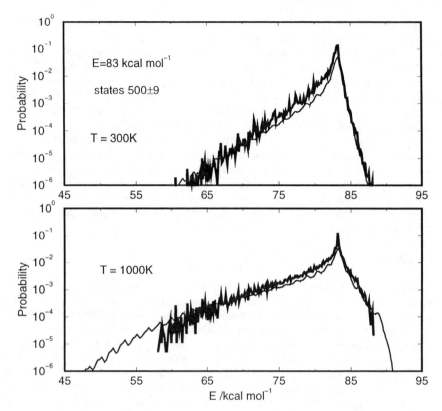

Figure 8.   Comparison of classical (thick line) and quantum (thin line) distribution
                 functions P(E, E') for E = 83 kcal/mol, T = 300K and 1000K. The
                 quantum distribution function is based on δ = 0.3 kcal/mol and states
                 500±9.

polyatomic molecules, and is responsible for the biexponential distributions that are often
used in characterizing the shape of P(E, E'). The present results show that this small
$|\Delta E|$ spike is even contained in single-mode progressions, so it appears that this is not
special to polyatomic molecules. It will be important therefore to study the variation of
this result with the nature of the intermolecular potential to see if that is responsible. The
faster than exponential fall-off for large $|\Delta E|$ has not been seen in 3D trajectory
simulations, although the low probabilities associated with this part of the distribution
makes its observation difficult. The fact that it shows up only at large $|\Delta E|$ points to its
origin in the anharmonicity of the intramolecular potential or in nonlinear intermolecular
interactions, but again it will be important to work through the sensitivity of these results
to potential parameters to establish its origin.

Acknowledgments

This research was supported by NSF Grant CHE-9527677, by the Hungarian National Research Fund grant OTKA-T7428, and by the US-Hungarian Joint Fund grant JF411.

Literature Cited

1.   Weston, R. E.; Flynn, G. W., *Annu. Rev. Phys. Chem.* **1992**, *43*, 559.
2.   *Vibrational Energy Transfer Involving Large and Small Molecules*; J. A. Barker, Ed.; Advances in Chemical Kinetics and Dynamics; JAI Press: Greenwich, Connecticut, 1995; Vols. 2A,2B.
3.   Hippler, H.; Troe, J. In *Bimolecular Reactions*; Baggott, J. E; Ashfold, M. N. R., Eds.; The Chemical Society: London, 1989.
4.   Barker, J. R.; Toselli, B. M. in *Photothermal Investigations of Solids and Fluids*; J. A. Sell, Ed.; Academic Press: New York, 1989.
5.   a) Bruehl, M.; Schatz, G. C., *J. Phys. Chem.* **1988**, *92*, 3190; b) Lendvay, G.; Schatz, G. C., *J. Chem. Phys.* **1992**, *96*, 4356; c) Lendvay, G.; Schatz, G. C. *J. Chem. Phys.*, **1993**, *98*, 1034; d) Lendvay, G.; Schatz, G. C. In Ref. 2, Vol. 2B, pp. 481-513.
6.   Gilbert, R. G. *Int. Rev. Phys. Chem.* **1991**, *10*, 319.
7.   Lenzer, T.; Luther, K.; Troe, J.; Gilbert, R. G.; Lim, K. F. *J. Chem. Phys.* **1995**, *103*, 626 and references therein.
8.   Nalewajski, R. F.; Wyatt, R. E. *Chem. Phys.* **1983**, *81*, 357; *85*, 117; *89*, 385.
9.   Clary, D. C.; Gilbert, R. G.; Bernshtein V.; Oref, I. *Faraday Disc. Chem. Soc.* **1995**, *102*, 423.
10.   (a) Clary, D. C. *J. Phys. Chem.* **1987**, *91*, 1718; (b) Clary, D. C. *J. Chem. Phys.* **1981**, *75*, 2023.
11.   Pan, B.; Bowman, J. M. *J. Chem. Phys.* **1995**, *103*, 9668; Bowman, J. M.; Padmavathi, D. A. *Mol. Phys.* **1996**, *88*, 21.
12.   Schatz, G. C. *Chem. Phys. Lett.* **1979**, *67*, 248.
13.   Rapp, D.; Kassal, T. *Chem.Rev.***1969**, *69*, 61; Schatz, G. C. *Mol. Phys.* **1978**, *35*, 477.
14.   Duff, J. W.; Truhlar, D. G. *Chem. Phys.* **1975**, *9*, 243.
15.   Schatz, G. C.; Lendvay, G. *J. Chem. Phys.*, in press.
16.   Carter, S.; Murrell, J. N. *Croat. Chim. Acta* **1984**, *57*, 355.
17.   (a) Park, T. J.; Light, J. C. *J. Chem. Phys.* **1986**, *85*, 5870; Gray, S. K.; Verosky, J. M. *J. Chem. Phys.* **1994**, *100*, 5011.
18.   Alexander, M. In *"Algorithms and Computer Codes for Atomic and Molecular Quantum Scattering Theory"*, L. Thomas, Ed.; NRCC Workshop Proceedings No. 5, Lawrence Berkeley Laboratory Report LBL-9501, **1979**. Vol I.
19.   Bruehl, M.; Schatz, G. C. *J. Chem. Phys.* **1990**, *92*, 6561.
20.   Lendvay, G.; Schatz, G. C. *J. Phys. Chem.* **1994**, *98*, 6530.

Chapter 15

Stalking the Step-Size Distribution: A Statistical–Dynamical Theory for Large-Molecule Collisional Energy Transfer

John R. Barker

Department of Atmospheric, Oceanic, and Space Sciences and Department of Chemistry, University of Michigan, Ann Arbor, MI 48109–2143

A theory of large molecule energy transfer based on state-to-state transition probabilities is derived and a demonstration calculation is presented for cyclopropane deactivation by helium. The calculations are shown to be practical even for energies where the vibrational state densities exceed 10^{10} states/cm^{-1}. For demonstration purposes the state-to-state transition probabilities are calculated using SSH(T) Theory, and thus the present calculations are expected to provide only general trends (in future work, more accurate state-to-state theories will be employed). The predicted collision step size distributions resemble the exponential model, but with strong fluctuations about the mean. The fluctuations arise from energy transfer propensities, which persist even at high energies. The predicted average energy transferred in deactivation collisions ($<\Delta E>_{down}$) depends on vibrational energy, due to the changing fraction of inelastic collisions; the shape of the distribution function remains nearly unchanged.

Energy transfer is of fundamental importance in unimolecular reactions (*1-3*), but theory has lagged seriously behind experiment for energy transfer involving highly excited large molecules (*4-6*). For small molecules, the theory has advanced to the point that quantitative predictions are possible (*7-9*). Beginning with the Landau-Teller Theory (*10*), several simple theories have been developed which are distinguished by their relative complexity and accuracy (*4,8*). One of the most basic of these is the Schwartz, Slawsky, and Herzfeld Theory (*11,12*), as implemented by Tanczos (*13*): SSH(T) Theory. Other theories of varying complexity include the impulsive called "ITFITS" theory (*14*) and its "DECENT" (*15-17*) and "INDECENT" (*18*) extensions. The co-linear atom+diatomic SSH and ITFITS models have been extended to three dimensions and to polyatomic systems by utilizing the "breathing sphere" approximation (*13,19*). These theories describe energy transfer with varying degrees of fidelity, but their predictive capabilities are limited. Classical trajectory calculations have been used extensively to investigate small molecule energy transfer in recent years (*7,20,21*). More recently, quantum mechanical close coupling calculations using realistic potential energy surfaces have been used to investigate energy transfer involving small and medium-sized molecules at low vibrational energies (*9,22*) and model systems at high excitation energies (*23*).

Classical trajectory calculations currently provide the most generally useful theoretical approach for studying large molecule energy transfer (*24,25*). They have become practical for large molecule energy transfer only recently, however, because of their great demands on computer time (*25,26*). These studies provide qualitative insight and even semi-quantitative agreement with experiments. However, it is not clear whether the agreement with experiments is due to an accidental cancellation of errors in the potential energy surface and those inherent in the use of classical mechanics.

Classical trajectory calculations may suffer from several deficiencies, due to differences between classical and quantum mechanics, but the quantitative importance of these deficiencies is not yet known (*27,28*). The most obvious deficiency is that zero point energy is not conserved in classical mechanics. Non-conserved zero point energy may enhance energy transfer and it can even contribute to bond dissociation at moderate vibrational energies which are by themselves insufficient to break a bond. A more subtle deficiency is that, unlike classical systems, quantum systems do not exhibit equipartition of energy (*29*). Therefore, quantum statistics are significantly different from classical statistics and it is not clear that classical trajectories are capable of properly reproducing effects which depend on the statistical amount of vibrational energy available in a given vibrational mode. It is often argued that the quasicontinuum of states in a large molecule at high vibrational energy validates the use of classical mechanics (*25*). Even at high densities of vibrational states, however, the average excitation per mode may be quite small and the zero point energy may be an important fraction of the total classical vibrational energy, conditions which tend to invalidate the use of classical mechanics. Quantitative comparisons between classical and quantum mechanics are currently being carried out on prototype energy transfer systems to determine the quantitative accuracy of classical mechanics (*28*).

In the present paper, we take a new approach, which has been described elsewhere in detail (*30*). The aim of this work is to develop a single semi-quantitative theory which can reliably describe both small and large molecule energy transfer over a very wide range of excitation energies. Such a theory is necessarily approximate and it may have only limited predictive capabilities, but it will provide an alternative to classical trajectory calculations. It also may provide insight into the important factors which govern large molecule energy transfer.

In this paper, we derive and demonstrate a statistical dynamical theory which conserves zero point energy, obeys detailed balance, and employs state-to-state energy transfer theories for separable degrees of freedom. The theory is modular and the state-to-state theoretical module can be selected according to the required accuracy and computational efficiency. For the purpose of demonstrating the statistical-dynamical theory, we have used SSH(T) breathing sphere theory, but work is currently underway in our laboratory to use ITFITS (14), which is more accurate.

Theory

Statistical-Dynamical Application of State-to-State Models. The derivation (30) is presented in this section. Our objective is to calculate the probability that a large molecule with internal energy E undergoes a transition to new energy in the range E' to E'+dE. For present purposes, we will treat this process as Vibration-to-Translation (V-T) energy transfer, but the theory can be extended easily to include Vibration-to-Vibration (V-V) energy transfer.

For a system with s separable degrees of freedom, the states α_s and β_s can be designated with sets of s quantum numbers (v_i, $1 \leq i \leq s$). A transition can be written in terms of α_s and β_s, or in terms of α_s and Δ_s, where the last represents the set of changes in quantum numbers:

$$\alpha_s = \{v_1, v_2, \ldots\}_s \tag{1}$$

$$\beta_s = \{v_1', v_2', \ldots\}_s \tag{2}$$

$$\Delta_s = \{(v_1'-v_1), (v_2'-v_2), \ldots\}_s \tag{3}$$

Equation (3) is convenient for considering transition probabilities, because it allows the quantum numbers to be partitioned into two groups: those which change during the transition and those which do not:

$$\Delta_s = \{(v_i'-v_i), (v_j'-v_j), \ldots\}_n + \{0, 0, \ldots\}_{s-n} \tag{4}$$

The states α_s and β_s can also be partitioned:

$$\alpha_s = \{v_i, v_j, \ldots\}_n + \{v_p, v_q, \ldots\}_{s-n} = \alpha_n + \alpha_{s-n} \tag{5}$$

$$\beta_s = \beta_n + \beta_{s-n} \tag{6}$$

where n of the quantum numbers change during the transition and the rest do not. The identities of the specific degrees of freedom (i, j, ...) in each subset are retained in sets η_n and η_{s-n} (i.e. the two sets record which modes change and which ones do not change during a transition). For separable degrees of freedom, the internal energy is also partitioned: $E = E_n + E_{s-n}$.

For a system with separable degrees of freedom, we assume that the state-to-state transition probabilities from state α_s to state β_s depend only on α_n and β_n:

$$P(\beta_s, \alpha_s) = P(\eta_n; \beta_n, \alpha_n) = P(\eta_n; \Delta_n, \alpha_n) \ . \tag{7}$$

Thus, every state with the same α_n (and η_n) has the same transition probability.

We are interested in the transition probability averaged over a small energy increment. The total number of states in an energy increment dE is $\rho_s(\eta_s; E)$ dE, where $\rho_s(\eta_s; E)$ is the total density of states for all s degrees of freedom at energy E. The functional form and magnitude of the density of states is determined by the identities, retained in η_s, of the s degrees of freedom. The average transition probability from a state in the energy range E to E+dE to all states in the energy range E' to E'+δE is equal to the sum of all such transition probabilities divided by the number of initial states. Thus, the average transition probability can be expressed

$$<P_c(E',E)> = \frac{1}{\delta E} \int_{E'}^{E'+\delta E} P_c(x,E) \, dx \ = \frac{1}{\delta E} \frac{1}{\rho_s(\eta_s; E) \, \delta E} \sum_{\alpha(E)}^{\alpha(E+dE)} \sum_{\beta(E')}^{\beta(E'+\delta E)} P(\beta_s, \alpha_s) \ , \tag{8}$$

Here, δE is a finite increment and the brackets around $P_c(E',E)$ emphasize the average over the finite interval; $P_c(E',E)$ is obtained in the limit as $\delta E \to 0$:

$$P_c(E',E) = \lim_{\delta E \to 0} \frac{1}{\delta E} \int_{E'}^{E'+\delta E} P_c(x,E) \, dx \ . \tag{9}$$

The summations range over all α_s states with energies in the range E to E+dE and β_s states in the energy range E' to E'+δE.

Not all of the quantum numbers necessarily change in a given transition. Consider the case when n of the s quantum numbers change during a transition. The remaining s-n quantum numbers may take any value, as long as the energy of the state is within the range E to E+dE. The number of states in this energy range which have n quantum numbers that change in the transition is the product of the number of ways of distributing energy E_n among n degrees of freedom multiplied by the number of ways of distributing energy E_{s-n} among the s-n remaining degrees of freedom. For a particular realization, the n quantum numbers are specified and there is only one way of distributing energy E_n. The remaining energy can be distributed in any fashion among the remaining states, as long as it is in the range E_{s-n} to E_{s-n}+dE, and the total number of ways this can be done is given by $\rho_{s-n}(\eta_{s-n}; E_{s-n})$ dE. Since the state-to-state transition probability depends only on α_n, all $\rho_{s-n}(\eta_{s-n}; E_{s-n})$ dE transitions have the same transition probability $P(\eta_n; \beta_n, \alpha_n)$. Thus, Equation (8) can be written

$$<P_c(E',E)> = \frac{1}{\delta E} \int_{E'}^{E'+dE'} P_c(x,E)\, dx \tag{10a}$$

$$= \frac{1}{\rho_s(\eta_s;E)\, dE\, \delta E} \sum_{n=1}^{s} \sum_{\alpha(E)}^{\alpha(E+dE)} \sum_{\beta(E')}^{\beta(E'+\delta E)} \rho(\eta_{s-n};E)\, P(\eta_n;\beta_n,\alpha_n)\, dE \tag{10b}$$

For a given transition, α_n is specified and E_n is exact, ranging from 0 to E; the residual energy E_{s-n} concurrently ranges from E to 0. The summation over initial states in the energy range E to E+dE is thus carried out by summing over all α_n states with $0 \le E_n \le E$:

$$<P_c(E',E)> = \frac{1}{\delta E} \sum_{n=1}^{s} \sum_{\alpha_n(0)}^{\alpha_n(E)} \sum_{\beta(E')}^{\beta(E'+\delta E)} \gamma(\eta_{s-n};E,E_{s-n})\, P(\eta_n;\beta_n,\alpha_n) \tag{11a}$$

where

$$\gamma(\eta_{s-n};E,E_{s-n}) = \frac{\rho_{s-n}(\eta_{s-n};E_{s-n})}{\rho_s(\eta_s;E)} \quad . \tag{11b}$$

When the summation is truncated at $n < s$, the energy differential dE does not appear explicitly in the final expression.

It is convenient to write Equation (11) as a sum of terms which are indexed by the number of quantum numbers which change during a transition:

$$<P_c(E',E)> = \sum_{n=1}^{s} P_n(E',E) \quad . \tag{12}$$

The use of this form is motivated by the notion that the series may converge as n increases. At high values of n, state-to-state transition probabilities are quite small, but many combinations of quantum number changes can contribute to the summations. Because of these two opposite tendencies, the magnitude of $P_n(E',E)$ may exhibit a maximum at low values of n and the series may converge at intermediate values. Truncation of the series results in a significant savings in computer time. This behavior is indeed observed, as discussed below.

SSH(T) Theory. The state-to-state transition probability $P(\beta_n,\alpha_n)$ must be obtained from a suitable theory. The choice should be based on the desired accuracy and the available computer resources. The present theory can be applied to V-T, or to V-V energy transfer by using a suitable state-to-state theoretical module.

For the purpose of this demonstration, we have used SSH(T) Theory (*11-13,32*), which includes most of the relevant physics; its strengths and weaknesses are well known ($8,31,33$). It is derived from the Landau-Teller model (10), which consists of a free atom colliding co-linearly with a harmonic oscillator. The free atom interacts with the nearer of the two atoms in the harmonic oscillator according to an exponential repulsion (with steepness characterized by length λ). In the limit of small vibrational amplitude and elastic collisions, an approximate analytical classical trajectory can be found, which is used to calculate an approximate time-dependent perturbation on the oscillator. The distorted wave Born (DWB) approximation is then used with the perturbation to obtain the transition probability. Detailed balance is imposed by using the approximate classical trajectory corresponding to the average of the incident and final (after the energy transfer) relative velocities. Various *ad hoc* assumptions have been introduced in attempts to match a Lennard-Jones potential to an exponential (12). Tanczos extended the co-linear SSH theory to multi-oscillator molecules by imposing a central potential approximation: the breathing sphere model (13).

Because the SSH(T) theory is based on an over-simplified potential energy function and approximate dynamics, it is does not give accurate quantitative predictions. The DWB approximation is not accurate for large transition probabilities and it does not satisfy unitarity, which means that the sum of the probabilities of transitions from a single initial state does not equal unity (8) Nevertheless, the SSH(T) Theory provides useful semi-quantitative descriptions of V-T and V-V energy transfer involving small molecules. Furthermore, it is computationally simple.

The transition probability $P(\eta_n;\beta_n,\alpha_n)$ is obtained by averaging $<P_c(E',E)>$ over the thermal distribution of initial relative speeds (31). The thermal energy transfer rate constant k_c is obtained from

$$k(T) = k_c <P_c(E',E)> = \pi\sigma^2 <u> <P_c(E',E)>, \tag{13}$$

where $<u>$ is the average relative speed and σ is the Lennard-Jones collision radius.

For a V-T energy transfer process involving a polyatomic, the SSH(T) Theory transition probability ($32,33$) depends sensitively on three sets of parameters: steric factors $P_0(\beta_n,\alpha_n)$, the characteristic length for the repulsive potential λ, and the squares of the vibrational amplitude coefficients A_i^2. For predictive calculations, the vibrational amplitude coefficients can be obtained from a normal coordinate analysis and the exponential repulsion characteristic length (λ) can be estimated on the basis of Lennard-Jones parameters ($12,13$). The steric factors ($P_0(\beta_n,\alpha_n)$) are usually taken as equal to 2/3 for each vibrational quantum number which changes during a transition (33) : $P_0(\beta_n,\alpha_n) = (2/3)^n$. For the purpose of fitting experimental data, or the results of higher order theories, the amplitude coefficients, steric factors, and length parameter may be adjusted. In the present work, anharmonicities have been introduced on an *ad hoc* basis by using anharmonic transition energies in place of the harmonic values.

Implementation

For computational purposes, note that each $P(\eta_n;\beta_n,\alpha_n)$ in Equation (7) is approximately a delta function at $E'-E = \Delta E_{\beta\alpha}$, the energy difference in the transition. By using a finite δE, one obtains a "binned" average transition probability $<P_c(E',E)>$, which approaches $P_c(E',E)$ as $\delta E \to 0$. Operationally, one can test for convergence of the sum over n after evaluating each P_n term in Equation (12) and the summation can be truncated when satisfactory convergence is achieved. Since the summations over α_n and β_n have many more terms for large n, truncation saves considerable computer time. It is also convenient for computational purposes to calculate all possible transitions in a relatively large energy range and then "bin" the results according to $\Delta E_{\beta\alpha}$. In this way, $<P_c(E',E)>$ can be obtained efficiently as a function of E'.

For the purpose of demonstrating the statistical-dynamical theory, we have applied it to the deactivation of cyclopropane by helium (see Tables I and II for parameters) (30). Cyclopropane is a famous test case for unimolecular reaction rate theory (34) and it is a moderately large molecule. The low-lying vibrational states (35) are widely separated and are in the small molecule limit. At high energies, however, the vibrational density of states is high. Thus cyclopropane deactivation provides a good qualitative demonstration of the transition between small and large molecule behavior.

In the present work, the thermal average for the SSH(T) transition probability was evaluated out by Gauss-Laguerre numerical integration (36) of order n=5, which was found by trial and error to give convergence to within <5%. Vibrational densities of states were calculated with the Whitten-Rabinovitch approximation (37). Although feasible, exact state counts are more time-consuming and not warranted for the present demonstration. We have retained the steric factors as equal to 2/3, but we have chosen to adjust the length parameter (λ) and the vibrational amplitudes (A_i^2) in order that state-to-state probabilities calculated using SSH(T) theory agree with those calculated using the VCC-IOS method. Since this work is only intended to demonstrate the statistical-dynamical theory and is not intended to provide quantitative predictions, or to fit experimental data at high energies, the parameterization is rather arbitrary. The results, however, are plausible.

Clary (38) used the VCC-IOS method to calculate $\Delta v = 0 \leftarrow 1$ rate constants for deactivation of eight of the 14 cyclopropane vibrational modes (some are degenerate) in thermal collisions with helium. In the present work (30), Clary's calculations were used as a reference in order to adjust the parameterization of the SSH(T) theory. Table I summarizes the collision parameters and Table II summarizes cyclopropane vibrational frequencies and the adjusted vibrational amplitude coefficients. The C-H stretching and wagging modes of cyclopropane are expected to have vibrational amplitude coefficients of about 1 amu^{-1} (an "amu" is one atomic mass unit expressed as gram mole^{-1}). State-to-state SSH(T) calculations

Table I. Collision Parameters

	σ (Å)	ε/k
Cyclopropane	4.807	248.9
Helium	2.551	10.22

Collision rate coefficient at 300 K:	$k_c = 4.8 \times 10^{-10}$ cm^3 s^{-1}
Repulsion length:	$\lambda = 0.35$ Å [c]

[c] Adjusted to fit VCC-IOSA calculations for 1-0 transitions.

Table II. Cyclopropane Properties

		ω (cm^{-1}) [a]	ω_x (cm^{-1}) [b]	A_i^2 (amu^{-1})
a_1'	1	3038	-60	1.00 [b]
	2	1479	-3	1.00 [b]
	3	1188	-3	0.86 [c]
a_1''	4	1126	-3	0.88 [c]
a_2'	5	1070	-3	0.51 [c]
a_2''	6	3103	-60	1.00 [b]
	7	854	-3	0.42 [c]
e'	8	3025	-60	1.00 [b]
	9	1438	-3	1.00 [b]
	10	1029	-3	0.51 [c]
	11	866	-3	0.25 [c]
e''	12	3082	-60	1.00 [b]
	13	1188	-3	0.98 [c]
	14	739	-3	0.29 [c]

[a] 1-0 transition frequencies from ref. 35.
[b] Estimated
[c] Adjusted to fit VCC-IOSA calculations for 1-0 transitions.

were carried out carried out for assumed values of λ and by assuming amplitudes of 1 amu^{-1} for all modes involving C-H bonds and the results were compared with Clary's calculations at 300 K. By varying λ, the SSH(T) calculations for each mode could be made to agree exactly with Clary's results. Finally, a length parameter λ = 0.35 Å was selected which gave A_{13}^2 = 0.98 amu^{-1} for v_{13} and smaller amplitudes for the remaining modes, as summarized in Table II. Vibrational amplitudes of 1 amu^{-1} were assigned arbitrarily to the modes not considered by Clary. Diagonal quadratic anharmonicities were estimated crudely on the basis of other hydrocarbons. The resulting parameters were employed with the full statistical-dynamical theory to obtain estimates of the collision step-size distribution at high vibrational energies.

Results and Discussion

Calculated step size distributions based on $\Delta v=0,\pm1$ transitions are shown in Figure 1, where $\delta E=10$ cm^{-1}. Calculated distributions with $\Delta v=0,\pm1,\pm2$ transitions differ by less than 5%, as shown in Figure 2, but they required approximately seven times the computer time (for E=20,000 cm^{-1}). All of these calculations were carried out as follows: at each vibrational energy, the $P_n(E',E;\delta E)$ terms were calculated for n=1 and 2. If the n=2 term was less than 10% of the sum, the summation was truncated at n=2. Otherwise, the calculation was continued, until the last calculated term was less than 10% of the total. The magnitudes of the thermally averaged $P_n(E',E;\delta E)$ terms are shown in Figure 3, where it is apparent that the terms are tending to converge, although they are not converging monotonically. Unless stated otherwise, the summations were truncated when the last $P_n(E',E;\delta E)$ term contributed less than 10% of the sum of the terms evaluated up to that point. With this condition, the results have converged to within about 6% for E = 10,000 cm^{-1} and about 10-15% at the higher energies, which is sufficient for the purposes of this demonstration.

By truncating the summation, the calculations become tractable using workstation computers. At E = 40,000 cm^{-1}, the calculations (for n=3) required the evaluation of only 59×10^6 individual state-to-state transition probabilities to calculate the step size distribution from E' = 30,000 cm^{-1} to 50,000 cm^{-1}. A full state-to-state calculation of every transition from a single state at 40,000 cm^{-1} to every state in the 30,000 to 50,000 cm^{-1} range would have required $>10^{15}$ evaluations. Even more evaluations would be needed in order to average over several initial states.

The calculated step size distributions at high vibrational energy exhibit one of the limitations of SSH(T) theory: the summed probabilities exceed unity by up to a factor of ~20 at E = 40,000 cm^{-1}. The ITFITS theory (14) obeys unitarity much more closely and will be used in future work. Nonetheless, the qualitative properties of the step-size distribution calculated using SSH(T) theory are still of considerable interest.

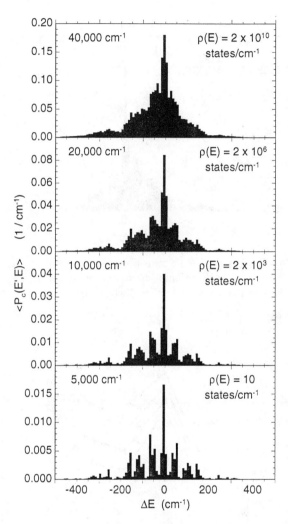

Figure 1. Collision step size distribution for cyclopropane deactivation by helium. Each panel shows the distribution calculated for a 300 K thermal speed distribution and 10 cm^{-1} bins. Results for 5000 cm^{-1}, 10000 cm^{-1}, and 20,000 cm^{-1} were obtained for $\Delta v = 0, \pm 1, \pm 2$ transitions, and those for 40,000 cm^{-1} were obtained for $\Delta v = 0, \pm 1$ transitions. In all cases, the summations were truncated when the last term was <10% of the sum.

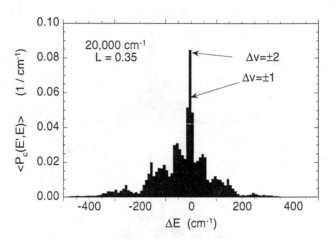

Figure 2. Contribution of $\Delta v = \pm 2$ transitions (10 cm^{-1} bins). The small black segment at the top of each column shows the contribution due to $\Delta v = \pm 2$ transitions. The summations were truncated when the last term was <10% of the sum.

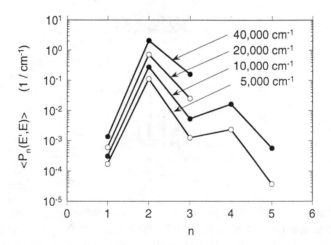

Figure 3. Convergence of series for $\Delta v = 0, \pm 1$ transitions. The magnitude of each term is shown.

One of the most striking properties of the calculated step size distributions is the presence of energy transfer "propensities", in which transitions with almost equal values of $\Delta E_{\beta\alpha}$ have significantly different probabilities. Such propensities have been measured in large molecule energy transfer at $E \leq 5000$ cm^{-1}, where the states can be optically resolved (*4,39*). The optical measurements are described adequately by SSH(T) Theory and thus the present calculations using SSH(T) theory may retain some validity at higher energies. The calculated step-size distributions indicate that the energy transfer propensities are at least partially preserved even at energies as high as 40,000 cm^{-1}, although they are "smeared out" by anharmonicity. The present theory is based on a model which assumes separable vibrational modes and it is therefore not known whether the propensities would be as distinct if inter-mode interactions could be included.

A second striking property is that the envelope of the transition probability distribution is approximately described by an exponential model, but the width of the distribution does not change with energy. Instead, the probability amplitude varies, which indicates that the inelastic cross section varies with vibrational energy.

From energy transfer measurements, the average energy transferred per collision is known to vary approximately linearly with vibrational energy (*40-42*). The conventional assumption is that every collision is inelastic and thus the collision step-size distribution is normalized to unity (*34*). For conventional modeling, the exponential model includes an energy-dependent parameter a(E):

$$P_{down}(\Delta E) = \frac{1}{N(E)} \quad \exp\left[\Delta E / a(E)\right] , \qquad \Delta E < 0 \qquad (14)$$

where N(E) is a normalization factor; the model for up-steps is determined by using $P_{down}(\Delta E)$ and detailed balance (*34*). Bi-exponential models (*21*) provide better descriptions of experimental data (*5,6*), but the average energy transferred per collision still depends approximately linearly on energy (*41,43*). The present calculations indicate, however, that the assumption that every collision is inelastic is probably in error. Instead, it is likely that the proportion of inelastic collisions varies with energy; this is certainly the case at low vibrational energy, where the density of states is sparse and individual state-to-state transitions are dominant.

The average energy transferred in down-steps was calculated from the theoretical results and a plot of $<\Delta E>_{down}$ as a function of E is presented in Figure 4. From the plot it is found that $<\Delta E>_{down}$ is approximately proportional to $E^{3/2}$, which agrees qualitatively with experimental measurements. In future calculations, more accurate state-to-state theories will be employed and perhaps the agreement with experiment will be better.

When the logarithm of the calculated collision step size distribution is plotted as a function of ΔE (Figure 5), the similarity to the exponential model becomes apparent. The envelope of the step-size distribution is approximately exponential, but there are large deviations which reflect the propensities. Note that a(E) in Figure 5 is much smaller than $<\Delta E>_{down}$ in Figure 4, because the step-size distribution does not sum to unity. It is not likely that this calculated step size

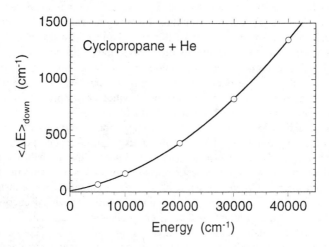

Figure 4. Calculated values of $\langle\Delta E\rangle_{down}$ as a function of cyclopropane vibrational energy.

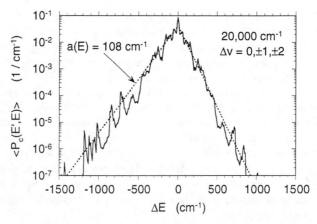

Figure 5. Comparison of exponential model and calculated collision step size distribution (10 cm^{-1} bins) calculated for $\Delta v=0,\pm1,\pm2$ transitions.

distribution can reproduce the behavior ascribed to "super collisions", in which surprisingly large amounts of energy are transferred (*44*). The asymmetry between slopes of the "exponential" envelopes for activating and deactivating collisions reflects the influence of detailed balance, which is obeyed in these calculations.

By varying the parameters in the calculations, it was found that the collision efficiency and the slope of the exponential envelope do not depend significantly on the vibrational frequencies. Instead, they are almost entirely controlled by the length parameter λ. This parameter describes the steepness of the repulsive wall encountered by the free atom and it determines, to a large extent, whether the collision is impulsive, or not. Impulsive collisions, in which the collision duration is short compared with the vibrational period, are highly efficient (*31*). The parameter adjustment procedure described above produced $\lambda = 0.35$ Å, which is considerably larger than the value of $\lambda = 0.19$ Å estimated from the Lennard-Jones parameters according to "Method B" (*12*). The smaller value of λ produces much higher transition probabilities than calculated by Clary (*38*) and much larger values of $<\Delta E>_{down}$ than typically determined in experiments. The larger value of λ is needed to agree with Clary's calculations; the resulting $<\Delta E>_{down}$ values (for higher excitation energies) are of the right order of magnitude. This agreement may indicate that the repulsive wall is "softer" than described by a Lennard-Jones potential. Accurate molecular structure calculations can address whether the actual repulsive walls are this "soft", or whether the parameter λ is compensating for other deficiencies in the SSH(T) theory. Classical trajectory calculations have demonstrated that energy transfer is quantitatively affected by the details of the repulsive wall (*45-48*) and they have also shown that the collision trajectories are far more complicated (*25*) than the simple analytical model which is the basis of SSH(T) Theory.

The sensitivity of the results to the parameter λ also suggests that this property of the potential energy function strongly affects the relative efficiencies of various collider gases. It has been known for many years that there is an ordering of collision efficiencies for the deactivation of excited species, but despite decades of work by numerous researchers, no single molecular property has emerged as the dominant property in governing this behavior (*5,6,34,49*). It is possible that the steepness of the repulsive wall as characterized by λ may be a dominant factor in determining relative collision efficiencies. Detailed molecular structure calculations will be needed to test this hypothesis.

As noted above, the convergence of the $P_n(E',E)$ terms is not monotonic. Instead, the odd-numbered and even-numbered terms appear to converge separately. Furthermore, the even-numbered terms are of greater magnitude. For even-numbered terms, the quantum number changes can produce $\Delta E_{\beta\alpha} \approx 0$ with high statistical probability, because of the vibrational mode degeneracies. When $\Delta E_{\beta\alpha}$ is near zero, the SSH(T) transition probabilities are higher. The $<P_n(E',E)>$ terms are plotted vs. E'-E in Figure 6, where it is apparent that the odd-numbered terms (solid lines) make smaller contributions than the even-numbered terms (broken lines) near $\Delta E_{\beta\alpha} \approx 0$.

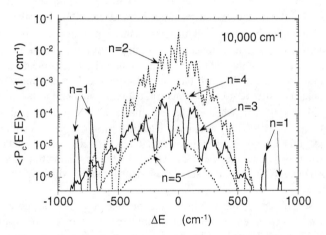

Figure 6. $\langle P_n(E',E)\rangle$ step-size distributions.

In summary, the statistical-dynamical theory presented here shows promise for calculating the collision step size distribution function needed for master equation calculations and predicting unimolecular reaction rate constants in the fall-off regime. Because SSH(T) theory is not suitable for predictions, the parameterization was carried out empirically by comparisons with accurate calculations for low vibrational energies. The present theory is useful even at energies where densities of states are greater than 10^{10} states/cm^{-1}. The results show that vibrational energy transfer propensities may persist to very high energy and that the collision step size distribution is energy dependent, in qualitative agreement with experiment. It is important to note that the calculated energy dependence arises from the energy-dependent fraction of inelastic collisions, rather than from a change in the shape of the collision step size distribution.

The present demonstration serves to raise questions about the details of energy transfer, but more accurate calculations are needed before any quantitative conclusions can be reached. We are currently writing the computer codes to use the ITFITS impulsive theory (*14*), which should provide more accurate results. If such calculations tend to produce results similar to those of trajectory methods, we will gain more confidence in both approaches.

Acknowledgments. Thanks go to the United States Department of Energy (Office of Basic Energy Sciences) for partial financial support of this work.

Literature Cited

1. Robinson, P. J.; Holbrook, K. A. *Unimolecular Reactions;* Wiley-Interscience: London, 1972.
2. Forst, W. *Theory of Unimolecular Reactions;* Academic Press: New York, NY, 1973.
3. Baer, T.; Hase, W. L. *Unimolecular Reaction Dynamics: Theory and Experiments;* Oxford University Press: Oxford, 1996.
4. Krajnovich, D. J.; Parmenter, C.; Catlett Jr., D. L. *Chem. Rev.* **1987**, *87*, 237.
5. Hippler, H.; Troe, J. In *Bimolecular Collisions;* Baggott, J. E.; Ashfold, M. N., Eds.; The Royal Society of Chemistry: London, 1989; p. 209.
6. Oref, I.; Tardy, D.C. *Chem. Rev.* **1990**, *90*, 1407.
7. Secrest, D. *Ann. Rev. Phys. Chem.* **1973**, *24*, 379-406.
8. Child, M. S. *Molecular Collision Theory;* Academic: New York, NY, 1974.
9. Clary, D.C.; Kroes, G.J. *Adv. Chem. Kin. Dyn.*, **1995**, *2A*, 135.
10. Landau, L.; Teller, E. *Phys. Z. Sowj. Un.*, **1936**, *10*, 34.
11. Schwartz, R. N.; Slawsky, Z. I.; Herzfeld, K. F. *J. Chem. Phys.*, **1952**, *20*, 1591.
12. Herzfeld, K. F.; Litovitz, T. A. *Absorption and Dispersion of Ultrasonic Waves;* Academic Press: New York, NY, 1959.
13. Tanczos, F. I. *J. Chem. Phys.*, **1959**, *30*, 1119.
14. Heidrich, F. E.; Wilson, K. R.; Rapp, D. *J. Chem. Phys.*, **1971**, *54*, 3885-3897.
15. Giese, C. F.; Gentry, W. R. *Phys. Rev. A*, **1974**, *10*, 2156.

16. Gentry, W. R.; Giese, C. F. *Phys. Rev. A*, **1975**, *11*, 90.
17. Gentry, W. R.; Giese, C. F. *J. Chem. Phys.*, **1975**, *62*, 1364.
18. Gentry, W. R.; Giese, C. F. *J. Chem. Phys.*, **1975**, *63*, 3144-3155.
19. Shobatake, K.; Rice, S. A.; Lee, Y. T. *J. Chem. Phys.*, **1973**, *59*, 2483-2489.
20. Stace, A. J.; Murrell, J. N. *J. Chem. Phys.*, **1978**, *68*, 3028.
21. Brown, N.; Miller, J. A. *J. Chem. Phys.*, **1980**, *80*, 5568-5580.
22. Clary, D. C. *J. Phys. Chem.*, **1987**, *91*, 1718-1727.
23. Clary, D. C.; Gilbert, R. G.; Bernshtein, V.; Oref, I. *Faraday Discuss.*, **1996**, *102*, in press.
24. Gilbert, R.G.; Smith, S.C. *Theory of Unimolecular and Recombination Reactions;* Blackwell Scientific Publications: Oxford, 1990.
25. Gilbert, R. G. *Int. Rev. Phys. Chem.*, **1991**, *10*, 319.
26. Lendvay, G.; Schatz, G.C. *Adv. Chem. Kin. Dyn.*, **1995**, *2B*, 481.
27. Guo, Y.; Thompson, D. L.; Sewell, T. D. *J. Chem. Phys.*, **1996**, *104*, 576-582.
28. Lendvay, G.; Schatz, G. C. *J. Chem. Phys.*, **1996**, submitted for publication.
29. Toselli, B. M.; Barker, J. R. *Chem. Phys. Letters*, **1990**, *174*, 304.
30. Barker, J. R. *Ber. Bunsenges. Phys. Chem.*, **1996**, in press.
31. Yardley, J.T. *Introduction to Molecular Energy Transfer;* Academic Press: New York, NY, 1980.
32. Stretton, J. L. *Trans. Faraday Soc.*, **1965**, *61*, 1053.
33. Lambert, J.D. *Vibrational and Rotational Relaxation in Gases;* Clarendon Press: Oxford, 1977.
34. Tardy, D.C.; Rabinovitch, B. S. *Chem. Rev.*, **1977**, *77*, 369.
35. Shimanouchi, T. *Nat. Stand. Ref. Ser., Nat. Bur.Stand. (U.S.)*, **1972**, *39*.
36. Press, W. H.; Teukolsky, S. A.; Vetterling, W. T.; Flannery, B. P. *Numerical Recipes in FORTRAN*; Cambridge University Press: Cambridge, 1992.
37. Whitten, G.Z.; Rabinovitch, B. S. *J. Chem. Phys.*, **1964**, *41*, 1883.
38. Clary, D. C. *J. Am. Chem. Soc.*, **1984**, *106*, 970.
39. Flynn, G. W.; Parmenter, C. S.; Wodtke, A. M. *J. Phys. Chem.*, **1996**, *100*, 12817-38.
40. Barker, J. R. *J. Phys. Chem.*, **1984**, *88*, 11.
41. Barker, J. R.; Toselli, B. M. *Int. Rev. Phys. Chem.*, **1993**, *12*, 305.
42. Barker, J. R.; Brenner, J. D.; Toselli, B. M. *Adv. Chem. Kin. Dyn.*, **1995**, *2B*, 135.
43. Barker, J. R.; King, K.D. *J. Chem. Phys*, **1995**, *103*, 4953.
44. Oref, I. *Adv. Chem. Kin. Dyn.*, **1995**, *2B*, 285.
45. Bruehl, M.; Schatz, J. C. *J. Phys. Chem.*, **1988**, *92*, 7223.
46. Lim, K. F; Gilbert, R. G. *J. Phys. Chem.*, **1990**, *94*, 72, 77.
47. Lim, K. F. *J. Chem. Phys.*, **1994**, *101*, 8756.
48. Lenzer, T.; Luther, K.; Troe, J; Gilbert, R. G.; and Lim, K. F. *J. Chem. Phys.*, **1995**, *103*, 626, and references cited therein.
49. Quack, M.; Troe, J. *Gas Kinetics and Energy Transfer;* Chem. Soc.: London, 1977; Vol. 2.

Chapter 16

Crossed-Molecular-Beam Observations of Vibrational Energy Transfer in Only Moderately Excited Glyoxal Molecules

Samuel M. Clegg, Brian D. Gilbert[1], Shao-ping Lu[2], and Charles S. Parmenter

Department of Chemistry, Indiana University, Bloomington, IN 47405

Crossed molecular beams were used to study rotational and rovibrational inelastic scattering of S_1 glyoxal (CHO-CHO) from D_2, N_2, CO and C_2H_4. A laser is used to prepare glyoxal to the 0^0, $K' = 0$ state. The rotational and rovibrational states excited by inelastic scattering were monitored by dispersed fluorescence. The collision partners were chosen to investigate the relative influence of the reduced mass versus the interaction potential of the collision pair. The glyoxal + D_2 and glyoxal + He inelastic spectra were nearly identical while the glyoxal + D_2 and glyoxal + H_2 spectra were clearly different. This indicates that the inelastic scattering is dominated by the kinematics rather than by the details of the interaction potentials of the various collision partners. Glyoxal collisions with N_2, CO and C_2H_4, each with a mass of 28 amu, were also nearly identical, further evidence that inelastic scattering is controlled by the reduced mass of the collision pair. The study of the relative rotational and rovibrational channel competition has also been extended to include the glyoxal + N_2, CO and C_2H_4 relative cross sections. These new data continue to demonstrate that the rotational versus rovibrational channel competition strongly depends on the reduced mass of the collision pair.

Crossed molecular beams with a laser pump-dispersed fluorescence probe has been shown to be an effective method of monitoring state-to-state vibrational energy transfer (VET). This method was initially used to study inelastic scattering from the BO_u^+ state of iodine (*1-4*). An already extensive study from the ^1A_u state of *trans*-glyoxal is still ongoing (*5-8*). Inelastic scattering from glyoxal has become a benchmark for fully quantal three-dimensional scattering calculations (*9-12*).

In this report, we describe the results of experiments designed to investigate the relative influence of the reduced mass of the collision pair versus the interaction potential in controlling the rotational and rovibrational energy transfer. This was

[1]Current address: Department of Chemistry and Physics, Coastal Carolina University, P.O. Box 261954, Conway, SC 29528–6054

[2]Current address: Innovative Lasers Corporation, 3280 East Hemisphere Loop #120, Tucson, AZ 85706

examined by adding rotationally resolved inelastic scattering spectra from glyoxal collisions with deuterium (D_2, 4 amu), nitrogen (N_2, 28 amu), carbon monoxide (CO, 28 amu) and ethylene (C_2H_4, 28 amu) to results already published.

Crossed molecular beams offer many advantages for VET study that are not possible in room temperature beam experiments. A primary asset is the relatively precise definition of collision energies. Additionally, the cold rotational temperatures from the beam expansion allows laser preparation of a specific initial rovibrational state, as well as monitoring with S_1 - S_0 dispersed fluorescence of all important inelastic scattering channels with rotational resolution.

Details of the VET channel selectivity and competition in glyoxal may be seen with the vibrational energy level diagram of Figure 1 and in its accompanying 0^0 and 7^1 rotational energy level diagram. The diagram shows all 27 S_1 vibrational levels through 1200 cm^{-1}, many of which may be pumped selectively. Initial state preparation in previous studies selected zero or little angular momentum (Kħ = 0) about the "a" axis in each of the vibrational energy levels shown in bold, namely 0^0, 7^2, 5^1 and 8^1 (5-8). All of the principal inelastic scattering channels were observed by collecting fluorescence from the rotational and rovibrational K states populated by inelastic scattering. Regardless of the initial vibrational level, VET involving $\Delta \upsilon_7 = \pm 1$ was the only channel observed even though many other energy levels are energetically accessible.

The rotational energy level diagram shows that the 0^0 and 7^1 rotational manifolds overlap at $\Delta E > 233$ cm^{-1} resulting in competition between rotational and rovibrational inelastic scattering. That competition was initially studied from glyoxal pumped to the level 0^0, K′ = 0, a state we shall call 0^0K^0, in collision with five target gases, namely H_2 (2 amu), D_2 (4 amu), He (4 amu), Kr (84 amu) and cyclohexane (84 amu) (5-8).

Gilbert *et al.* reported that the competition among rotational and rovibrational channels depended primarily on the mass of the target gas (6). Their conclusion was based on the observation of fluorescence spectra where rotational resolution is not achieved. The inelastic scattering spectra from glyoxal 0^0K^0 + He (4 amu) and D_2 (4 amu) were qualitatively similar, while qualitative differences were observed between the glyoxal 0^0K^0 + H_2 (2 amu) and D_2 (4 amu) lower resolution spectra. These results indicated that the mass of the target gas has dominant control over the VET channel competition while the interaction potential is much less important (6).

A unique aspect of our new work is to do the more difficult experiment of using higher resolution where the individual rotational channels are resolved. We have compared rotationally resolved glyoxal 0^0K^0 + D_2 inelastic scattering spectrum with the published glyoxal 0^0K^0 + H_2 and He spectra. Inelastic scattering spectra from glyoxal 0^0K^0 + N_2, CO and C_2H_4, all with a mass of 28 amu, are also used for a more definitive probe of the mass control of VET. All three inelastic scattering spectra should be similar if the mass truly exerts the dominant influence over the relative cross sections.

These new data will also better characterize the rotational versus rovibrational competition as the target gas mass changes. The rotational and rovibrational relative cross sections for glyoxal 0^0K^0 + N_2, CO and C_2H_4 are now added to the published H_2, He and Kr relative cross sections (5-8).

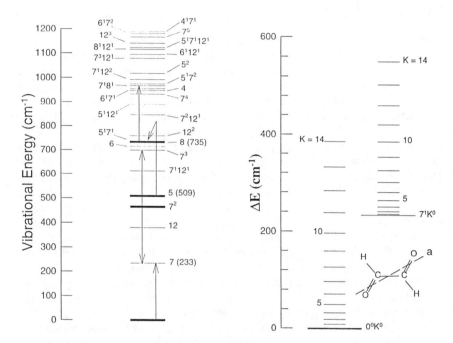

FIG. 1. A vibrational energy level diagram (left) of S_1 glyoxal with a rotational diagram (right) showing the K = J rotational energy levels that may be observed in inelastic scattering from the 0^0, K′ = 0 state of S_1 glyoxal. The bold levels are those so far pumped in published glyoxal experiments. The arrows show the principal vibrationally inelastic scattering channels from each initial levels. Glyoxal is almost a symmetric top with the top "a" axis passing near the oxygen atoms (insert).

Results and Discussion

Detailed descriptions of the experimental setup and data analysis used in obtaining the data presented here have been published (2,8). The relative cross sections are extracted by recreating the experimental spectrum with a simulation program. A description of the simulation program has been published (8). Figure 2 contains both simulated (line) and experimental spectra (circles) from glyoxal + D_2. Each of the peaks to the blue of the dip in the spectra indicates emission from a rotational or rovibrational state that has been excited by the inelastic collision.

Mass Has Dominant Control of the Relative Inelastic Scattering Cross Sections. The reduced mass of the collision pair and the interaction potential between the collision pair are important parameters in control of inelastic scattering. Inelastic scattering from glyoxal collisions with H_2, D_2 and He offers a particularly interesting opportunity to investigate the relative influence of the reduced mass versus the interaction potential. The interaction potentials between glyoxal + H_2 and glyoxal + D_2 are identical whereas the masses of the target gases are different by a factor of two. In contrast, the glyoxal + D_2 and glyoxal + He interaction potentials are different and the target gas masses are the same. Rotationally resolved inelastic scattering spectra from glyoxal + H_2 and glyoxal + He along with their relative cross sections have been published (8). A new rotationally resolved glyoxal + D_2 inelastic scattering spectrum, shown in Figure 2, can be compared to the glyoxal + H_2 and glyoxal + He rotationally resolved inelastic scattering spectra.

Figure 3 contains the superimposed glyoxal + H_2 and glyoxal + D_2 spectra (top) and the glyoxal + D_2 and glyoxal + He spectra (bottom). It is obvious that the H_2 and D_2 spectra begin to diverge for emission from $0^0K > 9$. Emission from $0^0K = 15\text{-}17$ can be observed in the glyoxal + D_2 spectrum where rotational structure from glyoxal + H_2 cannot be resolved past K^{14}. In contrast, the inelastic scattering spectra for D_2 and He, shown at the bottom of Figure 3, are nearly identical.

This mass effect was tested further in K state resolved experiments with three 28 amu collision partners, namely N_2, CO and C_2H_4. The interaction potential between glyoxal and each of the target gases is different. The inelastic scattering spectra are compared in Figure 4. Close inspection of the three spectra shows how well the rotational structure overlaps. The three spectra are again nearly identical, thus the extracted relative cross sections generally match to within experimental error.

Further indication of the insensitivity to the interaction potential comes from theoretical studies of H_2 and He interactions with glyoxal (9-12). The calculated inelastic cross sections for both H_2 and He agree with the experimental relative cross sections even though several interaction potential approximations were made (8,10,13). For example, the same glyoxal + He interaction potential was used in both calculations (9-12). Furthermore, the interaction potential for S_1 glyoxal + He was really a *ground electronic state* potential constructed from atom-atom pair potentials derived from the *ab-initio* formaldehyde + He system (9-12).

Competition Between Rotational and Rovibrational Inelastic Scattering. As described elsewhere, experimental relative cross sections can be obtained from the

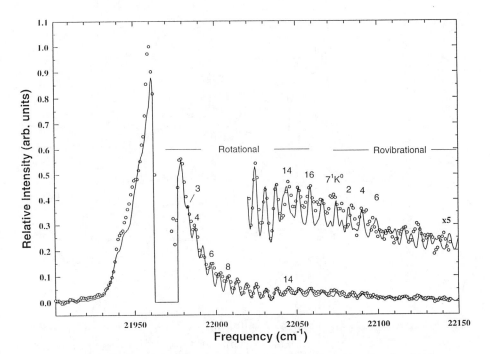

Fig. 2. The 0_0^0 and 7_1^1 bands of S_1 - S_0 fluorescence from glyoxal molecules that have been scattered from the S_1 (0^0, $K' = 0$) state by collisions with D_2. Circles show the experimental spectrum while the line is a computer simulation. The maxima are subbands from the K' states reached by rotationally inelastic scattering and are so labeled. Rovibrationally inelastic scattering produces emission from K' states within the 7^1 level that emits in the 7_1^1 band. This emission is also shown at 5x and displaced upward for clarity.

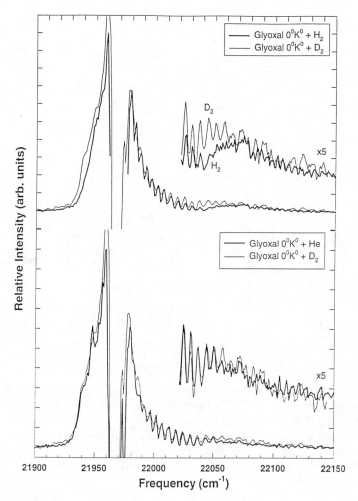

FIG. 3. A comparison of the experimental fluorescence spectra from S_1 glyoxal (0^0, $K' = 0$) that is inelastically scattered by collision with H_2, D_2 or He.

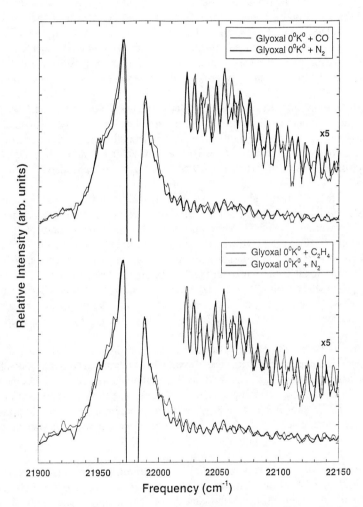

FIG. 4. A comparison of the experimental fluorescence spectra from S_1 glyoxal (0^0, $K' = 0$) that is inelastically scattered by collisions with N_2, CO or C_2H_4.

computer simulation of the scattering spectra. Those for glyoxal + D_2 from the present study are shown in Figure 5, where they are compared with H_2 and He cross sections obtained earlier (5,7,8). As can be seen in the display of cross sections against the energy ΔE transferred (T - R,V), the relative cross sections for D_2 and He are almost indistinguishable, whereas both sets are clearly different from the H_2 cross sections.

Figure 6 contains a plot of the relative cross sections for each mass set, H_2 (2 amu), He and D_2 (4 amu), N_2, CO and C_2H_4 (28 amu) and Kr and C_6H_{12} (84 amu). The relative cross sections for gases with the same mass are essentially identical, falling within their respective error bars. For convenience, the mass sets listed above will be referred to as H_2, He, N_2 and Kr where the other gases in each set are *implied*.

Rotational and Rovibrational Relative Cross Sections versus Target Gas Mass. Figure 7 contains a plot where the relative rotational cross sections are compared. The cross sections decrease more or less exponentially with increasing rotational excitation through the transition 0^0 K' = 10 (we shall call it $\sigma(0^0K^{10})$). Beyond that, the relative cross sections $\sigma(0^0K^{11})$ through $\sigma(0^0K^{14})$ have a strong dependence on the target gas mass.

The fall off in the glyoxal + H_2 and glyoxal + He relative rotational cross sections from K^{11} through K^{14} indicates that rotational excitation may be limited by angular momentum (5-8). The maximum initial orbital angular momentum can be estimated classically as L = μvb where μ is the reduced mass of the collision pair, v is the relative velocity of the collision pair and b is the impact parameter (5,13). The maximum rotational excitation about the "a" axis from collisions with H_2, He, N_2 and Kr is approximately 14ħ, 22ħ, 74ħ and 122ħ, respectively. The smaller H_2 and He relative cross sections may, therefore, be the result of angular momentum restrictions. Angular momentum limitations with N_2 and Kr are much too large to be observed in our spectra.

All of the mass sets of relative rovibrational cross sections show a strong dependence on the target gas mass. The glyoxal + H_2 rovibrational cross sections follow the same exponential scaling against ΔE as the rotational cross sections. As the target gas mass increases, the 7^1 rovibrational excitation becomes less competitive from 7^1K^0 through 7^1K^8, as shown in Figure 8. Angular momentum again limits the rotational excitation beginning near 7^1K^{10}. The relative value of the cross sections increases with mass for rovibrational excitation from 7^1K^{10} through 7^1K^{14}.

Rotational versus Rovibrational Competition. The rotational energy level diagram in Figure 1 indicates that the amount of energy, ΔE, required for rovibrational excitation to 7^1K^0 is approximately the same as for rotational excitation to 0^0K^{11}. Here, at $\Delta E \approx 230$ cm^{-1}, the rotational versus rovibrational competition becomes particularly interesting. This competition can be observed in the separation of the relative rotational and rovibrational cross sections shown in Figure 6. The relative magnitude of the rotational and rovibrational cross sections depends on the mass of the target gas. The relative rotational and rovibrational cross sections for glyoxal + Kr and glyoxal + N_2 are well separated. The relative rotational and rovibrational cross sections for glyoxal + He are also well separated but the separation is not as large as that for Kr and N_2. The rotational and rovibrational cross sections for glyoxal + H_2, however, are not separated.

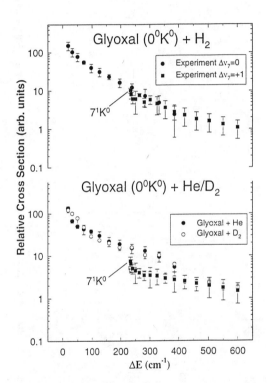

FIG. 5. The relative cross sections plotted against the energy transfer ΔE (T - R,V) for inelastic scattering from S_1 glyoxal (0^0, K′ = 0) by H_2 (top) and He (bottom, closed symbols) and D_2 (bottom, open symbols). The circles and squares are the rotational and 7^1 rovibrational relative cross sections, respectively.

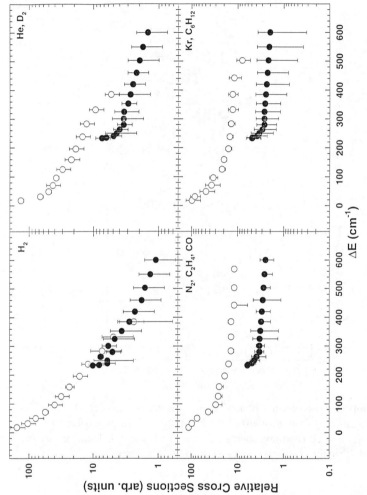

FIG. 6. The relative cross sections plotted against the energy transfer ΔE (T - R,V) for inelastic scattered S_1 glyoxal (0^0, K' = 0) with various target gases. Cross sections from different gases of the same mass are coincident to within the error bars. The open and closed circles are rotational and 7^1 rovibrational relative cross sections, respectively.

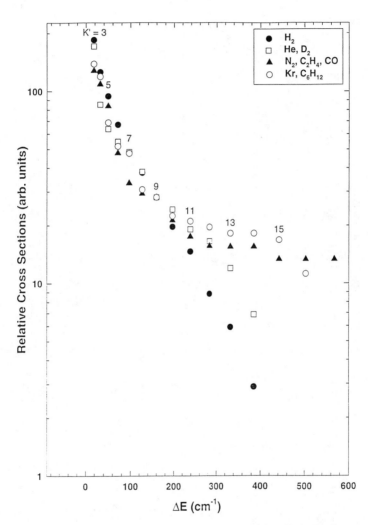

FIG. 7. A superposition of relative cross sections for rotationally inelastic scattering from S_1 glyoxal (0^0, $K' = 0$) by various target gases. Some K' identities of the final 0^0 state are noted.

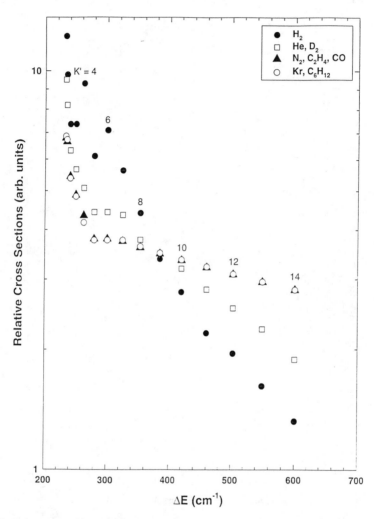

FIG. 8. A superposition of relative cross sections for rotationally inelastic scattering from S_1 glyoxal (0^0, $K' = 0$) by various target gases. Some K' identities of the final 7^1 states are noted.

Rotational versus rovibrational competition can be summarized by taking the ratio of the sum of the rotational, $\sigma(rot)$, and rovibrational, $\sigma(vib)$, relative cross sections. Figure 9 contains a plot of the experimental $\sigma(vib)/\sigma(rot)$ ratio plotted against target gas mass. The plot shows that the $\sigma(vib)/\sigma(rot)$ ratio decreases smoothly as the target mass increases. The value of each of the ratios is on the order of 0.1.

The relative nature of the experimental cross sections does not describe how each mass set changes relative to the others, whereas the absolute theoretical cross sections can. The calculations for H_2 and He, at our experimental $E_{c.m.}$, indicate that both $\sigma(rot)$ and $\sigma(vib)$ increase but that $\sigma(rot)$ increases by a larger amount.

Conclusion

Crossed molecular beams and a laser pump-dispersed fluorescence probe were used to study rotational and rovibrational inelastic scattering in S_1 glyoxal. Experiments were designed to characterize the influence of the reduced mass versus the interaction potential in setting the relative cross sections. This characterization was accomplished by comparing the K state resolved inelastic scattering spectra as well as the relative cross sections from glyoxal collisions with different target gases where the interaction potential or the target gas mass were similar. When the mass of the target gases were similar, namely glyoxal + D_2 and glyoxal + He, the inelastic scattering spectra and relative cross section were qualitatively similar.

When the interaction potentials between the collision pair are identical, namely glyoxal + H_2 and glyoxal + D_2, the spectra were clearly different. This indicates that the reduced mass of the collision pair is far more important than the interaction potential in controlling inelastic scattering.

Rotationally resolved inelastic scattering spectra from glyoxal collisions with N_2, CO and C_2H_4 further indicated that the target gas mass has far more control than the interaction potential in setting the relative cross sections. The inelastic scattering spectra and relative cross sections from collisions with N_2, CO and C_2H_4 were nearly identical.

The rotational energy level diagram in Figure 1 shows that the 0^0 and 7^1 rotational manifolds overlap. The most interesting part of the rotational and rovibrational competition was observed when there were approximately equal amounts of energy going into rotational and rovibrational excitation. Details of this competition between rotational and rovibrational energy transfer have been extended by the addition of the new N_2, CO and C_2H_4 data to the published H_2, He and Kr relative cross sections. The competition between rotational and rovibrational energy transfer has been shown to depend on the mass of the target gas. As the mass of the target gas increases, the separation between the relative rotational and rovibrational cross sections where equal amounts of energy are exchanged become larger.

Acknowledgment

We are grateful for the financial support of this work provided by the National Science Foundation and by the Petroleum Research Foundation administered by the American Chemical Society.

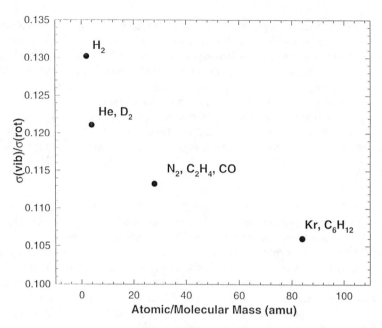

FIG. 9. A plot of $\sigma(\text{vib})/\sigma(\text{rot})$ against ΔE for the inelastic scattering of S_1 glyoxal $(0^0, K' = 0)$ by various target gases.

Literature Cited

1. Krajnovich, D.J.; Butz, K.W.; Du, H.; Parmenter, C.S. *J. Phys. Chem.* **1988,** 92, 1388.
2. Krajnovich, D.J.; Butz, K.W.; Du, H.; Parmenter, C.S. *J. Chem. Phys.* **1989,** 91, 7705.
3. Krajnovich, D.J.; Butz, K.W.; Du, H.; Parmenter, C.S. *J. Chem. Phys.* **1989,** 91, 7725.
4. Du, H.; Krajnovich, D.J.; Parmenter, C.S. *J. Phys. Chem.* **1989,** 95, 2104.
5. Butz, K.W.; Du, H.; Krajnovich, D.J.; Parmenter, C.S. *J. Chem. Phys.* **1988,** 89, 4680.
6. Gilbert, B.D.; Parmenter, C.S.; Krajnovich, D.J. *J. Chem. Phys.* **1994,** 101, 7440.
7. Gilbert, B.D.; Parmenter, C.S.; Krajnovich, D.J. *J. Phys. Chem.* **1994,** 98, 7116.
8. Gilbert, B.D.; Parmenter, C.S.; Krajnovich, D.J. *J. Chem. Phys.* **1994,** 101, 7423.
9. Clary, D.C.; Dateo, C.E. *Chem. Phys. Lett.* **1989,** 154, 62.
10. Kroes, G.-J.; Rettschnick, R.P.H.; Clary, D.C. *Chem. Phys.* **1990,** 145, 359.
11. Kroes, G.-J.; Rettschnick, R.P.H.; Dateo, C.E.; Clary, D.C. *J. Chem. Phys.* **1990,** 93, 287.
12. Kroes, G.-J.; Rettschnick, R.P.H. *J. Chem. Phys.* **1991,** 94, 360.
13. Levine, R.D.; Bernstein, R.B. *Molecular Reaction Dynamics;* Oxford University: New York, NY, 1974, pp 32-33.

Chapter 17

The Contribution of Wide-Angle Motions to Collisional Energy Transfer Between Benzene and Argon

V. Bernshtein and I. Oref[1]

Department of Chemistry, Technion-Israel Institute of Technology, Haifa 32000, Israel

Trajectory calculations of collisional energy transfer between an excited benzene molecule and an Ar bath atom are reported. Calculations were made for three groups of initial conditions of the excited molecule. a. Fully rotating and vibrating molecule. b. The overall rotations are frozen prior to the start of the collisions. c. The out-of-plane vibrations are frozen prior to the start of the collision. Trajectories that lead to large values of energy transfer, supercollisions, are analyzed individually. It is found that these collisions occur in a narrow cone of angular approach around 90^0. That the energy transfer in down collisions is either from rotation to translation or vibration to translation. In up collisions the reverse process takes place. The overall rotations and out-of-plane vibrations are the vehicle by which energy is transferred. The actual energy transfer process is shorter than intramolecular energy redistribution time. Therefore, only a small moiety participates in the actual energy transfer. Measurements of C-Ar and H-Ar van der Waals distances in the energy transferring moiety show that C is the prominent atom in the mechanism of energy transfer.

Collisional energy transfer, CET, plays a major role in reactive and non-reactive gas phase processes (*1*). In photophysical processes energy transfer at low and high levels of excitation provide an understanding of the preferred routes of vibration to vibration/rotation/translation energy exchange (*2*). In photochemical and thermal processes, CET is part of the reactive mechanism. It provides the mechanism by which energy is pumped up and down the energy ladder (*1b,3*). Thus, collisions cool highly excited molecules and excite molecules which are located in the part of the population distribution which is far away from the threshold value for reaction.

[1]Corresponding author

The energy transfer probability function from initial energy state E to final state E', P(E,E'), is an elusive quantity which is extremely hard to measure (*1b*). Most experiments do not give the whole functional form but only the first moment of the distribution (*1a*). In the rarest of cases the second moment can be obtained as well. Recent physical experiments which utilize spectroscopic means give greater insight into the shape of the function but even here, a priori assumptions must be made in order to fit the data to an assumed function (*2*). This unhappy situation led to the development of many empirical energy transfer models which were used in master equation calculations of unimolecular rate coefficients (*1b,4*). Without such models the great progress, due mainly to Marcus and Rabinovitch and their co-workers, in understanding unimolecular reactions which was made since the late 1950's would have come to nothing.

Lately, quasi-classical trajectory calculations have provided new insights into the collisional process (*5*). The ability to selectively probe the parameters which govern the energy transfer process provided an important technique by which the principal elements which effect the energy exchange can be studied. The effects of the intermolecular potential, of the mass, internal excitation, temperature, collision duration and internal modes were studied in a systematic way. An important feature of trajectory calculations is the ability, in principle, to evaluate P(E,E') (*6*). Regretfully, there is no one to one correspondence between P(E,E') and the trajectory results. The unknown elastic peak is an obstacle and P(E,E') is not fully determined. Exactly as in the physical experiments of energy transfer, also here some a priori assumptions must be made in order to obtain the complete form of P(E,E'). Nevertheless, trajectory calculations provide an important tool for understanding energy transfer. Quantum and classical calculations show that there is a general agreement between the two types of calculations (*5s,7*) lending important theoretical support to studying collisional energy transfer by quasiclassical calculations.

Previous work has provided information on the nature of the energy exchange process. Calculations of the duration of a collision have indicated that practically all collisions are very fast. The average collision duration is ~ 680 fs at 500 K and ~230 fs at 1500 K (*5o*). Probing the center-of-mass velocity during the collision show that most collisions are impulsive (*5t*). There is a fraction of "chattering" collisions in which the bath atom spends between 300 fs to 600 fs near the excited polyatomic molecule. However, the energy transferred does not seem to result from an accumulation of small quantities of energy in succession. Rather, the bath atom hovers over the molecule and the energy transfer event occurs during a short time at the end of the collision. It was also found by quantum mechanical calculation that the out-of-plane modes play a major role in the energy transfer process (*5s*). The role of the wide angle motions in the energy transfer process was confirmed by trajectory calculations (*8*). These calculations show that the out-of-plane modes and over all rotations play a major role in the CET of excited benzene colliding with Ar as a bath gas.

A sub-group within the inelastic collisions is those collisions which transfer very large quantities of energy in one collision, supercollision (9). Previous work on benzene/Ar and toluene/Ar has shown that the large values of ΔE encountered in those collisions are due to dynamical effects (5s,8). That is to say, the incoming atom has to be in a perpendicular orientation relative to the molecule and it has to be in phase with an out-of-plane motion or overall rotation. There are basically two types of mechanisms. The first is impulsive and the collision is of a very short duration. In the second, the atom hits the molecule once and transfers energy to it. As a consequence it is slowed down. After a very short time, it is hit by a wide angle motion and departs energy rich. These energy rich collisions were not found to depend on the relative translational energy. Thus, a slow moving atom with the right orientation can obtain a large value ΔE in a down-collision. In what follows we probe the dynamics of the collisions and the dependence of CET on orientation especially for supercollisions. We do so by monitoring the angle and the distances between the incoming atom and the atoms in the moiety which defines the collision site.

Theory

The numerical methods used in the present work are reported in refs. 5o and 5p. The equations of motion were integrated by using a modified public domain program Venus (10). The intermolecular potential was a pairwise Lennard-Jones potential. Its parameters are given in ref. 5s. The intramolecular potential includes all the normal modes contributions, stretching, bending and wagging. The values of the parameters of this potential were obtained from modified valance force field calculations by Draeger (11) and are also given in references 5o and 5p. The initial translational and rotational energies were chosen from the appropriate thermal energy distributions at 300K. The initial impact parameter was chosen randomly between 0 and its maximum value b_m. The internal energy was 51762 cm^{-1}. The beginning and the end of a collision were determined by the Forward and Backward Sensing (FOBS) method (5o,5p). In this method each trajectory is scanned forward and the moment that, for the first time, a change ε is observed in the internal energy of the hot molecule in a period τ is noted. Then, the trajectory is scanned backward and again, when a change ε is detected the time is noted again. These two times bracket the collisional event. The value of ε used in our calculations was 70 cm^{-1} and the value of τ was 20 fs. This value was obtained after a careful study in which ε was changed systematically and the value chosen such that a small variation in ε did not change the initial time or duration of the collision (5o,5p). For each trajectory the change in energy was noted as well as its duration as determined by FOBS.

The value of the maximum impact parameter b_m was determined separately (5o,5p) A value of 1.1 nm was used in the present calculations. A systematic study of post collision distances between Ar and the closest atom of toluene has shown that for distances larger than 0.8 nm, no changes in the values of the integral of $<\Delta E>_{all}$ were observed, therefore, this value was taken as the terminal distance for all trajectories and the end point from which the back sensing of FOBS was initiated. A convergence of $<\Delta E>_{up}$ and $<\Delta E>_d$ as a function of the number of trajectories was obtained well below the number of 12000 trajectories which was used in the present study.

The values of ΔE obtained from the calculations were multiplied by the factor $(b_m/b_{ref})^2$ in order to bring them in line with experimental data and previous calculations. b_{ref} is the product of σ_{LJ}, the Lennard-Jones collision radius, and $\Omega^{2,2*}$, the collision integral. The value calculated and used in all the calculations is 0.520 nm.

Supercollisions, SUC, are the tail of P(E',E) with unknown functional form which differs from a simple exponential function. In order to study the various facets of and interpret experimental result a biexponential function is customarily being used. Based on such a definition of P(E',E), it is possible to define a threshold value above which any contribution of the tail of P(E',E) is considered as coming from a SUC. For the microcanonical system studied here, this threshold was taken to be $5<\Delta E>_d$ which in the present work is equal to 1500 cm^{-1}. For a thermal reactive system a good criterion is using the value of the bulk average energy transferred per up collision $<<\Delta E>>_{up}$ (12). This quantity is a multiplication of two terms, $<\Delta E>_{up}$ and f(E) which is the non-equilibrium population at the energy of the molecule at which energy transfer is taking place. Supercollisions are rare events and normal trajectory calculations are very inefficient in producing a statistically meaningful sample. We have found in previous calculations that most collisions are produced when the impact parameter is between 0.2 and 0.6 nm. Therefore we have used an impact parameter "window" of 0.2 to 0.6 nm which gave us a factor of 4 in the number of high energy events.

In the present study we wish to study the effect of rotations and out of plane vibration on the mechanism of energy transfer. We do so by first "freezing" the rotational degrees of freedom by choosing an initial rotational temperature in which the molecule does not rotate **prior** to colliding with the Ar atom. Next we "freeze" the out-of-plane vibrations in the initial conditions allowing them to relax during the collision by direct CET excitation or by coupling with the rotations via the rare gas atom. The "freezing out" of the out-of-plane vibrations was done by dividing the initial internal energy among all the modes with the out-of-plane ones excluded. Animation programs were used for analyzing each SUC trajectory separately. The results of the calculations are presented in the next section.

Results and Discussion

The aim of the present work is to analyze the mechanism of supercollision and to determine the motions that lead to large CET. To achieve that, three "banks" of supercollision trajectories were created. In one were collected all trajectories which were created from normal initial conditions at 300 K. That is to say, the vibrations and rotations of the polyatomic molecule were fully active. The second was a collection of trajectories which describe a collision of Ar with non-rotating (frozen) excited benzene molecule. The third set of trajectories describes collisions of an Ar atom with frozen out-of-plane vibrations of the benzene molecule. Once a collision with a frozen molecule starts, there is excitation of the frozen modes via energy exchange between the atom and the molecule and a distortion of the molecular potential surface which leads to coupling of the various modes and they all become excited.

Previous work reported on the effects of freezing certain modes on average quantities (8). Thus, $<\Delta E>_d$ was found to increase from -347 cm^{-1} for normal collisions to -220 cm^{-1} for collisions with frozen rotations and to -282 cm^{-1} for collisions with frozen out-of-plane vibrations. $<\Delta E>_{up}$ was found to decrease from 323 to 286 cm^{-1} and to 307 cm^{-1} going from normal to frozen rotations and frozen out-of-plane vibrations respectively. It was also found that under normal conditions one trajectory out of 800 represented supercollisions while with frozen rotations the number dropped to one in 4000. For collisions with frozen out-of-plane vibrations the number was one in 1500. The present work reports on the mechanistic details of the energy transfer on a frame-by-frame basis. The procedure was as follows. a. The trajectory was animated and the frame (frames) at which energy transfer took place was marked. b. The distances between the bath atom and the various atoms in the molecule were measured. c. The angles between the Ar atom and pairs of atoms with the shortest distance to the Ar were measured. It was found that a C-H group in the molecule is the active moiety in the energy transfer process. The C, which is part of an out-of-plane motion or an over all rotation, is, together with the hydrogen atom, the vehicle for the energy transfer.

Measurements of the Ar-H-C angles and the center-of-mass distances between Ar and C and Ar and H do not provide the whole picture of what atoms are the major players in the energy transfer process. To know that, we have to take in consideration the atomic radii. Table I gives the values of center-of-mass interatomic distances before and after corrections for van der Waals radii. Up and down collisions as well as normal and frozen rotations are represented in the Table. The Ar-C center-of-mass distances are longer than the Ar-H ones. However, when the van der Waals radii are taken into consideration the situation changes and, for vibration to translation energy transfer, Ar-C distances are shorter by far. For rotation to translation energy transfer, there are very few cases where the Ar-H distance is about the same as the Ar-C one or even slightly shorter. The results of the calculations for each type of initial conditions are given below.

Table I Center-of-mass (CM) and van der Waals (vdW) distances[a] between Ar and C and Ar and H and the Ar-H-C angle for up and down supercollisions of normal and frozen overall rotations and out-of-plane vibrations

| Collision Duration | R(Ar-C) | | R(Ar-H) | | Ar-H-C angle | ΔE | ΔE_{vib} | ΔE_{rot} |
	CM	vdW[b]	CM	vdW				
ps	nm				Deg.	cm^{-1}		
	Normal collisions							
0.10	0.333	-0.043	0.283	-0.028	107	-2905	-2403	-562
4.82	0.339	-0.037	0.315	0.004	94	-2171	-1708	-463
0.12	0.315	-0.061	0.306	-0.005	85	-1858	-1676	-182
0.38	0.323	-0.053	0.294	-0.017	93	1696	73	1623
0.32	0.342	-0.034	0.306	-0.005	100	-1508	-941	-568
0.06	0.298	-0.078	0.339	0.028	55	-1458	-380	-1077
	Initially frozen rotations							
0.52	0.342	-0.034	0.313	0.002	98	-2102	-2307	205
0.38	0.324	-0.052	0.293	-0.018	81	-1846	-2508	662
0.38	0.317	-0.059	0.303	-0.008	86	-1591	-1894	304
0.44	0.298	-0.078	0.315	0.004	84	-1583	-2713	1134
0.34	0.322	-0.054	0.320	0.009	82	-1427	-1577	-150
0.14	0.363	-0.013	0.335	0.002	96	1444	654	790
	Initially frozen out-of-plane vibrations							
0.06	0.313	-0.063	0.273	-0.038	105	-2530	-132	-2398
0.12	0.314	-0.062	0.279	-0.033	98	-2361	-394	-1967
2.06	0.305	-0.071	0.304	-0.007	80	-2215	-1853	-362
0.10	0.315	-0.062	0.290	-0.021	94	-2176	-51	-2125
4.60	0.342	-0.034	0.304	-0.007	102	-1572	-1635	63
0.10	0.303	-0.073	0.267	-0.044	99	-1554	-166	-1388

a. Van der Walls radii (nm) : r(Ar) = 0.191, r(C) = 0.185, r(H) = 0.120 taken from ref. 13.
b. This quantity is calculated from $R_{CM} - (r_{vdW}(Ar) + r_{vdW}(C))$ and $R_{CM} - (r_{vdW}(Ar) + r_{vdW}(H))$.

a. Collisions with Normal Initial Conditions. Figures 1 and 2 show side and front views respectively of the configuration of a molecule/atom collision at the instance of the energy transfer as determined by monitoring the frame-by-frame energy content of the molecule. For this configuration the interatomic distances and angles were taken. It can be seen that the Ar is located above the C-H moiety. The angle Ar-H-C for few SUC trajectories of normal initial conditions as a function of frame (time) is shown in Figures 3-5. The arrow on each figure indicates the instant of energy transfer and the * sign indicates the beginning and end of the collision as determined by FOBS. Each figure has broad peaks and fine structure. The latter represents the fast out-of-plane bends where the H atom moves fast above and below the molecular plane which is itself distorted due to the skeletal out-of-plane movement of the C atoms. The broad peaks indicate the overall rotation of the molecule during the trajectory. In all 3 figures, the collision duration is short as can be seen by the * signs. Figures 3 and 4 which represent down collisions show that the collision event, as determined by FOBS, takes place rather late in the trajectory, i.e. the initial relative velocity is low. Figure 3 shows a case where the energy removed was from vibrational modes with almost no change in the rotational energy. This is seen from the width of the peaks which hardly change before and after the collision. Figure 4 represents a case where a big fraction of the rotational energy was removed as well. There is no peak after the collision is over. Figure 5 represents a case of up-collision where the excitation is only translation to rotation. The early collision indicates a fast atom and fast oscillations indicate that rotational excitation occurred. An important thing to notice in all the figures is that **the angle Ar-C-H is ~90⁰**. The perpendicular approach is a prerequisite for a supercollision to occur.

b. Collisions with Frozen Rotations. Rotations play a major role in the energy transfer process. They provide, together with the out-of-plane vibrations, the wide angle motions which are the vehicle by which large quantities of energy are transferred. Thus, it is expected that freezing those out will effect the energy transfer process and indeed it does. Figures 6 and 7 show what happens in a collision of Ar atom with an excited benzene molecule with frozen rotations. Prior to the beginning of the collision event, marked by the first *, the angle does not change as function of time since the molecule is not rotating. After the onset of the collision, it starts to rotate. The initial contact was not done necessarily with the same moiety which is involved in the final energy transfer event, hence, the odd angle. The final energy transfer in the two steps episode is done as in a. at ~90⁰. Figure 6 shows that after the collision is over, the molecule rotates only a little because in this particular trajectory the molecule is left rotationally cold. In this down-supercollision energy is removed only from internal energy. Figure 7 represents an up-collision. It is an early occurring collision since the relative velocity is high. The collision is very short, some of the energy goes into vibration and some of the energy goes into rotation as can be seen from the post-collision peak.

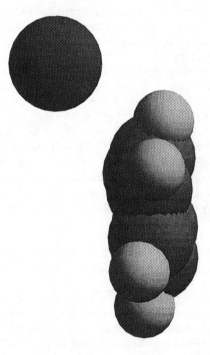

Figure 1. Side view of an Ar-benzene collision at the instance of energy transfer.

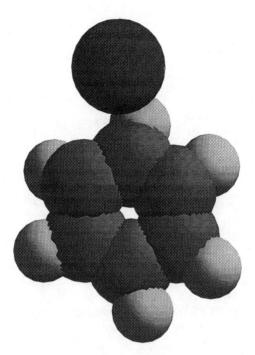

Figure 2. Front view of the Ar-benzene collision shown in Figure 1 at the instance of energy transfer.

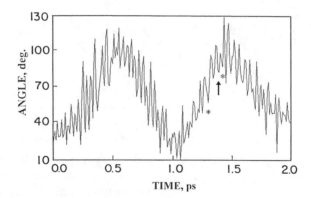

Figure 3. The angle Ar-H-C at the instance the energy is transferred vs. the time step (10 fs) in the trajectory for a collision with normal initial condition. This is an example of a collision of a short duration where there is only vibration to translation exchange. The arrow indicates the instance of energy exchange. The * indicate the beginning and the end of the collision.

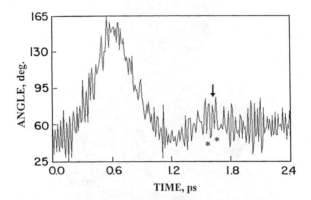

Figure 4. The angle Ar-H-C at the instance the energy is transferred vs. the time step (10 fs) in the trajectory for a collision with normal initial condition. This is an example of a collision of a short duration where there is mainly rotation to translation exchange. The arrow indicates the instance of energy exchange. The * indicate the beginning and the end of the collision.

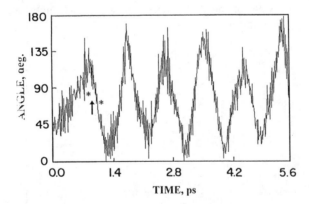

Figure 5. The angle Ar-H-C at the instance the energy is transferred vs. the time step (10 fs) in the trajectory for a collision with normal initial condition. This is an example of an up collision of a short duration where there is only translation to rotation exchange. The arrow indicates the instance of energy exchange. The * indicate the beginning and the end of the collision.

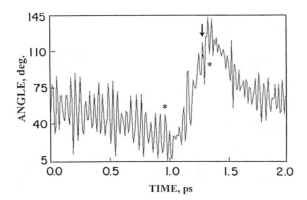

Figure 6. The angle Ar-H-C at the instance the energy is transferred vs. the time step (10 fs) in the trajectory for a collision with initially frozen overall rotations. This is an example of a down collision of a short duration where there is a large vibration to translation energy lose accompanied by rotational excitation of the molecule. The arrow indicates the instance of energy exchange. The * indicate the beginning and the end of the collision.

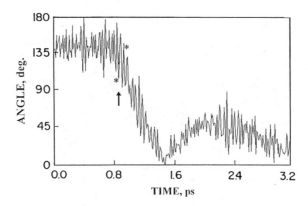

Figure 7. The angle Ar-H-C at the instance the energy is transferred vs. the time step (10 fs) in the trajectory for a collision with initially frozen overall rotations. This is an example of an up collision of a short duration where there is a large translation to vibration and rotation energy exchange. The arrow indicates the instance of energy exchange. The * indicate the beginning and the end of the collision.

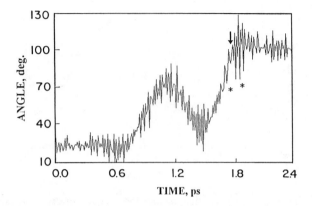

Figure 8. The angle Ar-H-C at the instance the energy is transferred vs. the time step (10 fs) in the trajectory for a collision with frozen out-of-plane vibrations. This is an example of a down collision of a short duration where there is only rotation to translation exchange. The arrow indicates the instance of energy exchange. The * indicate the beginning and the end of the collision.

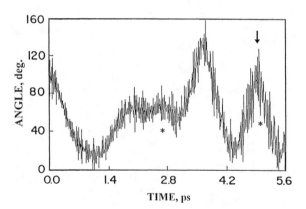

Figure 9. The angle Ar-H-C at the instance the energy is transferred vs. the time step (10 fs) in the trajectory for a collision with frozen out-of-plane vibrations. This is an example of a down collision of a long duration where there is only vibration to translation exchange. The arrow indicates the instance of energy exchange. The * indicate the beginning and the end of the collision.

c. Collisions with Frozen Out-of-Plane Vibrations. The energy transfer pattern which unfolds here is different from the mechanism of energy transfer in frozen rotations. There are two distinct cases to be considered. Short duration collisions lead to energy loss only from rotations, Figure 8. This is so because the collision duration is shorter than IVR times and energy from other excited modes can not flow into the out-of-plane mode which participate in the energy transfer process. The other case is where the collision duration is, very untypically, very long, ~2-5 ps, as in Figure 9. Here energy has time to equilibrate, the out-of-plane vibrations become excited and they in turn, transfer the energy to the bath atom. The data in Table I show the two types of trajectories. The division into the two types is readily seen as is the fact that the energy transferring kick occurrs at ~ 90^0. The van der Waals distance between Ar and the C atom is always shorter than the Ar-H distance indicating that the carbon atoms in the out-of-plane vibrations play the major role in the energy exchange process.

Conclusions. The major conclusions which are supported by the data given above are:

a. Energy transfer in a supercollision event occurs when the bath atom is nearly perpendicular to the C-H bond from which energy is transferred.

b. The energy transfer occurs in a kick of short duration. ~ 40 fs. Therefore IVR does not occur during that time.

c. The energy transfer happens when the bath atom collides with a moiety in the molecule, a C-H group, when it is in phase with the wide angle motion such as an energy rich (*14*) out-of-plane vibration and/or overall rotation.

d. The down energy transfer is either rotation-to-translation or vibration-to-translation. Only rarely would large quantities of energy be transferred simultaneously from vibration and rotation to translation.

e. In all cases studied, the van der Waals distances between the bath atom and the carbon atom in the collision moiety were shorter than between it and the hydrogen atom.

Acknowledgments. This work is supported by the fund for promotion of research at the Technion to (IO) and by the Ministry of Science and the Arts.

Literature Cited.
(1)   a. Tardy, D. C.; Rabinovitch, B. S. *Chem. Rev.* **1977**, *77*, 369.
      b. Oref, I.; Tardy, D. C. *Chem. Rev.* **1990**, *90*, 1407.
(2)   a. Chapters in: *Vibrational Energy Transfer Involving Large and Small Molecules*, Barker, J. R., Ed. JAI Press, Greenwich, Connecticut, 1995.
      b. Barker, J. R.; Toselli, B. M. *Int. Rev. Phys. Chem.* **1993**, *12*, 305.
      c. Luther, K; Reihs, K. *Ber. Bunzen-ges. Phys. Chem.* **1988**, *92*, 442.
      d. Hippler, H.; Troe, J. in *Bimolecular collisions;* Baggott, J. E., Ashfold, M. N. Eds.; *The Royal Soc. of Chemistry*: London, **1989**.
      e. Mullin, A. S., Park, J, Chou, J. Z., Flynn, G. W., Weston, R. E., *J. Chem. Phys.* **1993**, *53*, 175.

f. Mullin, A. S.; Michaels, C. A.; Flynn, G. W. J. *Chem. Phys.* **1995**, *102*, 6032.

(3) a. *Unimolecular Reactions* ; Robinson P. J. ; Holbrook, K. A. Wiley-Interscience, New York, N.Y. 1972.

b. *Theory of Unimolecular and Recombination Reactions;* Gilbert, R. G.; Smith, S. C. Blackwell Scientific Publications, Oxford, 1990.

(4) a. Lim, K. F.; Gilbert, R. G. *J. Chem. Phys.* **1986**, *84*, 6129; **1990**, 92, 1819.

b. Gilbert, R. G.; Oref, I. *J. Phys. Chem.* **1991**, *95*, 5007.

c. Dashevskaya, E.; Nikitin, E. E.; Oref, I. *J. Phys. Chem.* **1993**, *97*, 9397.

d. Dashevskaya, E.; Nikitin, E. E.; Oref, I. *J. Phys. Chem.* **1995**, *99*, 10797.

e. Tardy, D. C. *J. Chem. Phys.* **1993**, *99*, 963.

f. Tardy, D. C.; Song, B. H. *J. Phys Chem.* **1993**, *97*, 5628.

g. Ming, L.; Sewell, T. D.; Nordholm, S. *Chem. Phys.* **1995**, *83*, 199.

h. Borjesson, L. E. B.; Nordholm, S. *Chem. Phys.* in press.

(5) a. Date, N.; Hase, W. L.; Gilbert, R. G. *J. Phys. Chem.* **1984**, *88*, 5135.

b. Brown, N. J.; Miller, J. A. *J. Chem. Phys.* **1984**, *80*, 5568.

c. Bruehl, M.; Schatz, G. C. *J. Chem. Phys.* **1988**, *89*, 770.

d. Bruehl, M.; Schatz, G. C. *J. Phys. Chem.* **1988**, *92, 7223.*

e. Lim, K. F.; Gilbert, R. C. *J. Phys. Chem.* **1990**, *94*, 72; **1990**, *94*, 77.

f. Lendvay, G.; Schatz, G. C. *J. Phys. Chem.* **1990**, *94*, 8864.

g. Clarke, D. L.; Thompson, K. G.; Gilbert, R. G. *Chem. Phys. Lett.* **1991**, *182*, 357.

h. Lendvay, G.; Schatz, G. C. *J. Phys. Chem.* **1992**, *96*, 3752.

i. Clarke, D. L.; Oref, I.; Gilbert, R. G.; Lim, K. F. *J. Chem. Phys.* **1992**, *96*, 5983.

j. Lendvay, G.; Schatz, G. C. *J. Chem. Phys.* **1992**, *96*, 4356.

k. Bernshtein, V.; Oref, I. *J. Phys. Chem.* **1993**, *97*, 12811.

l. Lendvay, G.; Schatz, G. C. *J. Chem. Phys.* **1993**, *98*, 1034.

m. Bernshtein, V.; Oref, I. *J. Phys. Chem.* **1994**, *98*, 3782.

n. Lim, K. F. *J. Chem. Phys.* **1994**, *100*, 7385; **1994**, *101*, 8756.

o. Bernshtein, V.; Lim, K. F.; Oref, I. *J. Phys. Chem.* **1995**, *99*, 4531.

p. Bernshtein, V.; Oref, I. *Chem. Phys. Lett.* **1995**, *233*, 173.

q. Lenzer, T.; Luther, K.; Troe, J.; Gilbert, R.G.; Lim, K.F. *J. Chem. Phys.* **1995**, *103*, 626.

r. Koifman, I.; Dashevskaya, E. I.; Nikitin, E.; Troe, J. *J. Phys. Chem.* **1995**, *99,* 15348

s. Clary, D. C.; Gilbert, R. G.; Bernshtein, V.; Oref, I. *Faraday Discussions* **1995**, 102.

t. Bernshtein, V.; Oref, I. *J. Chem. Phys.* **1996**, *1958*, 104.

u. Lendvay, G.; Schatz, G. C. *J. Phys. Chem.* **1994**, *98*, 6530.

(6) Bernshtein, V.; Oref, I.; Lendvay, G. *J. Phys. Chem.* 1996, *100* 9738.

(7) Schatz, G.C.; Lendvay, G. private communication.

(8) Bernshtein, V.; Oref, I. In preparation.

(9)   a. I. Oref, "Supercollisions" in *Vibrational Energy Transfer Involving Small and Large Molecules*" Volume 2 of Advances in Chemical Kinetics and Dynamics, Barker, J. R. Ed. JAI Press 1995.
       b. Pashutzki, A; Oref, I. *J. Phys. Chem.* **1988**, *92*, 1978.
       c. Hassoon, S.; Oref, I.; Steel, C. *J. Chem. Phys.* **1988**, *89*, 1743
       d. Morgulis, I. M.; Sapers, S. S.; Steel, C.; Oref, I. *J. Chem. Phys.* **1989**, *90*, 923.
       e. Miller, L. A.; Cooks, C. D.; Barker, J. D. *J. Chem. Phys.* **1996**, *105*, 3012.
(10)   Venus, Quantum Chemistry Program Exchange by Hase, W. L.; Duchovic, R. J.; Hu, X..; Lim, K. F.; Lu, D. H.; Pesherbe, G.; Swamy, K. N.; Vande-Linde, S. R.; Rolf, R. J. *Quantum Chemistry Program. Exchange Bull.*
(11)   Draeger, J. A. *Spectrochimia Acta* **1985**, *41a*, 607.
(12)   Bernshtein, V.; Oref, I.; Lendvay, G. in preparation.
(13)   WebElements, http://www.shef.ac.uk/~c...elements/chem-nt
(14)   Oref, I. *Chem. Phys.* **1994**, *187*, 163.

Chapter 18

Collisional Energy Transfer of Highly Vibrationally Excited Molecules: The Role of Long-Range Interaction and Intramolecular Vibronic Coupling

Hai-Lung Dai

Department of Chemistry, University of Pennsylvania,
Philadelphia, PA 19104–6323

In the experimental study of collisional energy transfer of highly vibrationally excited molecules (*1*, *2*), several techniques have been developed and applied to monitoring the energy content of either the highly excited energy donor or the energy receiving bath molecules. The highly excited molecules are often prepared with electronic excitation followed by internal conversion (*3*) or with multiple resonance techniques such as Stimulated Emission Pumping (*4*). The probe techniques include thermal lensing (*5*), photoacoustic (*6*), Doppler width measurement, diode laser transient absorption spectroscopy (*2*), optical absorption (*7, 8*) and ionization (*9*), laser induced fluorescence (*4*), sensitized chemical reaction (*10*) and IR emission (*11*). The IR emission from vibrationally excited molecules can in principle be used to reveal the rovibrational energy content of both the energy donating and receiving molecules. This can be best achieved if the IR emission is detected with sufficient frequency and time resolution so that the energy content of the emitting molecules during a collisional relaxation process can be continuously monitored. The average energy and energy distribution of the highly excited molecules can then be directly extracted from the time-resolved IR emission spectra.

Time-Resolved FTIR Emission Spectroscopy

We have recently developed a Time-Resolved Fourier Transform Emission Spectroscopy (TR-FTIRES) technique based on a step-scan FTIR spectrometer (*12*). This technique affords the ability for acquiring emission spectra in the IR and near-IR region with 50 ns (presently limited by detector response time and excitation laser pulse width) time-resolution for the study of collisional deactivation of highly vibrationally excited molecules. In TR-FTIRES, a Fourier transform spectrometer is coupled to a fast IR detector/transient digitizer system to temporally and spectrally resolve IR emission from a pulsed laser excited sample. The attributes of TR-FTIRES, in addition to the fast time resolution, are fine spectral resolution (≈ 0.1 cm^{-1} in principle), wide spectral range covering the mid- and near-IR regions, and high sensitivity due to the multiplex detection and throughput advantages of FT spectrometers. By using TR-FTIRES the complete time evolution of all IR active vibrational modes can be monitored at high spectral resolution in a single experiment (*13*).

Using the TR-FTIRES technique we have examined the collisional energy transfer of NO$_2$ (excited to as high as 22,000 cm^{-1}) (*13-15* and G.V. Hartland et al., *J.*

Chem. Phys. submitted), SO_2 (32,000 cm^{-1}) (*15, 16,* and D. Qin et al., *J. Phys. Chem.,* submitted), CS_2 (32,000 cm^{-1}) (*15, 16*), and pyrazine (40,000 cm^{-1}) (*16*) to a variety of collisional partners, ranging from inert gases to aromatic molecules. In the experiments, all the highly excited molecules are prepared through photoexcitation of an electronic transition followed by internal conversion, which proceeds with or without the assistance of collisions.

The time resolved FTIR emission spectra following the (25 ns FWHM) laser pulse excitation of NO_2 at 21,050 cm^{-1} is shown in Figure 1. Emission from all the IR active modes has been detected. All emission peaks are shifting towards blue with time (by as much as 300 cm^{-1} for the NO_2 v_3 peak), indicating that the emitting molecules are relaxing from the highly excited, anharmonic levels to the lower energy, more harmonic ones during collisional quenching. Through spectral simulation (see below), the profile of the energy distribution is extracted. Figure 2 shows that the distribution of the ensemble of excited molecules during collisional deactivation can be well described by a Gaussian function as stipulated by many workers in this field. The average energy, <E>, as a function of time can be extracted, and the energy transferred per Lennard-Jones collision, <ΔE>, of the excited molecules can then be determined.

Analysis of the IR Emission Spectra

Simulation of the emission spectra from the highly excited molecules was done by using the following assumptions/principles: 1) The energy of all the vibrational levels are calculated using the experimentally determined vibrational constants. The NO_2 vibrational constants, for example, have been determined by Jost and coworkers (*17*) from assigned vibrational levels up to about 10,000 cm^{-1}. 2) Emission intensity from levels containing IR active mode quantum numbers is calculated according to the experimentally measured absorption intensity with scaling to vibrational quantum number according to the harmonic oscillator model. 3) Isoenergetic levels are assumed to have the same population. This statistical model is justified because of the rapid intramolecular vibrational relaxation and the high vibrational level densities at high energies. Fast IVR ensures that nearby molecular eigen levels have similar vibrational characters. High density of the levels favors a rapid equilibration of population of nearby vibrational levels during collisional deactivation. 4) An arbitrary functional form is used for describing the excited molecules' energy distribution at any given time during the deactivation process. It is shown that this distribution can be best described by a Gaussian function (C. Pibel, E. Sirota, and H.L. Dai, to be published). A second Gausian function is used to describe the molecules at lower vibrational levels populated by V-V transfer or Franck-Condon pumping. An example of simulated spectrum of NO_2 emission is shown in Figure 3. The average energy <E> of the excited Gaussian population as a function of time, which is directly proportional to the number of collisions, is obtained. The derivative of this curve gives the average amount of energy removed per collision, <ΔE>, as a function of <E>, which is shown for the case of NO_2 quenching by CO_2 in Figure 4, and NO_2 and CS_2 by a variety of colliders in Figure 5.

Observations That Point to the Importance of Long Range Interaction and Intra-Molecular Vibronic Coupling

Among the many interesting observations are the ones that seem to arise because of intramolecular vibronic couplings occuring at high energies of the excited molecule. These particular observations, in combination with some previous works on this subject (*18*), appear to point to a somewhat surprising revelation that the mechanism

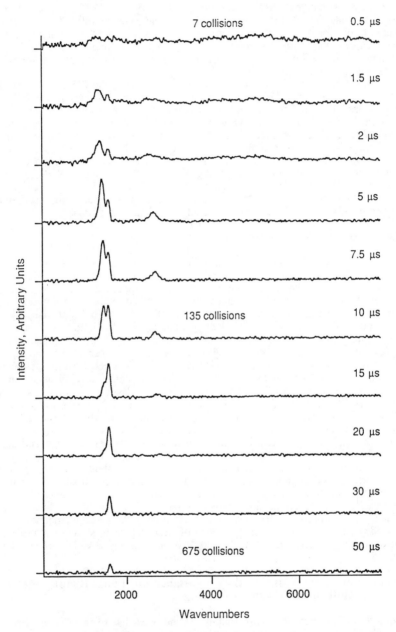

Figure 1. Selected FT emission spectra, recorded with a ~10 ns response time IR detector, from a sample of 1 Torr NO_2 excited by a 475 nm, ~1 mJ laser pulse. The baseline of each spectrum is marked on the abscisca.

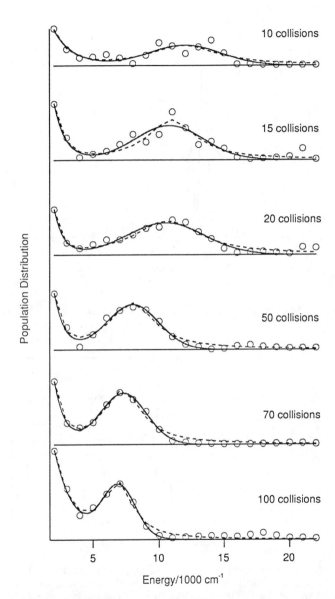

Figure 2. Energy distribution of excited NO_2 in collisional quenching with ambient NO_2. The points are obtained without assuming any functional form. The solid line is a Gaussian function and the dashed line Lorentzian.

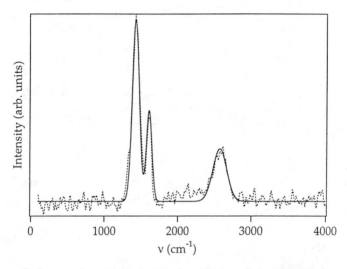

Figure 3. An experimental (dashed line) and its best fit (solid line) spectrum.

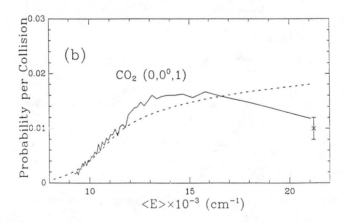

Figure 4. Energy transfer probability from NO_2 excited at $<E>$ to the $v_3=1$ CO_2 level. Solid line is calculated from the transition dipole model. Dashed line and the point are extracted from measurements by Flynn and coworkers (19). (Adapted from Ref. 14)

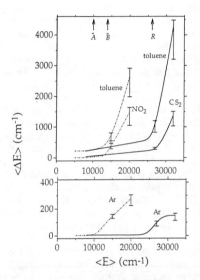

Figure 5. <ΔE>-<E> curves of excited NO_2 and CS_2 quenched by collision partners noted in the figure. The origin of the lowest, optically detected electronic excited states are marked; A and B states for NO_2 and the R state for CS_2. (Reproduced with permission from Ref. 15. Copyright 1995, American Institute of Physics.)

responsible for energy transfer of highly excited molecules, in contrast to those excited at lower energies, is dominated by long range, Coulombic interactions. These observations are:

Transition Dipole Coupling. Whenever we could quantitatively (in relative magnitude) test the V-V transfer probability, we have found that the transition-dipole coupling model works well. In this model, the donor molecule would make a downward transition while the receiver molecule makes an upward transition in energy during the collision. The probability of energy transfer is then proportional to the square of the downward and upward transition dipoles multiplied together. The receiver molecule defines the energy of the transition. Its transition dipole can be obtained from the conventional absorption spectrum of the receiver molecule. The downward transition dipole of the highly excited donor molecule, on the other hand, can be extracted (in relative magnitude as a function of the donor energy) from the emission intensity at the energy defined by the receiver. This emission intensity is registered in the emission spectra recorded in our experiment. So from our experimental emission spectra we could calculate the relative magnitude of energy transfer probability to a particular receiver vibrational mode as a function of the donor excitation energy.

One example of this test is shown in Figure 4. The solid line is the calculated probability of energy transfer per Lennard-Jones collision from highly excited NO_2 to the $v_3=1$ level of CO_2 as a function of the NO_2 energy. The calculation is done with the transition-dipole coupling model and the NO_2 IR emission spectra. An excellent comparison is made against the experimental measurements from Flynn and coworkers (19) who used the diode laser absorption spectroscopy to measure the CO_2 $v_3=1$ population. The importance of the transition dipole mechanism has been pointed out by several investigators on this subject previously (see for example Refs. 19-21). For energy transfer out of the vibrational levels in the low energy region, the dominant contribution of the transition dipole interaction has been alluded to for situations where the energy gap is small (21). For the highly excited molecules, the high density of states will relax the resonant conditions for V-V energy transfer (small energy gap), thus, it is consistent that the transition dipole coupling mechanism appears to work.

Threshold in <E> and Intramolecular Vibronic Coupling. For a few excited molecules such as NO_2 and CS_2 that have known strong vibronic coupling at high energies, <ΔE> appears to increase dramatically with <E> as the energy exceeds a threshold energy, Figure 5 (15). The threshold <E> for a specific excited molecule is always at the same value for all collision partners and correlates well with the beginning of vibronic coupling within the excited molecule. Similar threshold behavior has also been observed in V-T transfer studies by Barker and coworkers (22).

This threshold behavior may also be understood through the transition dipole coupling model (15, and G.V. Hartland et al., *J. Chem. Phys.*, submitted). In energy regions where strong vibronic coupling occurs, the highly excited vibrational levels of the electronic ground state are mixed in with the excited electronic states, i.e. the molecular eigenlevels at high vibrational energies contain excited electronic state characters. The downward transition from the highly excited donor molecule in this energy region may have its transition dipole enhanced due to the electronic character. The transition dipole may have electronic transition contribution in addition to the vibrational transition dipole. Consequently the energy transfer probability increases. The enhanced transition dipole in the post-threshold energy region is evidenced by the broad band emission peaks over wide wavelength range for both excited NO_2 and CS_2 (15).

For molecules that do not have strong vibronic coupling, such as pyrazine, the threshold behavior does not show up in the <ΔE>-<E> relationship. In terms of the magnitude of matrix element, vibronic coupling in pyrazine is a few orders of magnitude smaller than the triatomics. Here, energy transfer from pyrazine excited with 10,000 to 40,000 cm^{-1} does not show any threshold behaviour. The increase in <ΔE> with <E> appears to be linear (*16*) despite the fact that there are several excited electronic states within this energy range. But even here the transition dipole coupling model appears to be at least semi-quantitatively correct. In fact, in the pyrazine-pyrazine energy transfer the vibrational distribution of the energy receiving pyrazine molecules appears to be non-statistical which can be rationalized by the transition dipole coupling model (*16*).

Mass Effect vs. Polarizability. Even though the SSH theory (*23*) gives a qualitatively accurate description of the observed V-T transfer probabilities for molecules excited at lower vibrational energies, in the post threshold energy region, it does not appear to be adequate for understanding the observed energy transfer behaviour. For NO_2 excited below 10,000 cm^{-1} the lighter inert gas collider is more effective in relaxing the NO_2 vibrational energy. Here the <ΔE> per collision is on the order of 1 cm^{-1}. For NO_2 molecules excited at much higher energies, i.e. above 10,000 cm^{-1}, <ΔE> by inert gases increased to 10^2 cm^{-1}. Also, it seems that in this post threshold regime the efficiency of the collisional partner is no longer solely correlated with its mass (G.V. Hartland et al., *J. Chem. Phys.,* submitted and C. Pibel et al., to be punished). Ar is about as effective as Xe and both are much more effective than He for relaxing NO_2. Polarizability appears to play a role in collisional efficiency.

It has been suggested in the above that the transition dipole involved in the energy transfer mechnism for a highly excited molecule is much stronger in the post-threshold energy region because of intramolecular vibronic coupling. Based on this, we speculate that for a colliding partner with large polarizability, the non-repulsive contribution to V-T transfer can no longer be ignored. The SSH model is based primarily on the repulsive part of the inter-molecular potential. Its effectiveness is anticipated for V-T transfer involving low energy vibrational levels where the energy gap is large. But for the highly excited vibrational levels with high level density where the energy gap in V-T transfer is small and where the transition dipole is stronger, contributions from long range Coulombic interactions should be included in our consideration.

Concluding Remarks

The above observations, combined with observations made in other laboratories, have pointed to a seemingly non-intuitive mechanism for energy transfer out of a highly excited molecule: the large amount of energy transferred, either through V-V or V-T/R channels, to an ambient collider is through a long range, Coulombic type interaction. In this mechanism vibronic couplings at high energies will greatly accelerate the energy transfer, likely through an increase in the transition dipole of the excited molecule. With this possibility in mind, many questions now await to be answered. We would like to know in general how would the long range interaction channel compare quantitatively with the short range impact force induced energy transfer? Can the long range interaction mechanism involving vibronic coupling be able to provide a rationalization for the 'super collisions' where in one collisional encounter tens of thousands of wavenumber energy is transferred (*24*)? We are planning experiments using the TR-FTIRES on molecules with known vibronic coupling strength to further test our understanding on energy transfer from highly excited large molecules. On the theoretical front, in addition to classical trajectory calculations based on repulsive potentials (*25*), quantum or semi-classical calculations with realistic electrostatic coupling potentials involving the electronically excited states need to be performed for comparison.

Acknowledgments: This work is supported by the Basic Energy Sciences of the U.S. Department of Energy through Grant No. DE-FG02-86ER134584 and a U.S. Department of Energy University Instrumentation Grant. Support from the Department of Advanced Technology of the Brookhaven National Laboratory is also acknowledged. The author is grateful to Professor Gregory Hartland, Dr. Dong Qin, Professor Charles Pibel, and Mr. Egor Sirota for their contribution to this work.

Literature Cited

1. I. Oref and D. C. Tardy., *Chem. Rev.*, **1990**, 90, 140 .
2. R.E. Weston, Jr. and G.W. Flynn, *Annu. Rev. Phys. Chem.* **1992**, 559.
3. a). R. Atkinson, and B.A. Thrush, *Chem. Phys. Lett.*, **1969**, 3, 684.
   b). H. Hippler, K. Luther, J. Troe, and R. Walsh, *J. Chem. Phys.*,**1978**, 68, 323.
4. X. Yang, J.M. Price, J.A. Mack, C.G. Morgan, C.A. Rogaski, D. Mcguire, E. H. Kim, and A. M. Wodtke, *J. Phys. Chem.*,**1993**, 97, 3944.
5. a). D.R. Siebert, and G.W. Flynn, *J. Chem. Phys.*, **1974**, 60, 1564.
   b). B.M. Toselli, T.L. Walunas, and J. R. Barker, *J. Chem. Phys.*, **1990**, 92, 4793.
6. a). A. Karbach and P. Hess, *J. Chem. Phys.*, **1986**, 84, 2945;
   b). K.M. Beck and R.J. Gordon, *J. Chem. Phys.*, **1987**, 87, 5681.
7. H. Hippler and J. Troe, in *Bimolecular collisions*, Boggort, J.E. and Ashford, M.N.R., Eds.; The Royal Society of Chemistry, London, **1989**.
8. T.J. Bevilacqua and R.B. Weissman, *J. Chem. Phys.* **1993**, 98, 6316.
9. H.G. Lohmannsroben and K. Luther, *Chem. Phys. Lett.* **1988**, 144, 473.
10. S. Hassoon, I. Oref, and C. Steel, *J. Chem. Phys.* **1988**, 89, 1743.
11. J. R. Barker and B. M. Toselli, *Int. Rev. Phys. Chem.*, **1993**, 12, 305.
12. G. V. Hartland, W. Xie, and H.L. Dai, A. Simon, and M. J. Anderson, *Rev. Sci. Instrum.* **1992**, 63, 3261.
13. G.V. Hartland, D. Qin, and H.L. Dai, *J. Chem. Phys.*, **1994**, 100, 7832.
14. G.V. Hartland, D. Qin, and H.L. Dai, *J. Chem. Phys.* **1994**, 101, 8554.
15. G.V. Hartland, D. Qin, and H.L. Dai, *J. Chem. Phys.*, **1995**, 102, 8677.
16. D. Qin, Ph.D. Thesis, University of Pennsylvania **1996**.
17. a) A. Delon and R. Jost, *J. Chem. Phys.* **1991**, 95, 5686;
    b) A. Delon, R. Jost, and M. Lombardi, *J. Chem. Phys.* **1991**, 95, 5701.
18. a) A.S. Mullin, C.A. Michaels, G.W. Flynn., and R.E. Weston, *J. Chem. Phys.* **1993**, 175, 53.
    b) C.A. Michaels, A.S. Mullin., and G.W. Flynn., *J. Chem. Phys.*, **1995**, 102, 6682.
19. a) J.Z. Zhou, S.A. Hewitt, J.F. Hershberger, B.B. Brady, G.B. Spector, L. Chia, and G.W. Flynn, *J. Chem. Phys.* **1989**, 91, 5392.
    b) J.Z. Zhou, S.A. Hewitt, J.F. Hershberger, and G.W. Flynn, *J. Chem. Phys.* **1990**, 93, 8474.
    c) J.Z. Zhou and G.W. Flynn, *J. Chem. Phys.* **1990**, 93, 6099.
20. J.C. Stephenson and C.B. Moore, *J. Chem. Phys.* **1972**, 56, 1295.
21. a) R.D. Sharma and C.A. Brau, *J. Chem. Phys.* **1969**, 50, 924;
    b) R.D. Sharma and C.A. Brau, *Phys. Rev. Lett.* **1967**, 19, 1273.
22. B.M. Toselli, T.L. Walunas, and J.R. Barker, *J. Chem. Phys.* **1990**, 92, 4793.
23. a) R.N. Schwartz, Z.I. Slawsky and K. F. Herzfeld, *J. Chem. Phys.* **1952**, 20, 1591;
    b) R.N. Schwartz and K.F. Herzfeld, *J. Chem. Phys.* **1954**, 22, 767.
24. a) I.M. Margulis, S.S. Sapers, C. Steel, and I. Oref, *J. Chem. Phys.* **1989**, 90, 923.
    b) A. Pashutzki and I. Oref, *J. Phys. Chem.* **1988**, 92, 178.
25. See for example G. Lendvay, C. Schatz, and L.B. Harding, *Faraday Disc.* **1995**, 102.

INTERMOLECULAR DYNAMICS: REACTIVE COLLISIONS

Chapter 19

Simulations of Energy Transfer in the Collision-Induced Dissociation of $Al_6(O_h)$ Clusters by Rare-Gas Impact

W. L. Hase and P. de Sainte Claire

Department of Chemistry, Wayne State University, Detroit, MI 48202–3489

In collision-induced dissociation (CID) reagent relative translational energy is transferred to vibrational/rotational internal energy of the dissociating species. Classical trajectory simulations have been used to study the dynamics of this energy transfer process in collisions of $Al_6(O_h)$ with Ne, Ar, and Xe. The model Al_6 intramolecular potential and model Al_6-rare gas (Rg) intermolecular potentials, used in the simulations, were derived from *ab initio* calculations. The efficiency of transferring relative translational energy to Al_6 internal energy is studied as a function of cluster stiffness, relative translational energy, mass of the Rg-atom, and the repulsiveness of the Rg-Al_6 intermolecular potential. The results of these simulations are compared to those of an impulsive model (J. Chem. Phys. **1970**, *52*, 5221) for translation to vibration (i.e., T → V) energy transfer.

Experimental studies (1-16) of the collision-induced dissociation (CID) of semiconductor and metal atom clusters have provided information about the bond energies, dissociation pathways, and structures of the clusters. For the most part, ionic clusters have been studied, since specific cluster sizes can be selected by mass spectrometry. Experimental results have been reported for B_n^+ (10,11), Al_n^+ (8,9), C_n^+ (14-16), and Si_n^+ (12,13) and for first- and second-row transition metal clusters Ti_n^+ (4), V_n^+ (2), Cr_n^+ (1), Fe_n^+ (5), Ni_n^+ (6), Nb_n^+ (9), and Co_n^+ (3). In recent work CID has been used to study dissociation of the coordination complexes $Cr(CO)_n^+$ (17), $V(CO)_n^+$ (18), and $Li^+[O(CH_3)_2]_n$ (19).

Additional insight into the CID experiments would be gained by understanding the dynamics of the transfer of initial relative translational energy to cluster internal energy and the ensuing dynamics of cluster dissociation. Since the CID experiments are carried out at high energies, classical trajectory simulations (20-22) are expected to accurately describe the energy transfer and dissociation dynamics. In this chapter, recent classical trajectory simulations (23,24) of energy transfer in Rg + $Al_6(O_h)$ collisions are reviewed. An analysis is made of how the efficiency of energy transfer is affected by properties of the collision, i.e., the atomic masses and the relative translational energy, the Al_6 intramolecular potential, and the Rg-Al_6 intermolecular potential.

Model for T → V Energy Transfer

To describe T → V energy transfer in collinear A + BC collisions, Mahan (25-27) derived an *impulsive* model, in which Fourier components of the short-range A + BC repulsive forces are treated. The repulsive potential is represented by

$$V(r) = V_0 exp(-r/L) \tag{1}$$

where r is the A-B distance and L is the range parameter for the A-B interaction. For this impulsive model, the relationship between energy transferred to vibration ΔE_{int} and the initial relative translational energy E_{rel} is

where

$$\frac{\Delta E_{int}}{E_{rel}} = 4 \cos^2\beta \, \sin^2\beta \left(\frac{\xi}{2} cosech \frac{\xi}{2}\right)^2 \tag{2}$$

$$\cos^2 \beta = m_A m_C / [(m_A + m_B)(m_B + m_C)] \tag{3}$$

and ξ, often called the adiabaticity parameter (28), is

$$\xi = 4\pi^2 \nu L / v_{rel} \tag{4}$$

where ν is the BC vibrational frequency and v_{rel} is the A + BC initial relative velocity. For values of v_{rel} large compared to ν, so that $\xi < 1$, the collision time is short compared to the vibrational period, and the *sudden* limit is reached. In this limit equation 2 becomes

$$\Delta E_{sudden} / E_{rel} = 4 \cos^2 \beta \, \sin^2 \beta \tag{5}$$

For m_C much larger than either m_A or m_B, equation 5 reduces to

$$\Delta E_{sudden} / E_{rel} = 4 m_A m_B / (m_A + m_B)^2 \tag{6}$$

For large ξ the collision approaches the adiabatic limit (25-29).
    An important parameter for the above model is

$$m = \frac{m_A m_C}{m_B(m_A + m_B + m_C)} \tag{7}$$

since it determines whether A and B have a single collision or multiple collisions during the A + BC collision event. Secrest (30) has shown that for a harmonic BC oscillator and an exponential repulsive interaction between A and B multiple collisions between A and B occur if m > 0.697.

CID of first row transition metal clusters with xenon (1-5, 7-9) has been studied and it is of interest to consider what the above model predicts for energy transfer in these systems. In Figure 1, $\Delta E_{int}/E_{rel}$ from equation 2 is plotted versus ξ for Xe colliding with a cobalt dimer. For small ξ the collisions are in the sudden limit and $\Delta E_{int}/E_{rel}$ reaches its maximum value given by equation 5. For Xe + Co_2 collisions $\Delta E_{sudden}/E_{rel}$ equals 0.904. Increasing the adiabaticity parameter ξ decreases the efficiency of energy transfer (25-29).

The above model suggests that in atom-cluster collisions all of E_{rel} may not be transferred to the cluster and, as a result, the CID threshold may exceed the true threshold. However, the above model is for collinear atom-diatom collisions and its applicability to atom-cluster collisions can surely be questioned. For the latter, energy transfer from relative translation to cluster internal degrees of freedom may be enhanced for some collisions by the: (1) anisotropy of the atom-cluster intermolecular potential; (2) distribution of cluster frequencies; (3) range of collision impact parameters and different effective impact parameters for different collisional orientations; (4) range of an effective mass (31, 32) of the cluster, with which the colliding atom interacts; and (5) multiple atom-cluster impacts during a collision. It is of interest to determine whether such effects are sufficient to make the CID threshold equal the true threshold E_o.

Cluster Potentials

Lennard-Jones/Axilrod-Tellar Potential. A model analytic potential energy function for Al_n clusters has been developed from *ab initio* calculations (33). The potential function consists of two-body Lennard-Jones (L-J) terms:

$$V_{ij} = \frac{A}{r_{ij}^{12}} + \frac{B}{r_{ij}^6} \qquad (8)$$

and three-body Axilrod-Teller (A-T) terms (34):

$$V_{ijk} = C \frac{(1+3\cos\alpha_1 \cos\alpha_2 \cos\alpha_3)}{(r_{ij}r_{jk}r_{ki})^3} \qquad (9)$$

for which the parameters $A = 2975343.77$ kcal-$Å^{12}$/mol, $B = -17765.823$ kcal-$Å^6$/mol and $C = 81286.093$ kcal/mol. In equation 9 r_{ij}, r_{jk}, r_{ki} and α_1, α_2, α_3 represent the sides and angles, respectively, of the triangle formed by the three particles i, j and k. This L-J/A-T potential gives the following properties for the $Al_6(O_h)$ cluster: a nearest neighbor Al-Al distance of 2.834 Å, $Al_6 \rightarrow 6Al$ and $Al_6 \rightarrow Al + Al_5$ (C_{2v}) classical dissociation energies of 185.7 and 38.87 kcal/mol, respectively, and harmonic vibrational frequencies (and degeneracies) of 132(3), 162(2), 230(3), 330(3), and 475 cm^{-1}. For the L-J/A-T potential, the lowest energy Al_5 cluster has a planar C_{2v} geometry.

Morse Potential. To have a model potential that is easy to vary, an analytic function consisting of two-body Morse terms was developed (23) for Al$_6$(O$_h$). The exponential β-parameter in the Morse term is written as a cubic function of r_{ij} (35); i.e.,

$$V_{ij} = D_e \{1 - \exp [-\beta (r_{ij} - r_e)]\}^2 \tag{10a}$$

$$\beta = \beta_e + c_2 \Delta r^2 + c_3 \Delta r^3 \tag{10b}$$

where $\Delta r = r_{ij} - r_e$. Values of the parameters for the Morse function called Morse*1 were chosen so that: (1) the Ar + Al$_6$ (O$_h$) energy transfer probability distribution and CID cross section versus E_{rel} are in near exact agreement with those for the above L-J/A-T potential; (2) the Al$_6$ (O$_h$) structure and an Al$_6$ (O$_h$) $\rightarrow$ Al$_5$ + Al dissociation energy are in exact agreement with the values for the L-J/A-T potential; and (3) the Al$_6$ (O$_h$) vibrational frequencies are in approximate agreement with those for the L-J/A-T potential. The parameters for Morse*1 are r_e = 2.834 Å, D_e = 12.933 kcal/mol, β_e = 2.1128 Å^{-1}, c_2 = 1.756 x 10^{-2} Å^{-3} and c_3 = 20.436 Å^{-4}. The Al$_6$ (O$_h$) vibrational frequencies (and degeneracies) are 227(3), 227(2), 321(3), 393(3) and 454 cm^{-1} for the Morse*1 potential. It was found that the cluster vibrational frequencies are proportional to uniform scaling of β_e, c_2 and c_3. Scaling these parameters varies both the short-range and long-range "stiffness" of the cluster. Increasing β_e, c_2 and c_3 from their Morse*1 value by a factor of two gives a potential function called Morse*2, while decreasing the parameters by a factor of two gives Morse/2. Thus, for the Morse*n and Morse/n potentials the frequencies are n times larger and smaller, respectively, than for the Morse*1 potential.

Rare Gas-Cluster Potential. The intermolecular potential between R$_g$ and Al$_6$ is written as a sum of two-body potentials between R$_g$ and the Al atoms of the cluster. Two-body potentials for Ar-Al and Ar-Al$^+$ were calculated (23) with the GAUSSIAN series of programs (36), and convergence was obtained at the UMP2(fc)/6-31G* level of theory. In this method, unrestricted perturbation theory, with a second-order energy correction, is applied to the ground state energy. Excitations out of the inner shell molecular orbitals are not included, i.e. the frozen core (fc) approximation. At long-range the Ar-Al$^+$ potential is more attractive than that for Ar-Al because of the former's ion-induced dipole interaction; see Figure 1 of Reference 23. However, for the high translational energies of CID it is the short-range potential that is important and, at short-range, the Ar-Al and Ar-Al$^+$ potentials are nearly identical. Thus, Ar-Al$_6$ and Ar-Al$_6^+$ are expected to have similar intermolecular potentials at the energies involved in CID.

The *ab initio* Ar-Al potential is fit by

$$V_i = \frac{a}{r_i^{12}} + \frac{b}{r_i^6} + c \exp(-d \, r_i) \tag{11}$$

where r_i is the distance between Ar and the ith Al atom. The values of the parameters in equation 11 are: a = 5212.13281 kcal-Å$_{12}$/mol, b = 884.959534 kcal-Å$_6$/mol, c =

10421.443 kcal/mol and d = 2.75301790 Å^{-1}. The UMP2(fc)/6-31G* level of theory was also used to calculate Ne-Al and Xe-Al two-body potentials (24).

The high-energy region of the aluminum-rare gas *ab initio* potentials are excellently fit by equation 1. This is illustrated in Figure 2. The V_0 and L parameters for the fits are: V_0 = 17642.938 kcal/mol and L = 0.2819 Å for Ne; V_0 = 17352.530 kcal/mol and L = 0.3503 for Ar; and V_0 = 18479.848 kcal/mol for L = 0.4019 for Xe. The Xe-Al and Ne-Al potentials are the most and least repulsive, respectively. This is expected from the relative sizes of the rare-gas atoms. For the classical trajectory simulations reviewed here, the Rg-Al$_6$ intermolecular potentials were constructed from the two-body potentials in equations 1 and 11.

Procedure for the Classical Trajectory Simulations

The classical trajectory calculations were performed with the general chemical dynamics computer program VENUS96 (37). The potential energy functions described above are standard options in VENUS96. Initial conditions for the trajectories were chosen to represent experiments of Al$_6$ CID (8, 9). A 138 K rotational energy of RT/2 was added to each of the principal rotation axes of the Al$_6$(O$_h$) cluster, to give a total initial cluster rotational energy E_{rot} of 0.4 kcal/mol. The clusters contained no initial vibrational zero-point energy. The energy transfer cross sections and probabilities of energy transfer are insensitive to whether the zero-point energy of 4.1 kcal/mol is added to Al$_6$(O$_h$) in the initial conditions (23). The reactants were initially separated by 15 Å, so that there was no cluster-rare gas intermolecular interaction. The cluster was then randomly oriented about its Euler angles. The impact parameter b was chosen randomly between 0 and b$_{max}$. The value for b$_{max}$ was chosen to be the minimum impact parameter at which not one of 500 trajectories transferred sufficient energy to dissociate Al$_6$(O$_h$).

The trajectories were terminated when the cluster-rare gas intermolecular interaction became sufficiently weak that the internal energy of the cluster changed by less than 10^{-7} kcal/mol between two successive iterations, as the rare gas scattered from the cluster. The final cluster internal energy was evaluated at this point. The cluster is assumed to have sufficient energy to dissociate if its final internal energy is larger than its dissociation threshold E_0, which is 38.8 kcal/mol for all the calculations reported here. The CID cross section is

$$\sigma_{CID} = \pi b_{max}^2 \, (N_d/N) \tag{12}$$

where N_d is the number of trajectories which excite the cluster above E_0 and N is the total number of trajectories. Sets of 600 or 3000 trajectories were calculated to determine σ_{CID} at a particular value of E_{rel}. The error bar for σ_{CID} reflects the number of trajectories calculated.

The cross sections calculated versus E_{rel} were fit with

$$\sigma_{CID} = A(E_{rel} + E_{rot} - E_0^{CID})^n / E_{rel} \tag{13}$$

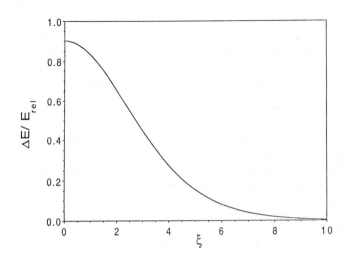

Figure 1. Plot of $\Delta E_{int}/E_{rel}$ versus the adiabaticity parameter ξ for Xe + CO$_2$ collisions.
(Reprinted from ref 24. Copyright 1996 American Chemical Society.)

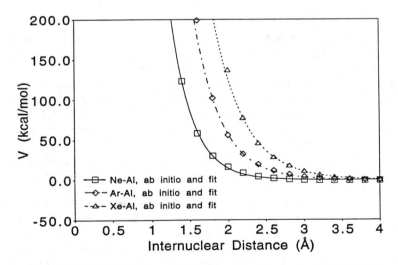

Figure 2. UMP2(fc)/6-31G* two-body intermolecular potentials. The fits are with equation 1.
(Reprinted from ref 24. Copyright 1996 American Chemical Society.)

where A, E_0^{CID}, and n are the fitting parameters, and E_{rot} is the cluster initial rotational energy of 0.4 kcal/mol. Equation 13 has been used to fit experimental measurements of cluster CID cross-sections and extract a value for E_0^{CID} (1-5, 7).

The energy transferred to $Al_6(O_h)$ is identified as ΔE_{int}. The probability of this energy transfer $P(\Delta E_{int})$ was studied by determining the number of trajectories which transferred energy into bins of width $\Delta(\Delta E_{int})$. The probability that the energy transfer falls into bin i and is greater than 38.4 kcal/mol, so that CID can occur, is

$$P(\Delta E_{int}) = \frac{N_i}{N_d \Delta(\Delta E_{int})} \qquad (14)$$

where N_i is the number of trajectories which fell into bin i and N_d is the number of trajectories with ΔE_{int} greater than 38.4 kcal/mol. Trajectories with ΔE_{int} in the range 38.4-40.0 kcal/mol are included in the first bin, which is thus normalized by a larger $\Delta(\Delta E_{int})$ than the fixed $\Delta(\Delta E_{int})$ for the remaining bins.

Energy Transfer Probabilities and CID Cross Sections for Rg + $Al_6(O_h)$ Collisions

Trajectory calculations are reviewed below in which the Rg + $Al_6(O_h)$ energy transfer probability and CID cross section are studied as a function of the cluster's stiffness, the initial relative translational energy E_{rel}, and the Rg-Al_6 intermolecular potential. CID cross sections and fitted E_0^{CID} values, equation 13, are compared for Ne, Ar, and Xe + $Al_6(O_h)$. The results of the trajectories are compared with the predictions of Mahan's model, equations 1-8.

Effect of Cluster Stiffness. Calculations with Ar as the colliding gas atom and for E_{rel} of 120.8 kcal/mol were performed to determine how variations in the $Al_6(O_h)$ vibrational frequencies affect energy transfer. The Morse*n and Morse/n potentials were used for Al_6 and equation 11 was used for the Ar-Al two-body potential. The resulting $P(\Delta E_{int})$ distributions are given in Figure 3. The energy transfer efficiency is strongly dependent on cluster stiffness, i.e., efficient for soft clusters and inefficient for stiff clusters. There appears to be soft and stiff asymptotic limits. The results for the Morse/4 and Morse/2 potentials (the soft limit) are nearly the same, as are the results for the Morse*6 and Morse*8 potentials (the stiff limit). $P(\Delta E_{int})$ for the Morse*8 potential is statistically the same as that for the Morse*6 potential and is not shown.

For the Morse*1 and softer clusters, energy transfer is efficient and the CID cross section is large, i.e., >20 Å². However, in going from the Morse*1 to Morse*2 cluster, there is a large drop in the energy transfer efficiency and the cross section decreases. The specific values for the CID cross sections, in units of Å², are: Morse/4 (26.3 ± 0.3), Morse/2 (25.5 ± 0.3), Morse*1 (23.1 ± 0.3), Morse*2 (13.5 ± 0.3), Morse*4 (6.9 ± 0.3), Morse*6 (4.2 ± 0.2) and Morse*8 (3.0 ± 0.2).

As the cluster stiffens, energy transfer to cluster rotation becomes more important. Calculating average ΔE_{vib} and ΔE_{rot} values from all the trajectories for a particular cluster gives in units of kcal/mol: $<\Delta E_{vib}> = 35.5$ and $<\Delta E_{rot}> = 4.8$ for

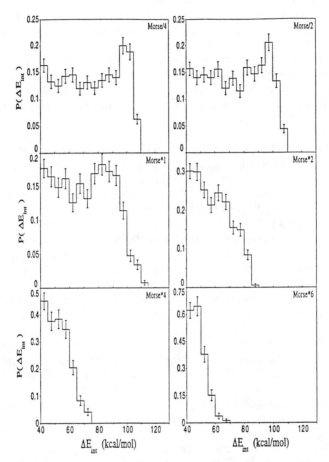

Figure 3. Energy transfer distributions for Ar + Al$_6$(O$_h$) versus the stiffness of the Al$_6$ intramolecular potential. E$_{rel}$ = 120.8 kcal/mol. P(ΔE$_{int}$) is in units of (kcal/mol)$^{-1}$, and the distributions are normalized so that the total probability is unity for transferring energy in excess of the true dissociation threshold E$_0$. The probabilities have been multiplied by a factor of 10.
(Reprinted from ref 23. Copyright 1995 American Chemical Society.)

Morse*1, $<\Delta E_{vib}> = 19.4$ and $<\Delta E_{rot}> = 10.9$ for Morse*2, $<\Delta E_{vib}> = 6.3$ and $<\Delta E_{rot}> = 13.6$ for Morse*4, and $<\Delta E_{vib}> = 1.5$ and $<\Delta E_{rot}> = 14.7$ for Morse*6. For the Morse*6 and Morse*8 potentials, 92% and 96%, respectively, of the energy transfer is to cluster rotation. In calculating the CID cross sections it is assumed that CID occurs when the cluster internal energy $E_{int} = E_{vib} + E_{rot}$ exceeds the dissociation threshold of 38.8 kcal/mol. This model assumes that E_{vib} and E_{rot} are equally effective in inducing dissociation. In RRKM theory these two energies are treated differently (38); i.e., E_{vib} is active, while E_{rot} is adiabatic.

The increased probability for extensive $T \rightarrow V$ energy transfer and larger CID cross sections, observed in the trajectories as the Al_6 cluster is softened, is consistent with Mahan's model for energy transfer. According to equation 4, lowering the cluster's vibrational frequencies decreases ξ, which predicts more efficient energy transfer.

Effect of Initial Relative Translational Energy. Trajectories were calculated using the L-J/A-T potential for $Al_6(O_h)$ and Ar as the colliding gas atom to determine how the efficiency of energy transfer to the cluster varies with E_{rel}. Equation 11 was used for the Ar-Al two-body potential. For $E_{rel} = 120.8$ kcal/mol, the maximum ΔE_{int} divided by E_{rel} is ≈ 0.9 (Figure 3 has $P(\Delta E_{int})$ at 120.8 kcal/mol for the Morse*1 Al_6 potential). $P(\Delta E_{int})$ distributions for E_{rel} from 60 to 400 kcal/mol are plotted in Figure 4. There is a high probability of transferring large amounts of energy as E_{rel} is increased. For $E_{rel} = 400$ kcal/mol, there are collisions in which $\Delta E_{int}/E_{rel}$ equals unity. The shape of the $P(\Delta E_{int})$ distribution for the high E_{rel} of 400 kcal/mol is similar to that in Figure 3 for the soft Al_6 Morse/4 potential.

It is of interest to determine the average energy transfer $<\Delta E_{int}>$ for all trajectories calculated when choosing b randomly between 0 and b_{max}, not just those which lead to CID. Using the same b_{max} of 3.5 Å, the ratio $<\Delta E_{int}>/E_{rel}$ is 0.38, 0.39, 0.40, 0.41, 0.41, and 0.41 for E_{rel} of 60, 80, 120.8, 170, 200, and 400, respectively. Given the statistical uncertainties, there is no meaningful change in this ratio versus E_{rel}. Thus, increasing E_{rel} does not alter $<\Delta E_{int}>/E_{rel}$ but broadens the $P(\Delta E_{int})$ distribution so that there are trajectories which transfer all of E_{rel} to E_{int}. The broadening of $P(\Delta E_{int})$ with increasing E_{rel} is consistent with Mahan's model, since the adiabaticity parameter in equation 4 decreases as v_{rel} increases.

Effect of the Rg-$Al_6(O_h)$ Intermolecular Potential. Collisions between Xe and $Al_6(O_h)$ were simulated to study how variations in the cluster-rare gas intermolecular potential affect σ_{CID}. These calculations used the L-J/A-T cluster potential and equation 1 to represent the intermolecular potential. Values for L equal to 0.2, 0.33 and 0.4 Å were considered for the intermolecular potential with $V_0 = 20257.938$ kcal/mol. As discussed above, following equation 11, L equal to 0.33 and 0.4 Å are representative of the values for the Ar-Al and Xe-Al potentials, respectively.

From Figure 5 it is seen that both the high energy, asymptotic value of σ_{CID} and the CID threshold increase as L is increased. The former is expected, since the Xe + Al_6 collision radius increases as L is increased. The increase in E_0^{CID} with increase in L is consistent with the adiabaticity parameter ξ of equation 4 and Mahan's model for T

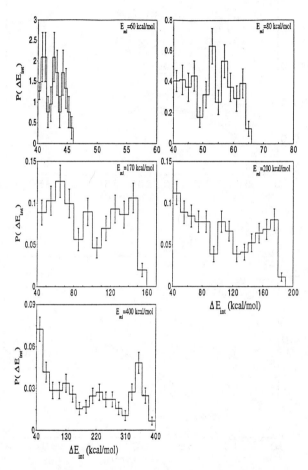

Figure 4. Energy transfer distributions for Ar + Al₆(Oₕ) on the L-J/A-T surface for different E_{rel}'s. The normalization and units for the distributions are described in the caption for Figure 3.

(Reprinted from ref 23. Copyright 1995 American Chemical Society.)

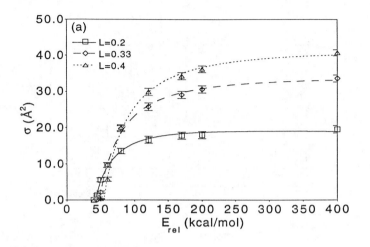

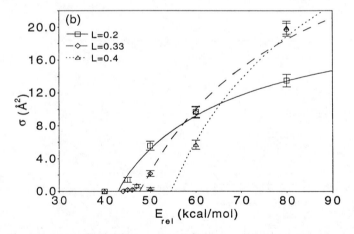

Figure 5. CID cross sections versus E_{rel} for Xe + $Al_6(O_h)$ with the L-J/A-T cluster potential and different model potentials based on equation 1. Part (a) contains all the calculated σ_{CID} and their fit, Part (b) illustrates the fit in (a) at low E_{rel}.
(Reprinted from ref 24. Copyright 1996 American Chemical Society.)

→ V energy transfer. Increasing L decreases the probability of energy transfer and as a result the CID threshold increases. The fits to the $\sigma_{CID}(E_{rel})$ curves give: A = 48.3 Å^2, E_0^{CID} = 43.4 kcal/mol, and n = 0.86 for L = 0.2; A = 63.3 Å^2; E_0^{CID} = 48.5 kcal/mol, and n = 0.91 for L = 0.33; and A = 81.0 Å^2; E_0^{CID} = 55.0 kcal/mol, and n = 0.90 for L = 0.4. For L = 0.2 the fitted E_0^{CID} value is only slightly in excess of the actual threshold of 38.8 kcal/mol.

For the Xe + Al_6 calculation with L = 0.33, σ_{CID} was calculated for a range of low E_{rel} values (24). The resulting σ_{CID} curve was fit by equation 13 to give A = 6.91 Å^2, E_0^{CID} = 44.9 kcal/mol, and n = 1.61. This fit gives an E_0^{CID} value only 3-4 kcal/mol lower than E_0^{CID} = 48.5 kcal/mol from the complete $\sigma_{CID}(E_{rel})$ curve, which suggests that meaningful E_0^{CID} values are obtained by fitting the complete $\sigma_{CID}(E_{rel})$ curves. On the other hand, a value for n significantly larger than unity is obtained from the fits to σ_{CID} at low E_{rel}. This is because near threshold the collisions are not impulsive (25-27). Values of n larger than unity have been observed in fits of experimental σ_{CID} curves to equation 13 (1-5, 7).

Ne, Ar, Xe + $Al_6(O_h)$ CID Cross Sections. A comparison was made between the CID cross sections for Ne, Ar, and Xe + $Al_6(O_h)$. The L-J/A-T function was used for the Al_6 potential. The cluster-rare gas intermolecular potentials used are those resulting from the fits of equation 1 to the Ne-Al, Ar-Al, and Xe-Al *ab initio*two-body potentials. The V_0 and L parameters are listed below equation 11. The calculated values of σ_{CID} versus E_{rel} are plotted in Figure 6, where they are compared with fits by equation 13. The fitting parameters are: A = 12.1 Å^2, E_0^{CID} = 47.2 kcal/mol, and n = 1.27 for Ne; A = 11.9 Å^2, E_0^{CID} = 50.0 kcal/mol, and n = 1.48 for Ar; and A = 1.64 Å^2; E_0^{CID} = 47.2 kcal/mol and n = 2.05 for Xe. The fitted E_0^{CID} values for the three curves are in the range of 47-49 kcal/mol and approximately 10 kcal/mol larger than the actual threshold of 38.8 kcal/mol.

The ξ parameter of equation 4 is smallest for Ne + Al_6 collisions, since the range parameter L is the smallest and the relative velocity v_{rel} is the largest. Inserting the values for ξ into equation 2 and using the mass of one Al-atom for the masses of B and C gives $\Delta E_{int}/E_{rel}$ of 0.50, 0.39, and 0.13 for Ne, Ar, and Xe, respectively, for E_{rel} = 38.4 kcal/mol. This is the value of E_{rel} needed if dissociation actually occurred at the true threshold; i.e. $E_{rel} = E_0 - E_{rot}$ = 38.4 kcal/mol. Thus, Mahan's model, equations 1-8, predicts significantly different $\Delta E_{int}/E_{rel}$ values for Ne, Ar, and Xe + $Al_6(O_h)$ at the true threshold and different E_0^{CID} values might have been expected from the trajectory simulations. In contrast, the trajectory E_0^{CID} values for Ne, Ar, and Xe are very similar. The conclusion one draws is that Mahan's model provides a useful qualitative picture, but does not incorporate all the dynamical details which may affect energy transfer to the cluster.

Comparison of Experimental and *Ab Initio* Rg-Al Interaction Energies

The UMP2(fc)/6-31G* *ab initio* calculations (23, 24) give the following classical well depths and minimum energy geometries for the Rg-Al (Al^+) potentials: Ne-Al, 0.6

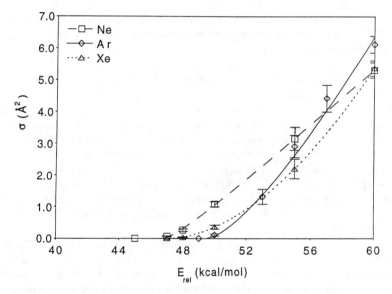

Figure 6. CID cross sections versus E_{rel} for Ne, Ar, Xe + Al$_6$ with the L-J/A-T cluster potential and intermolecular potentials determined from fits of equation 1 to the Ne-Al, Ar-Al, and Xe-Al *ab initio* potentials.
(Reprinted from ref 24. Copyright 1996 American Chemical Society.)

kcal/mol and 3.5 Å; Ar-Al, 0.2 kcal/mol and 4.4 Å; and Ar-Al$^+$, 1.1 kcal/mol and 3.6 Å. The Xe-Al *ab initio* potential is purely repulsive. In contrast to these *ab initio* results, one would expect a deeper potential well for Xe-Al than for Ne-Al. To determine accurate Rg-Al attractive potentials, higher-level treatments of electron correlation are required.

The well-depths which have been measured are with respect to diatom zero-point levels. The values in units of kcal/mol are: 0.38 (39), 0.39 ± 0.19 (40) and 0.47 [+0.11 and -0.03] (41) for Ar-Al; 0.52 ± 0.19 (40) for Kr-Al; 0.88 ± 0.17 (40) for Xe-Al; 2.81 ± 0.01 (42) for Ar-Al$^+$; and 4.37 ± 0.01 (42) for Kr-Al$^+$. The N$_2$-Al$^+$ experimental well depth is 1.33 kcal/mol (43). In future work, it would be of interest to use high-level *ab initio* calculations to derive Rg-Al potential curves that reproduce these experimental well depths as well as accurately representing the short-range repulsive interactions.

Acknowledgments

This research was funded by the National Science Foundation. M. A. Duncan is thanked for very insightful discussions regarding rare gas-Al two-body potentials.

Literature Cited

(1) Su, C.-X.; Armentrout, P.B. *J. Chem. Phys.* **1993**, *99*, 6506.

(2) Su, C.-X.; Hales, D. A.; Armentrout, P. B. *J. Chem. Phys.* **1993**, *99*, 6613.

(3) Hales, D. A.; Su, C.-X.; Lian, L.; Armentrout, P. B. *J. Chem. Phys.* **1994**, *100* 1049.

(4) Lian, L.; Su, C.-X.; Armentrout, P. B. *J. Chem. Phys.* **1992**, *97*, 4084.

(5) Lian, L.; Su, C.-X.; Armentrout, P. B. *J. Chem. Phys.* **1992**, *97*, 4072.

(6) Lian, L.; Su, C.-X.; Armentrout, P. B. *J. Chem. Phys.* **1992**, *96*, 7542.

(7) Hales, D. A.; Lian, L.; Armentrout, P. B. *Int. J. Mass. Spectrom. Ion Processes* **1990**, *102*, 269.

(8) Jarrold, M. F.; Bower, J. E.; Kraus, J. S. *J. Chem. Phys.* **1987**, *86*, 3876.

(9) Hanley, L.; Ruatta, S. A.; Anderson, S. L. *J. Chem. Phys.* **1987**, *87*, 260.

(10) Hanley, L.; Anderson, S. L. *J. Phys. Chem.* **1987**, *91*, 5161.

(11) Hanley, L.; Whitten, J. L.; Anderson, S. L. *J. Phys. Chem.* **1988**, *92*, 5803.

(12) Jarrold, M. F.; Bower, J. E. *J. Phys. Chem.* **1988**, *92*, 5702.

(13) Jarrold, M. F.; Honea, E. C. *J. Phys. Chem.* **1991**, *95*, 9181.

(14) von Helden, G.; Gotts, N. G.; Bowers, M. T. *Chem. Phys. Lett.* **1993**, *212*, 241.

(15) Hunter, J. M.; Fye, J. L.; Roskamp, E. J.; Jarrold, M. F. *J. Phys. Chem.* **1994**, *98*, 1810.

(16) Lifshitz, C.; Sandler, P.; Grutzmacher, H.-F.; Sun, J.; Weiske, T.; Schwarz, H. *J. Phys. Chem.* **1993**, *97*, 6592.

(17) Khan, F. A.; Clemmer, D. E.; Schultz, R. H.; Armentrout, P. B. *J. Phys. Chem.* **1993**, *97*, 7978.

(18) Sievers, M. R.; Armentrout, P. B. *J. Phys. Chem.* **1995**, *99*, 8135.

(19) More, M. B.; Glendening, E. D.; Ray, D.; Feller, D.; Armentrout, P. B. *??* **1996**, *100*, 1605.

(20) Bunker, D. L. *Methods Comput. Phys.* **1971**, *10*, 287.

(21) Porter, R. N.; Raff, L. M. In *Dynamics of Molecular Collisions;* Miller, W. H., Ed.; Plenum: New York, NY, 1976; Part B, Chapter 1.

(22) Raff, L. M.; Thompson, D. L. In *Theory of Chemical Reaction Dynamics*; Baer, M., Ed.; Chemical Rubber: Boca Raton, FL, 1985; Vol. III.

(23) de Sainte Claire, P.; Peslherbe, G. H.; Hase, W. L. *J. Phys. Chem.* **1995**, *99*, 8147.

(24) de Sainte Claire, P.; Hase, W. L. *J. Phys. Chem.* **1996**, *100*, 8190.

(25) Mahan, B. H. *J. Chem. Phys.* **1970**, *52*, 5221.

(26) Yardley, J. T. In *Introduction to Molecular Energy Transfer*, Academic Press: London, 1980; pp 95-129.

(27) Shin, H. K. In *Dynamics of Molecular Collisions*; Miller, W. H., Ed.; Plenum: New York, NY, 1976; Part A, Chapter 4.

(28) Levine, R. D.; Bernstein, R. B. In *Molecular Reactions Dynamics and Chemical Reactivity*; Oxford: New York, NY, 1987; pp 312-331.

(29) Rapp, D.; Kassal, T. *Chem. Rev.* **1969**, *69* 61.

(30) Secrest, D. *J. Chem. Phys.* **1969**, *51*, 421.

(31) Schulte, J.; Lucchese, R. R.; Marlow, W. H. *J. Chem. Phys.* **1993**, *99*, 1178.

(32) Grimmelmann, E. K.; Tully, J. C.; Cardillo, M. J. *J. Chem. Phys.* **1980**, *72*, 1039.

(33) Pettersson, L. G. M.; Bauschlicher, Jr., C. W.; Halicioglu, T. *J. Chem. Phys.* **1987**, *87*, 2205.

(34) Axilrod, B. M.; Teller, E. *J. Chem. Phys.* **1943**, *11*, 299.

(35) Hase, W. L.; Mondro, S. L.; Duchovic, R. J.; Hirst, D. M. *J. Am. Chem. Soc.* **1987**, *109*, 2916.

(36) Frisch, M. J.; Trucks, G. W.; Head-Gordon, M.; Gill, P. M. W.; Wong, M. W.; Foresman, J. B.; Johnson, B. G.; Schlegel, H. B.; Robb, M. A.; Resplogle, E. S.; Gomperts, R.; Andres, J. L.; Raghavachari, K.; Binkley, J. S.; Gonzales, C.; Martin, R. L.; Fox, D. J.; Defrees, D. J.; Baker, J.; Stewart, J. J. P.; Pople, J. A. *GAUSSIAN92*; Gaussian, Inc.: Pittsburgh, 1992.

(37) Hase, W. L.; Duchovic, R. J.; Hu, X.; Komornicki, A.; Lim, K. F.; Lu, D.-H.; Peslherbe, G. H.; Swamy, K. N.; Vande Linde, S. R.; Varandas, A.; Wang, H.; Wolf, R. J. *QCPE* **1996**, *16*, 671.

(38) Baer, T.; Hase, W. L. *Unimolecular Reaction Dynamics.   Theory and Experiments*; Oxford University Press: New York, NY, 1996.

(39) Gardner, J. M.; Lester, M. I. *Chem. Phys. Lett.* **1987**, *137*, 301.

(40) Callendar, C. L.; Mitchell, S. A.; Hackett, P. A. *J. Chem. Phys.* **1989**, *90* 5252.

(41) McQuaid, M. J.; Gole, J. L.; Heaven, M. C. *J. Chem. Phys.* **1990**, *92*, 2733.

(42) Heidecke, S. A.; Fu, Z.; Colt, J. R.; Morse, M. D. *J. Chem. Phys.* **1992**, *97*, 1692.

(43) Brock, L. R.; Duncan, M. A. *J. Phys. Chem.* **1995**, *99*, 16571.

Chapter 20

Collision-Induced Dissociation of Highly Excited NO_2 in the Gas Phase and on MgO (100) Surfaces

A. Sanov[1], D. W. Arnold[2], M. Korolik, H. Ferkel[3], C. R. Bieler[4], C. Capellos[5], C. Wittig, and Hanna Reisler

Department of Chemistry, University of Southern California, SSC 619, Los Angeles, CA 90089–0482

Collision-induced dissociation (CID) of NO_2 in highly excited mixed $^2A_1/^2B_2$ states is studied in crossed molecular beams at collision energies of ~2000 cm^{-1} and on crystalline MgO(100) at collision energies of ~2000 and 4400 cm^{-1}. The yield spectra obtained by scanning the excitation laser wavelength while monitoring NO fragments show features identical to those in the fluorescence excitation spectrum of NO_2, but the yield of CID decreases exponentially with the increase of the amount of energy required to reach the threshold for the monitored NO state. The results are discussed in terms of a mechanism in which highly excited NO_2 undergoes further activation by collisions, followed by unimolecular decomposition. The NO product spin-orbit excitations are sensitive to the chemical identity of the collider and bear the imprints of exit-channel interactions, which are more significant on the MgO(100) surface than in the gas-phase.

Studies of energy transfer are central to the understanding of unimolecular and bimolecular interactions. In collisional environments and at high temperatures, the high internal and/or translational energies that molecules acquire via collisions in the gas phase and with surfaces eventually lead to decomposition (1–3), but such collisions also lead to relaxation (4–6) and reactions (7–11). In the work described here, our approach has been to study the final decomposition step by exciting molecules to well defined internal and translational energies. By using molecular beams under single-collision conditions and state-selected laser detection of products,

[1]Present address: Joint Institute for Laboratory Astrophysics, University of Colorado, Boulder, CO 80309

[2]Present address: Sandia National Laboratories, Livermore, CA 94551

[3]Present address: Technische Universität, Clausthal, 38678 Clausthal-Zellerfeld, Germany

[4]Present address: Department of Chemistry, Albion College, Albion, MI 49224

[5]Permanent address: U.S. Army ARDEC, Picatinny Arsenal, AMSTA-AR-AEE, Dover, NJ 07801–5001

it is possible to investigate these processes with good energy and quantum state resolution.

We describe here our studies of CID of internally excited NO_2 (hereafter denoted NO_2^*) with atomic and molecular colliders and with crystalline surfaces. The studies involving activation by gas-phase colliders are compared to the results of Hartland and Dai who observed that in the *deactivating* collisions of NO_2^*, the average energy transferred per collision is much higher for triatomic colliders, compared with atomic and diatomic colliders (*12*). It is also known that the quenching rates of NO_2^* by molecular colliders such as NH_3 and H_2O are particularly high (*13, 14*).

An important issue concerns the influence of attractive forces and reactive pathways in collisions of highly excited molecules. For example, HCl has a permanent dipole moment which may enhance long-range interactions. Also, with HCl and NH_3, reactive channels producing HONO are energetically possible for the high excitation energies of NO_2^* used in this study. Would such attractive interactions manifest themselves in the CID channel? In reactions with surfaces, considerable energy transfer to the surface can slow down the departing molecules, which experience also long-range forces by the surface. Moreover, binding energies of the excited parent, as well as the fragments, to the surface may be significant. For example, will one or both of the fragments be captured (if only temporarily) by the surface? Alternatively, will the momentum of the rebounding parent succeed in carrying the products away from the surface, albeit with energy transfer in the exit channel? At the very least, a significant percentage of the NO product may experience forces during its escape that may alter its internal states.

The results obtained in this work show that the probability of CID is significant when NO_2 is excited to energies approaching its dissociation threshold (D_0), and that the average energy transferred per activating collision with polyatomic colliders is smaller than that obtained with the atomic and diatomic colliders studied so far. In addition, no participation of reactive pathways could be discerned, suggesting that the probability of inelastic scattering is much larger than that for reaction. Signatures of exit-channel interactions with the collider or the surface are revealed primarily in the spin-orbit population ratios of the NO fragment.

Experimental Methodologies

Gas-Phase Crossed Beams Experiments. The gas-phase experiments are performed in a crossed beams arrangement described in detail elsewhere (*15*). Briefly, it consists of a main collision chamber and two adjacent differentially pumped beam source chambers. A pulsed supersonic molecular beam of 2% NO_2 seeded in 1.5 atm of He or Ar carriers, and a beam of neat collider gas expanded under similar conditions are introduced into the collision chamber through skimmers. The rotational temperature of NO_2 in the beam (estimated from the rotational distribution of background $NO(^2\Pi_{1/2})$ always present in NO_2 samples) is ≤ 5 K. The relative arrangement of the molecular and laser beams in the collision chamber is shown schematically in Figure 1a. The molecular beams cross at 90° approximately 50 mm from the skimmers, creating an overlap region of ~ 1 cm^3. Number density estimates indicate that single-collision conditions prevail (*15*). The relative collision energies are evaluated under the assumption of fully expanded beams (*16*); these estimates are confirmed experimentally.

Jet-cooled NO_2 is excited into mixed $^2B_2/^2A_1$ molecular eigenstates using an excimer-laser pumped dye-laser system. It is crucial for laser excitation to precede collisions in order to minimize the probability of *photodissociating* collisionally - excited NO_2 molecules (*15*). Therefore, the excitation laser beam (15 ns; ~ 5 mJ; 396 – 414 nm) crosses the NO_2 beam 20-30 mm upstream from the collider beam overlap region (see Figure 1a). The frequency-doubled output from a second, similar laser system is used to probe NO produced in NO_2 CID. The probe beam (~ 226 nm;

15 ns; ~ 150 μJ) crosses the two molecular beams at the center of the overlap region. Both pump and probe laser beams are loosely focused with 1 m focal-length lenses. Since the laser excitation takes place outside the collision region, a pump-probe delay (typically 14-20 μs) is required to allow NO$_2^*$ to reach the collision/detection region. The long lifetime of NO$_2^*$ (~ 50 μs) (*17*) ensures that a significant fraction of molecules remain in the excited state upon arrival at the interaction region. NO is detected state-selectively by 1+1 (one-color) resonantly enhanced multiphoton ionization (REMPI) via the $A^2\Sigma^+ \leftarrow X^2\Pi$ transition using a microchannel plate detector located at the end of a time-of-flight mass spectrometer, whose flight-tube is mounted perpendicular to the plane of the molecular and laser beams.

Several competing processes and background signals must be accounted for in the data analysis, the most important arising from background NO in the NO$_2$ beam (*15*). Expansion-cooling ensures that only the lowest NO($^2\Pi_{1/2}$) rotational levels have significant populations; nonetheless, collisions excite background NO from low *J*'s into higher rotational states (*18, 19*), necessitating corrections to the monitored NO signal. Background signals of different origin are singled out by operating the collider nozzle and the pump laser in *on/off* sequences, and subtracted from the total signal on a shot-to-shot basis for each data acquisition cycle, thereby minimizing errors due to long-term drifts. Data processing also includes shot-to-shot normalization by the pump and probe pulse energies.

Beam-Surface Experiments. The surface CID experiments are carried out in a UHV apparatus that has been described in detail elsewhere (*20-22*). Pulsed supersonic expansion of an NO$_2$/O$_2$ mixture seeded in He or H$_2$ produces a molecular beam of cold NO$_2$ (T_{rot} < 10 K). The beam, generated in a source chamber, is skimmed and collimated before entering the UHV chamber (~ 10^{-10} Torr). In the UHV region, collisions occur with an *ex situ* cleaved MgO(100) crystal. The quality of the crystal surface is checked by He diffraction and Auger spectroscopy. Experiments conducted using two different NO$_2$ translational energies, E_{col} = 1940 cm^{-1} (θ_{inc} = 20° and 45°)(*23*) and 4400 cm^{-1} (θ_{inc} = 15°) are described.

The experimental arrangement is shown schematically in Figure 1b. NO$_2^*$ is prepared ~ 20 mm from the MgO crystal (before the collision) by laser excitation using an excimer-pumped dye laser (~ 400 nm) collimated to a ~ 2 mm line along the molecular beam. The NO fragments that result from dissociation are detected 3-5 mm from the surface via 1+1 (226 nm + 280 nm) REMPI through the $A^2\Sigma^+ \leftarrow X^2\Pi$ transition as described before (*20-22*). The probe beams are produced by a Nd:YAG pumped dye laser whose doubled output frequency (~ 280 nm) is Raman shifted to ~ 226 nm. After ionization, NO$^+$ is detected with a bare channeltron. The probe beams are propagated perpendicular to the scattering plane (Figure 1b). Angular distributions are investigated by measuring NO signal along an arc around the collision center. Other conditions and the treatment of spurious signals are similar to those in the gas-phase experiments.

Results

CID Yield Spectra. NO$_2$ CID yield spectra are obtained by scanning the excitation laser frequency $h\nu$ below D_0 of NO$_2$, while monitoring the production of NO in specific quantum states. Figure 2 shows the gas-phase CID yield spectra obtained in collisions of NO$_2^*$($E = h\nu$) with O$_2$, CO, CO$_2$, N$_2$O and HCl, respectively, by monitoring NO($^2\Pi_{1/2}$; v = 0; J = 5.5). In addition, the partial yield spectrum obtained in NO$_2^*$ + Ar CID by monitoring the same NO level is shown in Figure 3a. All CID yield spectra obtained with different colliders are rather similar, including those observed in collisions with MgO(100) at comparable E_{col} (not shown). Likewise, probing different NO product rotational states results in similar CID yield spectra. Each spectrum exhibits similar features throughout the wavelength region studied, the

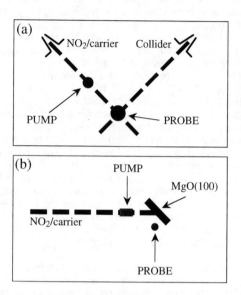

Figure 1. Schematics of the experimental arrangements used in the (a) gas-phase and (b) molecule-surface CID experiments. Sizes are not to scale.

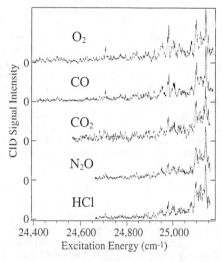

Figure 2. $NO(^2\Pi_{1/2}; v = 0; J = 5.5)$ yield from NO_2^* CID with indicated colliders as a function of NO_2^* excitation energy. D_0 of NO_2 corresponds to the position of the tallest peak in each curve at 25, 131 cm^{-1}.

only notable difference being the relative decrease in the CID signal intensity as the excitation laser energy is scanned away from D_0. The rate of decrease varies with the collider and the collision energy. For example, a more rapid decrease, compared to the spectrum shown in Figure 3a, was observed in NO$_2$ + Ar CID at smaller collision energy (E_{col} = 750 cm^{-1}).

A key observation in identifying the origin of the monitored NO signal is the structure in the yield spectra. For comparison, an NO$_2$ LIF spectrum taken in a companion molecular beam chamber under similar conditions is shown in Figure 3b. The similarities with the CID spectra shown in Figure 2 and Figure 3a are apparent. The observation that the CID spectra obtained by monitoring NO carry the fingerprints of the absorption spectrum of NO$_2$ is the primary indication that the observed signal is indeed due to photoexcitation of expansion-cooled NO$_2$ to levels *below D_0*, followed by CID.

An important difference between the NO$_2$ LIF and CID yield spectra, the scaling of their relative intensity with the excitation energy, reflects the decreasing efficacy of the energy transfer leading to CID as the amount of energy required to produce NO in the monitored level increases. The relative scaling of the CID and LIF spectra is reflected in Figure 3c, which shows a logarithmic plot of the point-by-point ratio of the CID yield spectrum in (a) and the LIF spectrum in (b). Similar plots were also obtained for CID with other colliders. The linear dependence (on average) of the $\ln(I_{CID}/I_{LIF})$ signals on the excitation energy in the investigated energy range signifies that the fraction of the NO$_2{}^*$ molecules that undergoes CID decreases exponentially with increasing energy deficiency, in accordance with the often observed exponential energy gap law for energy transfer (*24*). The sharp decrease in $\ln(I_{CID}/I_{LIF})$ with decreasing $h\nu$ just below D_0 (as seen for example in Figure 3c) arises in part from inelastic scattering of NO generated by photodissociation of rotationally excited NO$_2$. The contribution of this process is significant only near D_0, and this portion of the spectrum is ignored in the data analysis. Linear least square fits in the range ($D_0 - h\nu$) = 50 – 460 cm^{-1} are used to obtain quantitative estimates of the average energy $<\Delta E>$ transferred per activating collision; the values of $<\Delta E>$ for different gas-phase colliders are summarized in Table I.

Due to poor signal-to-noise ratio for the 1940 cm^{-1} data, a quantitative value for $<\Delta E>$ has not been extracted. For experiments at 4400 cm^{-1}, the transferred energy ($<\Delta E>$ = 100 cm^{-1}) is lower than would be expected in the gas phase at the same collision energy. Energy transfer to the bulk MgO makes the collision less hard-sphere-like. This will cause the $<\Delta E>$ value to be lower because not all of the collision energy is available for transfer into the NO$_2$ molecule.

NO Product State Distributions. NO product state distributions are obtained by fixing the excitation laser at a specific wavelength, below D_0, and scanning the probe laser over the NO $A \leftarrow X$ transition. Figure 4a shows a typical rotational distribution obtained in the NO$_2$/He + CO experiment by fixing the excitation laser on a prominent peak in the CID yield spectrum at 160 cm^{-1} below D_0. Note that for both spin-orbit states there appears to be no preference in the population of the Λ-doublet components. A Boltzmann plot of the distribution in Figure 4a is shown in (b). The distribution is Boltzmann-like, as are the distributions obtained with other colliders. In the NO$_2$ + Ar experiment, distributions measured at energies ranging from 60 to 440 cm^{-1} below D_0 are rather similar to each other. The rotational distributions for each NO spin-orbit state can be characterized by similar rotational temperatures, which are listed (in cm^{-1}) in Table I. Also summarized in Table I are the NO spin-orbit population ratios, i.e. [$^2\Pi_{3/2}$]:[$^2\Pi_{1/2}$], estimated from the product state distributions obtained in gas-phase CID. Note that in the crossed-beams experiment, increasing the NO$_2$ + Ar collision energy [e.g. NO$_2$(Ar seeded) + Ar *vs.* NO$_2$(He seeded) + Ar] leads to a significant increase in the CID product spin-orbit excitation. When He carrier is used for NO$_2$ and CID is effected by Ar, CO, or O$_2$, E_{col} does not

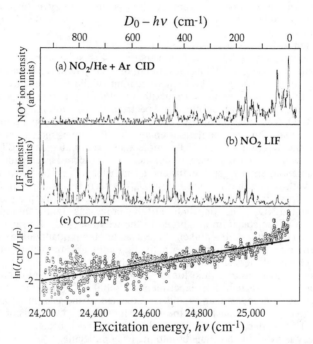

Figure 3. (a) NO($^2\Pi_{1/2}$; v = 0; J = 5.5) CID yield as a function of NO_2^* excitation energy $h\nu$ in the NO_2/He + Ar experiment. (b) Jet-cooled NO_2 LIF excitation spectrum. (c) Log plot of the point-by-point ratio of the CID and LIF spectra shown in (a) and (b), respectively. The top axis indicates energy below D_0.

Table I. E_{col} (cm⁻¹) for NO_2^* + M CID; $<\Delta E>$ (cm⁻¹) in CID versus relaxation; NO product rotational temperatures (cm⁻¹) and spin-orbit population ratios

M	E_{col}	$<\Delta E>$ in CID	$<\Delta E>$ in relaxation.[a]	$T_{rot}(^2\Pi_{1/2})$[b]	$T_{rot}(^2\Pi_{3/2})$[b]	$[^2\Pi_{3/2}]:[^2\Pi_{1/2}]$
Ar[c]	750	114 ± 5	—	103 ± 10	77 ± 10	0.11 ± 0.03
Ar[d]	2400	307 ± 20	278	256 ± 15	273 ± 25	0.18 ± 0.04
CO	2100	175 ± 8	560	183 ± 15	187 ± 15	0.23 ± 0.05
O_2	2200	207 ± 10	266	211 ± 15	173 ± 15	0.29 ± 0.06
CO_2	2600	196 ± 7	1130	—	—	—
N_2O	2600	158 ± 5	1600	—	—	—
HCl	2350	150 ± 5	—	—	—	—
NH_3	1850	135 ± 5	—	—	—	—

[a] From Ref. 12 and Hartland, G. V.; Dai, H.-L., personal communication.

[b] Characteristic rotational "temperatures" are determined from Boltzmann fits to the rotational distributions.

[c] NO_2 seeded in Ar.

[d] NO_2 seeded in He.

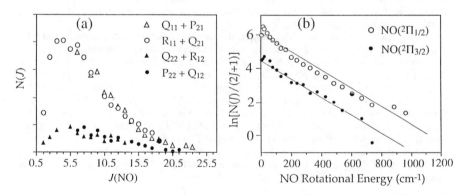

Figure 4. (a) $NO(^2\Pi_{1/2}; v = 0)$ rotational distribution measured in the NO_2^* + CO CID experiment at a fixed NO_2 excitation energy of $D_0 - h\nu = 160$ cm⁻¹. (b) Boltzmann plot of the NO rotational distribution shown in (a) with averaged contributions of the two Λ-doublet components.

vary markedly; nevertheless, the NO spin-orbit populations exhibit a marked dependence on the nature of the collider, but are colder than statistical.

NO spin-orbit ratios observed in the surface experiments vary as a function of the incident kinetic energy of the NO_2 molecules. At 1940 cm^{-1}, the rotational distributions are Boltzmann in nature and the spin orbit population ratio is nearly unity (23). On the other hand, at 4400 cm^{-1}, [$^2\Pi_{3/2}$]:[$^2\Pi_{1/2}$] ~ 0.5 and the NO rotational temperature is not well described by Boltzmann distribution, but rather appears bimodal.

NO Angular Distributions. In the gas-surface experiments, additional insight into the molecule-surface interactions can be gained by examining the angular dependence of the NO signal. While no marked variations in the CID yield spectra or the NO state distributions are observed while varying the detection angle around the specular angle, the NO signal intensities are largest near the specular angle, indicating that accommodation on the surface is not achieved.

Discussion

The CID Mechanism. The CID yield spectra obtained following gas-phase collisions or scattering from MgO(100) exhibit no state-specificity in the collisional energy transfer. Within our signal-to-noise level, the normalized CID yield spectra (i.e., the I_{CID}/I_{LIF} plots as a function of energy below D_0) are best described as unstructured, monotonically decaying exponentials (see Figure 3c). A phenomenological, limiting model was developed which was successful in describing the gas-phase observations. The model assumes that once collisional excitation of NO_2^* occurs, no further interactions with the collider exist. In other words, the activation step is impulsive and the collider plays a spectator role in the decomposition. The subsequent dissociation of collisionally excited $NO_2^{\ddagger}$ [where $\ddagger$ denotes collisionally activated states above D_0] can then be treated as a statistical unimolecular reaction. This mechanism is referred to as the unimolecular decomposition (UMD) model. However, the possibility of interaction with the collider in the exit channel cannot be ruled out *a priori* and may affect energy partitioning, as discussed below.

In the framework of the UMD model, the lack of structure in the normalized CID yield spectra is the result of averaging over all activated $NO_2^{\ddagger}$ states contributing to the population of the monitored NO level. Recall that collisions create a distribution $N_{NO2}(E^{\ddagger})$ of $NO_2^{\ddagger}$ levels above D_0, which is formally given by the population $N_{NO2}(h\nu)$ of NO_2^* prepared by laser excitation at energy $h\nu$, multiplied by the probability $P(h\nu, E^{\ddagger})$ of promoting the molecule from the energy hν to the energy $E^{\ddagger}$, which lies above D_0, in a single collision. We assume that each of these excited molecules decomposes statistically into a number of NO states, subject only to energy and angular momentum constraints (i.e., as per Phase Space Theory). Thus, unlike the photodissociation of NO_2 (25–27), every NO product state originates from a distribution of NO_2 excess energies, and the resulting NO population at specific energy E_{NO} can be obtained as

$$N(E_{NO}, h\nu) = C \cdot \int_{E_{NO}}^{h\nu + E_{col} - D_0} N_{NO2}(h\nu) \cdot P(h\nu, E^{\ddagger}) \cdot P(E^{\ddagger}, E_{NO}) \; dE^{\ddagger} , \qquad (1)$$

where C is a constant and $P(E^{\ddagger}, E_{NO})$ is the probability (calculated using Phase Space Theory) of forming a specific NO state.

Equation 1 describes in a formal way the major steps of the UMD model. The first term in the integrand describes the laser excitation step, the second term accounts for the collisional energy transfer probability, while the third reflects the subsequent unimolecular decay of the activated complex (under the implicit assumption of no post-collision collider–$NO_2^{\ddagger}$ interaction). The averaging over the ensemble of activated $NO_2^{\ddagger}$ molecules is described by integration in $E^{\ddagger}$ space. The only explicit assumption in the integral is that of statistical decomposition of the activated $NO_2^{\ddagger}$ complex. Checks were carried out to confirm that the laser excitation step can be well described by the LIF excitation spectrum of NO_2, i.e. $N_{NO2}(h\nu) \propto I_{LIF}(h\nu)$. Regarding the energy transfer probability $P(h\nu, E^{\ddagger})$, no definite conclusion can be drawn about its *state-to-state* behavior, because the distribution of collisionally excited $NO_2^{\ddagger}$ is not measured experimentally. The linearity, on average, of the $\ln(I_{CID}/I_{LIF})$ vs. $h\nu$ plots leads us to believe that the probability of collisional excitation decays exponentially with $\Delta E = E^{\ddagger} - h\nu$. Consider $P(h\nu, E^{\ddagger}) = c \bullet \exp(-\Delta E/\langle\Delta E\rangle)$, where c is a constant and $\langle\Delta E\rangle$ remains constant over the investigated energy range. Although $\langle\Delta E\rangle$ is defined at a fixed energy $h\nu$, experimentally the values of $\langle\Delta E\rangle$ are determined by changing $D_0 - h\nu$; this approach is justified since no significant variations of $\langle\Delta E\rangle$ with $h\nu$ are observed. We also point out that because of the integration in equation 1, state-to-state fluctuations, which are prominent in the photodissociation of NO_2 (25–27), are washed out, justifying the "smooth" exponential decay model of energy transfer (28). For the same reason, no fluctuations can be observed in the product state distributions.

Collisional Energy Transfer. Within the above assumptions, equation 1 describes both the linearity of the $\ln(I_{CID}/I_{LIF})$ vs. $h\nu$ plots and the Boltzmann-like NO rotational distributions when $\langle\Delta E\rangle \ll E_{col}$ (15). $\langle\Delta E\rangle$ is sensitive to the kinematic and microscopic properties of the collisional system. By comparing $\langle\Delta E\rangle$ for different colliders (Table I), we find that the efficiency of the activating collisions is largest for the atomic collider and smallest for polyatomic colliders. Although $\langle\Delta E\rangle$ depends also on E_{col}, this alone cannot explain the observed trend. Contrary to the activating collisions, studies of collisional relaxation of highly excited ($\sim 20,000$ cm^{-1}) NO_2^* by various colliders under thermal (300 K) conditions reveal that the NO_2^* average energy *loss* per collision is significantly larger with triatomic colliders than with atomic or diatomic colliders (see Table I) (12). For larger polyatomic colliders, this trend of larger deactivating values (SF$_6$, toluene) and smaller activating values (NH$_3$) continues.

The simplest explanation for the enhanced deactivation efficiency of molecular colliders is based on statistical considerations: as the number of degrees of freedom of the collider increases, deactivation channels other than V(NO$_2$) → T [i.e. V(NO$_2$) → V, R(collider)] become available, while the number of activation channels [T → V, R(NO$_2$)] remains the same, and the balance shifts to deactivation. The initial internal excitation of NO_2 ($\sim 20,000$ cm^{-1}) is larger than the collider vibrational quanta; thus, all internal degrees of freedom of the collider, including its vibration, may participate to some extent in the energy transfer (29, 30). In particular, with polyatomic colliders, rotational energy transfer may be quite efficient. There is strong evidence that translations and rotations track each other well in the relaxation of NO_2^* by triatomic colliders (31). The probability of energy transfer to vibrations of the collider is small but since vibrational frequencies are usually large, the collisions may affect the value of $\langle\Delta E\rangle$. In essence, molecular colliders look "softer" to NO_2^* due to their vibrational and rotational degrees of freedom, thereby affecting the efficiency of the T → V, R(NO$_2$) energy transfer. The "stiffness" of the collider has also been found important in previous surface CID experiments, where in high energy collisions (≤ 7 eV) CID of C$_3$F$_7$NO on a hard surface [MgO(100)] was more efficient than on "softer" surfaces, such as GaAs and Ag, which have lower Debye temperatures (20–22). Long-range chemical interactions may also enhance energy flow to the collider, and in such cases energy will be transferred primarily to the deactivation channels. Note however, that direct comparisons between the beam experiments and the thermal relaxation measurements of NO_2^* are not possible because of the significant difference in the experimental conditions (beam vs. bulk) and collision energies. For example, at 300 K (i.e. average collision energy of ~ 200 cm^{-1}), long-range mechanisms such as dipole-dipole couplings can be more important than at the high E_{col} of our CID studies where impulsive collisions should predominate (*30*).

Effect of Exit-Channel Interactions. Exit-channel interactions of the activated $NO_2^{\ddagger}$ with the collider may influence the outcome of CID. These interactions will not change the CID yield spectra significantly, since they do not affect the collisional excitation step. However, the partitioning of excess energy among the NO + O products and the collider can be affected.

The role of post-collision interactions in the gas-phase CID can be illustrated by the following rough estimates. Suppose an NO_2 molecule is collisionally excited to an energy level just above D_0. From separate studies (*32*), it is known that an isolated NO_2 molecule excited just above D_0 decomposes with a rate of ~ 0.2 × 10^{12} s^{-1}. The activated $NO_2^{\ddagger}$ molecule travels away from the collider with a speed of 15 Å/ps. This corresponds to a characteristic $NO_2^{\ddagger}$-collider separation of tens of Ångstroms during dissociation. Under these conditions, post-collision interactions are expected to be of minor importance. On the other hand, for those molecules with $E^{\ddagger}$ ~ 1000 cm^{-1}, the speed of separation with the collider will be reduced to ~ 10 Å/ps, while the dissociation rate will increase up to 2 × 10^{12} s^{-1} (*32*), yielding a characteristic $NO_2^{\ddagger}$-collider separation during dissociation of ~ 5 Å; thus, a significant fraction of these molecules will dissociate in the vicinity of the collider.

The NO rotational temperatures obtained by CID with different colliders are not significantly different and can be rationalized within the UMD model (*15*). On the other hand, the NO spin-orbit excitations exhibit a marked dependence on the chemical identity of the collider (see Table I), which cannot be explained by the UMD model. From the vast amount of data gathered in studies of photoinitiated NO_2 unimolecular decomposition in the gas phase, one signature emerges as particularly robust: the $[^2\Pi_{3/2}]/[^2\Pi_{1/2}]$ ratio of the NO product is always substantially smaller than the statistical value (*25-27*). This propensity is also present in the case of gas phase CID with small collision partners such as Ar, and is in qualitative agreement with our picture of this process as being essentially serial; i.e., the collision first energizes NO_2, which then decomposes. On the other hand, in the case of chemically active colliders, e.g., CO and O_2, the interaction with the departing CID products, NO and O, is likely to be stronger and may affect the energy partitioning in the exit channel, resulting in more equilibrated NO spin-orbit populations. By analogy with NO_2 photodissociation, it is suggested that in the decomposition of $NO_2^{\ddagger}$ rotational populations are fixed first, while the product spin-orbit ratio is determined at larger O–NO separation (*26*), and is therefore more sensitive to exit-channel interactions. Support for stronger exit-channel interactions with chemically active colliders can also be found in the results of inelastic scattering of NO, where the general trend in NO spin-orbit excitations is the same as in CID; the ratios are lowest for scattering by Ar, and significantly higher for collisions with O_2, CO (*19*) and with a MgO(100) surface (*32*).

In the CID of NO_2^* colliding with a crystalline MgO(100) surface, exit-channel interactions are more complex than in the case of gas phase collisions. Molecule-surface binding energies of the parent, as well as its fragments, are significant, and the surface-molecule forces are longer-range than are the weak attractive forces encountered in the gas-phase experiments. Thus, one cannot assume *a priori* that unimolecular decomposition proceeds in a way that parallels gas phase CID. However, by comparing the results of experiments performed at two different collision energies, we find that exit channel interactions influence the product state distributions of the NO products resulting from CID on MgO(100).

Suppose molecules with 2000 cm^{-1} of translational energy and internal energies near D_0 gain an additional 500 cm^{-1} by colliding with the MgO surface. This is reasonable when considering factors such as typical collision energies and the stiffness of the MgO crystal. The corresponding decomposition rate is ~ 10^{12} s^{-1} (*31*). Assuming that the velocity of the rebounding NO$_2$ molecules traveling away from the surface following T→V(NO$_2$) energy transfer is ~ 5 Å/ps, a large fraction of the dissociation events will occur near the surface, i.e., 50% dissociate within ~ 3.5 Å. However, at this distance there will be an attractive interaction between the fragments and the surface that can alter their internal energy distributions. For example, one or both of the fragments may be captured (if only temporarily) by the surface. Alternatively, the momentum of the rebounding parent may be sufficient to carry the products away from the surface, albeit with energy transfer in the exit channel.

In a reasonable scenario, at least some of the NO products are attracted to the surface, since the binding energy of NO on MgO(100) is ~ 0.2 eV and some of the energized NO$_2$ molecules will experience large O–NO bond extensions on a subpicosecond time scale, i.e., while the rebounding NO$_2$ is still close to the surface. In separate studies it was found that when expansion-cooled NO scatters from MgO(100), the [$^2\Pi_{3/2}$] / [$^2\Pi_{1/2}$] ratio is near unity (i.e., the statistical value) (*33*) which is similar to the corresponding value obtained in NO$_2$ CID on MgO(100) at 1940 cm^{-1} of translational energy. This sharply contrasts the NO product which derives from decomposition of isolated NO$_2$ (*25–27*). However, at the higher collision energy, we find a spin-orbit population ratio of ~ 0.5 which is closer to the value obtained in the collision-free decomposition of NO$_2$ and indicative of a faster escape from the surface after the collision. A smaller fraction of the NO fragments is generated near the surface, and thus the spin-orbit scrambling is less important. Thus, the extent of secondary interactions with the surface depends upon the incident energy of NO$_2$* which affects the molecule-surface interaction times during the decomposition.

The peaking of the NO product near the specular direction means that the detected NO molecules retain some memory of the momentum of the incident NO$_2$; i.e., they are not the consequence of trapping followed by desorption. Thus, all of the evidence collected to date is consistent with a dissociation mechanism in which NO$_2$ gains energy in the collision and some percentage of it decomposes before it can free itself from the environs of the surface.

Participation of Reactive Pathways. Finally, no indication of the existence of reactive pathways in collisions of NO$_2$* has been obtained despite the possible reactivity of some of the colliders (e.g., CO and O$_2$ yielding NO as a product, HCl and NH$_3$ giving HONO). For example, in the case of NO$_2$* + O$_2$ ↔ NO + O$_3$ the reverse reaction is known to produce NO$_2$* (*34, 35*), and thus one might expect vibrational excitation in NO$_2$ to enhance reactivity. Our inability to observe effects due to reaction may be due to the general consideration that inelastic scattering may have fewer geometrical constraints compared to an approach leading to reaction. It should also be recognized that not all vibrational motions are equally effective in coupling to the reaction coordinate and promoting reactivity. In this regard, a diatomic reactant such as vibrationally excited O$_2$ (which has only one vibrational mode) may exhibit a higher probability for reaction than a polyatomic collider with a comparable level of excitation.

Summary

Gas-phase CID of highly excited NO$_2$* has been studied with different gas-phase colliders as well as a MgO(100) surface, at well-defined hyperthermal collision energies. When the internal energies of the incident NO$_2$ lie just below D_0, the probability of dissociation is high, as expected, since the probabilities are comparable for adding energy to, or removing it from, the incident NO_2. The probability of CID decreases exponentially with the amount of energy needed to reach dissociation threshold.

CID can be described as a two-step process: collisional activation, creating a distribution of $NO_2^{\ddagger}$ above D_0, followed by unimolecular decomposition to yield NO + O. In the gas phase, the collider plays mostly a spectator role, and a statistical model based on the assumption that the fractional population of $NO_2^{\ddagger}$ above D_0 decreases exponentially with increasing energy reproduces the experimental observations.

The average energy transferred per activating collision decreases with the level of complexity of the collider. It is possible that the greater efficacy of T → V, R energy transfer involving internal degrees of freedom of the molecular colliders competes with conversion of translational energy to NO_2 internal energy.

Regarding the molecule-surface CID results, it appears that NO_2 is different than larger polyatomics, which are known to travel far from the surface before decomposition occurs. Depending upon the incident translational energy, (and thus the velocity of the departing activated $NO_2^{\ddagger}$ and/or NO fragments), exit-channel interactions with the surface result in variable NO spin-orbit state distributions.

Finally, in all cases, the CID yield spectra are very similar to the spectrum observed with inert Ar as the collider, and no signatures of chemical reactions are discerned. CID is thus the dominant channel at the collision and excitation energies employed.

It is noteworthy that the family of small polyatomics (3 or 4 atoms) undergoing potentially reactive molecule-surface and molecule-molecule scattering includes many technically important systems. Thus, we anticipate that the issues which have emerged in the present study have relevance in a large number of systems.

Acknowledgments

The authors wish to thank the U.S. Army Research Office, the National Science Foundation and Air Force Office of Scientific Research for supporting this research. We thank the other participants in these experiments, Martin Hunter, Lori Hodgson, and James Singleton for their significant contributions to this research and Professor George Flynn for his valuable comments.

Literature Cited

(1)   Forst, W. *Theory of Unimolecular Reactions*; Academic: New York, NY, 1973.

(2)   Robinson, P. J.; Holbrook, K. A. *Unimolecular Reactions*; Wiley: New York, NY, 1972.

(3)   Gilbert, R. G.; Smith, S. C. *Theory of Unimolecular and Recombination Reactions*; Blackwell Scientific Publ.: Oxford, UK, 1990.

(4)   Hippler, H.; Troe, J. In *Bimolecular Collisions*; Bagott, J. E.; Ashfold, M. N. R. Eds.; The Royal Society of Chemistry: London, UK, 1989.

(5)   Flynn, G. W.; Weston, R. E., Jr. *J. Phys. Chem.* **1993**, *97*, 8116.

(6)   Yardley, J. T. *Introduction to Molecular Energy Transfer*; Academic Press: New York, NY, 1980.

(7)   Tardy, D. C.; Rabinovitch, B. S. *Chem. Rev.* **1977**, *77*, 396.

(8)   Metz, R. B.; Pfeiffer, J. M.; Thoemke, J. D.; Crim, F. F. *Chem. Phys. Lett.* **1994**, *221*, 347.

(9) Metz, R. B.; Thoemke, J. D.; Pfeiffer, J. M.; Crim, F. F., *J. Chem. Phys.* **1993**, *99*, 1744.

(10) Sinha, A.; Thoemke, J. D.; Crim, F. F., *J. Chem. Phys.* **1992**, *96*, 372.

(11) Sinha, A.; Hsiao, M. C.; Crim, F. F., *J. Chem. Phys.* **1991**, *94*, 4928.

(12) Hartland, G. V.; Qin, D.; Dai, H.-L. *J. Chem. Phys.* **1995**, *102*, 8677.

(13) Donnelly, V. M.; Keil, D. G.; Kaufman, F. *J. Chem. Phys.* **1979**, *71*, 659.

(14) Capellos, C. *Proceedings of the Army Science Conference*; Orlando, FL, 1994; Assistant Secretary of the Army, Office of the Director of Research, Development and Acquisition, Contract number DAAH04-93-C10048: vol. 1, pp 51-58.

(15) Sanov, A.; Bieler, C. R.; Reisler, H. *J. Phys. Chem.* **1995**, *99*, 7339.

(16) Kolodney, E.; Amirav, A. *Chem. Phys.* **1983**, *82*, 269.

(17) Patten, Jr., K. O.; Burley, J. D.; Johnston, H. S. *J. Phys. Chem.* **1990**, *94*, 7960.

(18) Joswig, H.; Andresen, P.; Schinke, R. *J. Chem. Phys.* **1986**, *85*, 1904.

(19) Bieler, C. R.; Sanov, A.; Reisler, H. *Chem. Phys. Lett*. **1995**, *235*, 175.

(20) Kolodney, E.; Baugh, D.; Powers, P. S.; Reisler, H.; Wittig, C. *J. Chem. Phys.* **1989**, *90*, 3883.

(21) Kolodney, E.; Powers, P. S.; Hodgson, L.; Reisler, H.; Wittig, C. *J. Chem. Phys.* **1991**, *94*, 2330.

(22) Powers, P. S.; Kolodney, E.; Hodgson, L.; Ziegler, G.; Reisler, H.; Wittig, C. *J. Phys. Chem.* **1991**, *95*, 8387.

(23) Ferkel, H.; Singleton, J. T.; Reisler, H.; Wittig, C. *Chem. Phys. Lett.* **1994**, *221*, 447.

(24) Troe, J. *J. Chem. Phys.* **1977**, *66*, 4745.

(25) Hunter, M.; Reid, S. A.; Robie, D. C.; Reisler, H. *J. Chem. Phys.* **1993**, *99*, 1093.

(26) Reid, S. A.; Robie, D. C.; Reisler, H. *J. Chem. Phys.* **1994**, *100*, 4256.

(27) Reid, S. A.; Reisler, H. *J. Phys. Chem.* **1996**, *100*, 474, and references therein.

(28) Sanov, A. *Ph.D. Thesis*; University of Southern California, Los Angeles, CA, 1996.

(29) Chou, J. Z.; Hewitt, S. A.; Hershberger, J. F.; Flynn, G. W. *J. Chem. Phys.* **1990**, *93*, 8474.

(30) Hartland, G. V.; Qin, D.; Dai, H.-L. *J. Chem. Phys.* **1994**, *101*, 8554.

(31) Zheng, L.; Chou, J. Z.; Flynn, G. W. *J. Phys. Chem.* **1991**, *95*, 6759.

(32) Ionov, S. I.; Brucker, G. A.; Jaques, C.; Chen, Y.; Wittig, C. *J. Chem. Phys.* **1993**, *99*, 3420.

(33) Kolodney, E.; Baugh, D.; Powers, P. S.; Reisler, H.; Wittig, C. *Chem. Phys. Lett.* **1988**, *145*, 177.

(34) Kahler, C. C.; Ansell, E.; Upshur, C. M.; Green, Jr., W. H. *J. Chem. Phys.* **1984**, *80*, 3644.

(35) Redpath, A. E.; Menzinger, M.; Carrington, T. *Chem. Phys.* **1978**, *27*, 409.

Chapter 21

Direct Calculation of Reactive Thermal Rate Constants

Ward H. Thompson[1]

Department of Chemistry, University of California, and Chemical Sciences Division, Lawrence Berkeley National Laboratory, Berkeley, CA 94720

The thermal rate constant for a chemical reaction can be obtained *directly* (*i.e.*, without first obtaining state-selected or energy-dependent quantities) as the time integral of the flux-flux autocorrelation function, $C_{ff}(t)$. The correlation function, $C_{ff}(t)$, can be calculated efficiently by taking advantage of the low rank of the Boltzmannized flux operator. In addition, including absorbing potentials in the time propagation eliminates unphysical reflection from the edge of the finite basis; the basis can then be drawn in to represent a smaller area around the interaction region. Application of this method to the D + H_2 reaction is presented.

The last decade has seen great progress in theoretical methods for treating chemical reaction dynamics. Full quantum scattering calculations are now routinely carried out for numerous three atom systems (atom-diatom collisions). Indeed, full-dimensionality scattering calculations have been carried out for some four atom systems[1-4] (most notably the OH + H_2 reaction). However, such large systems still present a major challenge and methods are needed to make reactions involving many atoms (> 3) more amenable to theoretical calculations. To this end, our research group has worked on *directly* calculating the dynamical quantity of interest. While the S-matrix elements $(S_{n_p,n_r}(E))$ provide the most detailed information obtainable for a chemical reaction (resolution on the full state-to-state and amplitude level), frequently one is not interested in this extreme amount of detail. Often more averaged quantities are desired, such as the initial state-selected reaction probability,

[1]Current address: Department of Chemistry and Biochemistry, University of Colorado, Boulder, CO 80309

$$P_{n_r}(E) = \sum_{n_p} |S_{n_p,n_r}(E)|^2, \tag{1}$$

the cumulative reaction probability,

$$N(E) = \sum_{n_p,n_r} |S_{n_p,n_r}(E)|^2, \tag{2}$$

or the thermal rate constant,

$$k(T) = \frac{1}{2\pi\hbar Q_r(T)} \int_0^\infty e^{-\beta E} \sum_{n_p,n_r} |S_{n_p,n_r}(E)|^2 \, dE, \tag{3}$$

where $Q_r(T)$ is the reactant partition function and $\beta = 1/k_b T$. Significant progress has recently been achieved in the calculation of initial state-selected[5-7] and cumulative reaction probabilities[5,8-10] *directly* (*i.e.*, without first obtaining the state-to-state S-matrix elements), yet correctly (without introducing approximations). Methods for directly obtaining the thermal rate constant have existed for over 35 years[11] and have been the focus of much attention for the last two decades.[12-28] This paper describes what we believe is the latest (and best) technology for the direct (but correct) calculation of the thermal rate constant for a chemical reaction.

The exact thermal rate constant for a system obeying classical mechanics is given by

$$k(T) = \frac{1}{Q_r(T)h^F} \int d\mathbf{p} \int d\mathbf{q} \, e^{-\beta H(\mathbf{p},\mathbf{q})} \delta[f(\mathbf{q})] \frac{df(\mathbf{q})}{d\mathbf{q}} \cdot \frac{\mathbf{p}}{m} \chi(\mathbf{p},\mathbf{q}), \tag{4}$$

where $\mathbf{p}$ and $\mathbf{q}$ are the momenta and coordinates of the classical system of F degrees of freedom governed by the Hamiltonian $H(\mathbf{p},\mathbf{q})$. The equation $f(\mathbf{q}) = 0$ defines a dividing surface which separates reactants and products. $\chi(\mathbf{p},\mathbf{q})$ is known as the *characteristic function* and is equal to 1 for reactive trajectories and 0 for nonreactive trajectories. Practically, one runs a trajectory with initial conditions $\mathbf{p}$ and $\mathbf{q}$ (which must lie on the dividing surface by virtue of the $\delta[f(\mathbf{q})]$ factor) backward in time to determine if it started as reactants in the infinite past; starting with the same initial conditions the trajectory is run forward in time to determine if it ends up as products in the infinite future. (Here "infinite" refers to times long enough to ensure the trajectory will not return to the interaction region.) A trajectory which positively satisfies these two criteria is reactive. This is the point of departure for various dynamical approximations, most notably transition state theory[29-32] (TST). In TST the characteristic function is assumed to be given by $\chi(\mathbf{p},\mathbf{q}) = h(p_s)$ where $h(x)$ is the Heaviside step function equal to 1 if $x > 0$ and 0 if $x < 0$, and p_s is the momentum perpendicular to the dividing surface $f(\mathbf{q}) = 0$. Thus, trajectories crossing the dividing surface are assumed never to come back – there is *no recrossing* of the dividing surface.[30] Thus TST provides an upper bound to the classical rate constant and a variational principle exists in which the best dividing surface is that which gives the lowest rate (*i.e.*, the fewest recrossing trajectories).

Transition state theory is the most powerful model in chemical reaction rate theory. However, there is no unique *quantum mechanical* transition state theory. Part of the difficulty is that the fundamental TST assumption is stated in terms of (recrossing) trajectories; this cannot be straightforwardly interpreted in the context of quantum mechanics. In addition, a strict, meaningful upper bound to the exact quantum rate constant has never been provided by any quantum mechanical version. Many quite useful quantum mechanical transition state theories have been proposed, however. (One might more appropriately call them analogues to or generalizations of classical transition state theory.) All these approaches invoke, in some way, the spirit of TST: reactivity is determined at the transition state (or in the case of quantum mechanics and the uncertainty principle, in a small area around the transition state).

This paper presents a quantum mechanical method for obtaining rate constants which is evocative of this spirit. However, the goal here is to obtain the exact rate constant and therefore the current approach is more akin to the method described above for calculating exact classical rate constants. That is, if one is interested in obtaining the exact classical rate, a straightfoward approach would be to start trajectories with initial conditions as reactants and run them forward in time to see if they react. With the wisdom of TST in hand, we would follow the procedure described above and trajectories would only have to be followed for short(er) times (in practice) forward and backward. In addition, trajectories which never reach the transition state, and are thus destined not to react, would not be followed at all. Similarly, if the quantum mechanical rate constant is desired, it can be obtained by calculating all the reactive S-matrix elements and then averaging away all that detail as indicated in Eq. (4). However, the lesson of TST would tell us that a simpler, direct route to the rate constant would be to somehow measure the net reactive flux through the transition state.

Exactly such an approach is provided by flux correlation functions. Miller[12] has shown that the quantum mechanical rate constant is given by

$$k(T) = Q_r(T)^{-1} \operatorname{tr}[e^{-\beta \hat{H}} \hat{F} \hat{\wp}] \tag{5}$$

where $\hat{H}$ is the Hamiltonian of the system and $\hat{F}$ the symmetrized flux operator. Also, $\hat{\wp}$ is the quantum mechanical analogue to the characteristic function χ; where χ can be thought of as a projector onto reactive trajectories, $\hat{\wp}$ is the projection operator onto the reactive part of Hilbert space. (Thus $\hat{F}\hat{\wp}$ can be regarded as the reactive flux operator.) $\hat{\wp}$ can be written in numerous formally equivalent ways[12,13] which have different (and interesting) practical properties. For the present purpose we are interested in the form

$$\hat{\wp} = \lim_{t \to \infty} e^{i\hat{H}t/\hbar} h(\hat{s}) e^{-i\hat{H}t/\hbar}, \tag{6}$$

where here, for simplicity of notation, $s = 0$ defines the dividing surface. This gives the rate constant as

$$k(T) = Q_r(T)^{-1} \lim_{t \to \infty} \text{tr}[e^{-\beta \hat{H}/2} \hat{F} e^{-\beta \hat{H}/2} e^{i\hat{H}t/\hbar} h(\hat{s}) e^{-i\hat{H}t/\hbar}]$$
$$= Q_r(T)^{-1} \lim_{t \to \infty} C_{fs}(t), \tag{7}$$

where $C_{fs}(t)$ is the flux-side (sometimes flux-step) correlation function. Note that $\hat{H}$ commutes with $\hat{\rho}$ so any division of the Boltzmann operator can be used;[12] here we choose a symmetric apportionment for reasons that will be evident later. It has recently been shown how $C_{fs}(t)$ can be efficiently evaluated.[26,27] If this expression for the rate constant is manipulated by writing it as the integral of the derivative of the correlation function, one obtains[13]

$$k(T) = Q_r(T)^{-1} \int_0^\infty C_{ff}(t) \, dt, \tag{8}$$

where

$$C_{ff}(t) = \text{tr}[e^{-\beta \hat{H}/2} \hat{F} e^{-\beta \hat{H}/2} e^{i\hat{H}t/\hbar} \hat{F} e^{-i\hat{H}t/\hbar}], \tag{9}$$

is the flux-flux autocorrelation function. We have used the fact that the flux operator can be written as

$$e^{i\hat{H}t/\hbar} \hat{F} e^{-i\hat{H}t/\hbar} = e^{i\hat{H}t/\hbar} \frac{i}{\hbar} [\hat{H}, h(\hat{s})] e^{-i\hat{H}t/\hbar} = \frac{d}{dt} \left\{ e^{i\hat{H}t/\hbar} h(\hat{s}) e^{-i\hat{H}t/\hbar} \right\}. \tag{10}$$

The rate constant as the integral of the flux-flux autocorrelation function was first derived (in a different way) by Yamamoto in 1960.[11]

Section I reviews how the flux-flux correlation function can be obtained very efficiently by taking advantage of the low rank of the flux operator and by employing absorbing potentials in the real time propagation. Additionally, the relation of this approach to TST is discussed. Section II describes some of the details involved in carrying out a practical calculation, in this case for atom plus diatom scattering. Section III presents results for the D + H_2 reaction and Section IV concludes.

I. SUMMARY OF THEORY

The evaluation of the quantum mechanical trace to obtain $C_{ff}(t)$ is the computational bottleneck. This is because to obtain a matrix element of the operator inside the trace, propagation must be carried out in both real time and imaginary time (the Boltzmann operator). However, the trace can be calculated efficiently by evaluating it in the optimum basis (since the trace is independent of the basis in which it is computed). In the case of the flux-flux correlation function the operator inside the trace is of low rank. It has previously been shown that if the flux operator in one-dimension is diagonalized in a finite basis, there will be only two nonzero eigenvalues.[19,20,33] These two eigenvalues are of equal magnitude but opposite sign, corresponding to motion forward and backward across the dividing surface. In a multidimensional case the flux operator is *not* of low rank as this pair of positive and negative eigenvalues will be repeated for every state in the perpendicular degrees of freedom (the degrees of freedom that are parallel to the dividing surface). However, the Boltzmannized flux operator

$$\hat{F}(\beta) = e^{-\beta\hat{H}/2}\,\hat{F}\,e^{-\beta\hat{H}/2}, \tag{11}$$

is of low rank in many dimensions. This is because the Boltzmann factor restricts the number of states accessible in the perpendicular degrees of freedom (though this number naturally rises with temperature). Thus, the rank of this operator will be approximately equal to twice the number of states in the perpendicular degrees of freedom that are significantly populated thermally at temperature T. If the dividing surface is placed at the saddle point then this can be thought of as twice the number of states of the activated complex. (A slightly different version of the Boltzmannized flux operator was first defined by Park and Light.[20])

If we could somehow obtain the eigenvaules $\{f_n\}$ and eigenfunctions $\{|n\rangle\}$ of the Boltzmannized flux operator,

$$\hat{F}(\beta)|n\rangle = e^{-\beta\hat{H}/2}\,\hat{F}\,e^{-\beta\hat{H}/2}|n\rangle = f_n|n\rangle, \tag{12}$$

this would constitute the ideal basis in which to evaluate the trace in Eq. (9). In fact, there is a method which can do just that: the iterative Lanczos algorithm[34] obtains only the largest (in absolute value) eigenvalues (and the corresponding eigenfunctions) of an operator. The procedure is to first choose a random vector $|v_0\rangle$. (We continue to use bra-ket notation, though this procedure is carried out within a finite basis set. In a such a representation, the states become vectors and operators become matrices.) A Krylov basis is formed by repeated application of $\hat{F}(\beta)$ onto $|v_0\rangle$:

$$|v_1\rangle = \hat{F}(\beta)\,|v_0\rangle + S.O.$$
$$|v_2\rangle = \hat{F}(\beta)\,|v_1\rangle + S.O. \tag{13}$$
$$\vdots$$
$$|v_{M-1}\rangle = \hat{F}(\beta)\,|v_{M-2}\rangle + S.O. \tag{14}$$

where *S.O.* implies Schmidt orthogonalization to all previous vectors. In this process the matrix of $\hat{F}(\beta)$ in the Krylov basis $\{|v_0\rangle, |v_1\rangle, \ldots, |v_{M-1}\rangle\}$ is automatically obtained. Diagonalizing this Krylov space version gives the M largest (in absolute value) eigenvalues $\{f_n\}_{n=1}^{M}$ and eigenvectors $\{|n\rangle\}_{n=1}^{M}$ of $\hat{F}(\beta)$ in the full basis.

Once these eigenvalues and eigenvectors are in hand, the rate constant is given by

$$k(T) = Q_r(T)^{-1} \sum_{n=1}^{M} f_n \int_0^\infty \langle n| \, e^{i\hat{H}t/\hbar} \, \hat{F} \, e^{-i\hat{H}t/\hbar} \, |n\rangle \, dt. \tag{15}$$

Thus the cost of calculating the rate constant is propagation in imaginary time (applying $\hat{F}(\beta)$) M-1 times, where M is the rank of $\hat{F}(\beta)$, plus propagation in real time of the resulting M eigenfunctions. Note that in general M is much, much smaller than the total size of the basis since it is roughly equal to twice the number of thermally accessible states of the activated complex.

In carrying out the real time propagation, it is advantageous to introduce an imaginary absorbing potential:[35-40]

$$\hat{H} \to \hat{H} - i\hat{\epsilon}, \tag{16}$$

where $\hat{\epsilon} = \epsilon(\mathbf{q})$ is zero in the interaction region and "turns on" in the reactant and product valleys. By including ϵ the outgoing flux is absorbed before it can undergo unphysical reflection from the edge of the finite basis set. The absorbing potential actually has two desirable consequences: (1) Its inclusion guarantees that the flux-flux correlation function will go to zero at long times. And, (2) it allows the edge of the finite basis to be moved in closer to the interaction region in the reactant and product valleys, thus reducing the size of the basis. Absorbing potentials have previously been used by Brown and Light[22] in calculating the flux-flux autocorrelation function.

The rate constant now becomes

$$k(T) = Q_r(T)^{-1} \sum_{n=1}^{M} f_n \int_0^\infty dt \, \langle n| \, e^{i(\hat{H}+i\hat{\epsilon})t/\hbar} \, \hat{F} \, e^{-i(\hat{H}-i\hat{\epsilon})t/\hbar} \, |n\rangle. \tag{17}$$

Since the rate constant is given by the time integral of the flux-flux correlation function, the time propagator in Eq. (17) should be carried out with a method which gives the result at all intermediate times for the same effort as propagating to one long time. The split operator propagation scheme satisfies this requirement. For a small time, Δt, the propagator is approximated as

$$e^{-i(\hat{H}-i\hat{\epsilon})\Delta t/\hbar} \simeq e^{-i\hat{T}\Delta t/2\hbar} \, e^{-i(\hat{V}-i\hat{\epsilon})\Delta t/\hbar} \, e^{-i\hat{T}\Delta t/2\hbar}. \tag{18}$$

(See Section II for the precise form of the split operator propagator used in the present applications.) Thus, an eigenstate $|n\rangle$ is propagated in steps of Δt by successive application of this split operator:

$$|n(t + \Delta t)\rangle = e^{-i\hat{T}\Delta t/2\hbar} \, e^{-i(\hat{V}-i\hat{\epsilon})\Delta t/\hbar} \, e^{-i\hat{T}\Delta t/2\hbar} \, |n(t)\rangle. \tag{19}$$

Note that the $\langle n(t)|$ is given by the complex conjugate of $|n(t)\rangle$ so it is only necessary to propagate the ket state. The correlation function is thereby obtained at all the intermediate times (needed to perform the integral) for the same cost as finding it at the largest time. (The same propagation scheme is used in imaginary time for the Boltzmann operator.) Each eigenstate of the Boltzmannized flux operator is propagated and its contribution to the flux-flux correlation function, and hence the rate, is obtained. Writing out the expression for the rate constant explicitly with some integration scheme for the time integral gives

$$k(T) = Q_r(T)^{-1} \sum_{n=1}^{M} f_n \sum_{\ell=0} w_\ell \, \langle n(\ell\Delta t)| \, \hat{F} \, |n(\ell\Delta t)\rangle, \qquad (20)$$

where the w_ℓ are the weights in the time integration scheme.

II. DETAILS OF THE CALCULATION

Here we outline the procedure for the calculation of the rate constant for the D + H$_2$ reaction. We have used the reactant Jacobi coordinates, as shown in Figure 1, where R is the distance from D to the center of mass of H$_2$, r the H$_2$ bond distance, and γ the angle between $\mathbf{R}$ and $\mathbf{r}$. A discrete variable representation[41-44] has been chosen for the basis set. The radial sinc-function (DVR) of Colbert and Miller[44] is used for the coordinates r and R, while a symmetry-adapted Gauss-Legendre DVR is employed to describe the angular motion. Thus, the exchange symmetry of H$_2$ is taken into account and the total rate constant is obtained by performing both even and odd parity calculations and adding the resulting rate constants with the proper weights (1:3) from the nuclear spin statistics:

$$k(T) = k_{p=0}(T) + 3\, k_{p=1}(T). \qquad (21)$$

Here p is the parity quantum number and $p = 0$ and 1 refer to even and odd parity, respectively. The reactant partition function is given by

$$Q_r(T) = \left(\frac{\mu_R}{2\pi\hbar^2\beta}\right)^{3/2} \left[\sum_{v,j\,even} (2j+1)e^{-\beta\varepsilon_{v,j}} + 3\sum_{v,j\,odd}(2j+1)e^{-\beta\varepsilon_{v,j}}\right], \qquad (22)$$

where the $\{\varepsilon_{v,j}\}$ are the energy levels of the isolated H$_2$ molecule (calculated numerically).

The general procedure for a calculation is as follows. First, a direct product DVR grid is laid out in the three coordinates. The spacing of grid points in the radial coordinates is determined by specifying the grid constant, N_B, where

$$\Delta x = \frac{2\pi}{N_B}\left(\frac{2\mu_x k_b T}{\hbar^2}\right)^{-1/2}, \qquad (23)$$

with N_B typically between 10-13. [Note that in the symmetry-adpated Gauss-Legendre DVR only half of the grid points in an unsymmetrized basis in γ are used. Thus, the points lie only in the interval $(\pi/2, \pi)$ rather than $(0, \pi)$; the information in the other half of the range $(0, \pi/2)$ is redundant due to the symmetry.] The grid is then truncated according to two criteria: (1) A grid point in the reactant or product valley is discarded if the translational Jacobi coordinate for that arrangement, R_τ, (where τ is an arrangement index) is greater than some value, R_{max}. (Note that R_{max} can vary with arrangement.) (2) If the potential energy at a given grid point is larger than a specified cuttoff energy, V_{cut}, that grid point is discarded. The assumption being that at such a point the wavefunction will be vanishingly small and can be taken to be zero.

Once the truncated grid has been formed, the DVR matrix elements of the one-dimensional propagators are calculated and stored. For these systems the Hamiltonian for $J = 0$ is given by

$$\hat{H} = -\frac{\hbar^2}{2\mu_R}\frac{\partial^2}{\partial R^2} - \frac{\hbar^2}{2\mu_r}\frac{\partial^2}{\partial r^2} + \left(\frac{\hbar^2}{2\mu_R R^2} + \frac{\hbar^2}{2\mu_r r^2}\right)\hat{\jmath}^2 + V(R, r, \gamma) \qquad (24)$$
$$= \hat{T}_R + \hat{T}_r + \hat{T}_\gamma + \hat{V},$$

where $\hat{\jmath}^2$ is the angular momentum operator associated with the diatom rotation. Note that $\hat{T}_\gamma$ does not commute with $\hat{T}_R$ and $\hat{T}_r$. Thus, we form the split operator in two steps, first separating out the radial kinetic energy terms,

$$e^{-i(\hat{H}-i\epsilon)\Delta t/\hbar} \simeq e^{-i\hat{T}_R\Delta t/2\hbar}e^{-i\hat{T}_r\Delta t/2\hbar}e^{-i(\hat{T}_\gamma+\hat{V}-i\epsilon)\Delta t/\hbar}e^{-i\hat{T}_r\Delta t/2\hbar}e^{-i\hat{T}_R\Delta t/2\hbar}, \quad (25)$$

and then the middle term is split once again:

$$e^{-i(\hat{T}_\gamma+\hat{V}-i\epsilon)\Delta t/\hbar} \simeq e^{-i\hat{T}_\gamma\Delta t/2\hbar}e^{-i(\hat{V}-i\epsilon)\Delta t/\hbar}e^{-i\hat{T}_\gamma\Delta t/2\hbar}. \qquad (26)$$

Thus we need to obtain the matrices of the three one-dimensional kinetic energy propagators and the full-dimensional propagator containing the potential terms. The latter is simple since the potentials only depend on position and in a DVR are thus approximated as diagonal:

$$\langle j' | e^{-i(\hat{V}-i\epsilon)\Delta t/\hbar} | j \rangle = e^{-i[V(\mathbf{q}_j)-i\epsilon(\mathbf{q}_j)]\Delta t/\hbar}\delta_{j',j}, \qquad (27)$$

where $\mathbf{q}_j$ represents the coordinates R, r, and γ for DVR grid point j. For the kinetic energy propagator matrix elements in the radial sinc-function DVR an analytical form can be obtained in terms of error functions (which can be evaluated numerically). Finally, the angular kinetic energy propagator is obtained by transforming to the corresponding finite basis representation (of Legendre polynomials) in which the propagator is diagonal and then transforming back:

$$\mathbf{P}_\gamma^{DVR} = \mathbf{U} \cdot \mathbf{P}_\gamma^{FBR} \cdot \mathbf{U}^\dagger, \qquad (28)$$

where $\mathbf{P}_\gamma^{DVR}$ is the matrix of the angular kinetic energy propagator in the DVR, and similarly for $\mathbf{P}_\gamma^{FBR}$. The transformation matrix is

$$(\mathbf{U})_{k,l} = \sqrt{w_k}\sqrt{1 + (-1)^{p+l}}\,\tilde{P}_l(\cos\gamma_k), \qquad (29)$$

where $k = 0, 1, \ldots, N/2$ is the DVR grid point index, $l = 0, 1, \ldots, N-1$ the Legendre polynomial index, w_k the weight for the Gauss-Legendre quadrature,

$\tilde{P}_l$ the normalized Legendre polynomial, and p the parity quantum number. Also,

$$\left(\mathbf{P}_\gamma^{FBR}\right)_{l,l'} = \exp\left\{-\frac{i\Delta t}{\hbar} l(l+1)\hbar^2\right\} \delta_{l,l'}, \tag{30}$$

is the propagator in the finite basis representation. The split operator described above is applied using sparse matrix multiplication routines.

Once the propagator is known the Lanczos algorithm is carried out for the Boltzmannized flux operator to obtain the eigenvalues and eigenfunctions. Each eigenfunction is then propagated in real time to obtain the flux-flux correlation function which is integrated to obtain the thermal rate constant.

Two additional notes are necessary on the form of the absorbing potential and the flux operator. We have used a quartic function of the translational Jacobi coordinate of each arrangement for $\epsilon(\mathbf{q})$. Namely, $\epsilon(\mathbf{q}) = \epsilon_p(R_p) + \epsilon_r(R_r)$ where

$$\epsilon_\tau(R_\tau) = \lambda_\tau \left(\frac{R_\tau - R_{max,\tau}}{R_{max,\tau} - R_{0,\tau}}\right)^4, \tag{31}$$

where $\tau = r, p$ is the arrangement label, $R_{0,\tau}$ is the point where the absorbing potential "turns on," $R_{max,\tau}$ is the end of the absorbing potential (and the truncated grid), and λ_τ is the strength parameter. There are multiple formally equivalent expressions for the flux operator which have different numerical properties. We use the commutator form given in Eq. (10) as it is straightforwardly applied to multidimensional problems. Within a DVR, the matrix elements of the flux operator are given by,

$$\mathbf{F}_{j,j'} = \frac{i}{\hbar} \mathbf{T}_{j,j'} [h(s_{j'}) - h(s_j)], \tag{32}$$

where $\mathbf{T}$ is the DVR kinetic energy matrix and $h(s_j)$ is the step function evaluated at the jth DVR point. This form of the flux operator can easily be applied with a sparse matrix multiplication routine.

III. RESULTS AND DISCUSSION

The $D + H_2$ reaction provides an ideal system for application of a new method for calculating rate constants. An accurate potential energy surface exists and there are recent exact[21,45] calculations of the thermal rate constant available for comparison.

We have carried out calculations of the thermal rate constant for the $D + H_2$ reaction for total angular momentum $J = 0$ on the LSTH (Liu-Siegbahn-Truhlar-Horowitz) potential energy surface.[46,47] This surface is an analytical fit by Truhlar and Horowitz[47] to the ab initio calculations of Siegbahn and Liu.[46] The potential is thermoneutral with a symmetric barrier of 0.425 eV and a collinear transition state.

The flux-flux autocorrelation function for the $D + H_2$ reaction is shown in Figure 2 for (a) T = 300 K, (b) T = 500K, and (c) T = 1000 K. Fig. 2(a) compares the correlation functions calculated with (solid line) and without (dashed line) an absorbing potential included in the real time propagation. The two correlation functions are virtually identical for the first 35 fs. At larger times, however, the absence of the absorbing potential results in unphysical reflection from the edge of the finite DVR grid and the correlation function diverges. Incorporating an absorbing potential eliminates this spurious reflection and the correlation function converges smoothly to zero at long times.

The correlation functions also illustrate the TST-like nature of this method. The propagation time required for the correlation function to decay to zero is quite short (< 25 fs for all the temperatures shown here). This is analogous to the short times one would need to follow trajectories starting at the transition state forward and backward in time to obtain the exact classical rate constant. The present method obtains the exact quantum rate constant from short time dynamics in the region of the transition state. The time for the correlation function to decay to zero for a direct reaction over a barrier (such as this one) is expected to be on the order of $\hbar\beta$; naturally the time decreases with increasing temperature. Indeed, for $D + H_2$ we find the correlation function converges to zero in 22 fs at 300 K for which $\hbar\beta \simeq 25$ fs, in 15 fs at 500 K ($\hbar\beta \simeq 15$ fs), and in 10 fs at 1000 K ($\hbar\beta \simeq 8$ fs).

The rate constant is plotted against the number of Lanczos iterations (number of eigenstates of the Boltzmannized flux operator used to calculate the rate) is plotted in Figure 3 for (a) T = 300 K, (b) T = 500 K, and (c) T =1000 K. At all temperatures a reasonable approximation (< 10% error) to the rate can be obtained with just 6 eigenvalues; after 8 Lanczos iterations the error is less than 3%. The rate constant is converged after only 12 eigenvalues of the Boltzmannized flux operator (corresponding to approximately 6 states of the activated complex contributing to the rate). This small number can be compared with the size of the DVR basis used for these calculations: about 690 grid points at 300 K, 950 at 500 K, and 1750 at 1000 K. Thus, while the size of the basis has increased by more than a factor of two, the number of eigenstates of the Boltzmannized flux operator needed to obtain the rate has not changed.

An Arrhenius plot of the rate constant vs. 1000/T is shown in Figure 4. The present calculated rate constants for $J = 0$ are compared with previous exact calculations by Park and Light[21] and Mielke *et al.*[45] (The calculations of Park and Light also obtained the rate constant from the flux-flux autocorrelation function.) Excellent agreement is observed between the present results and the previous calculations over the temperature range 300 - 1500 K.

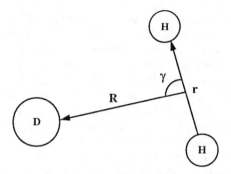

Figure 1. The Jacobi coordinates of the D + H$_2$ arrangement.

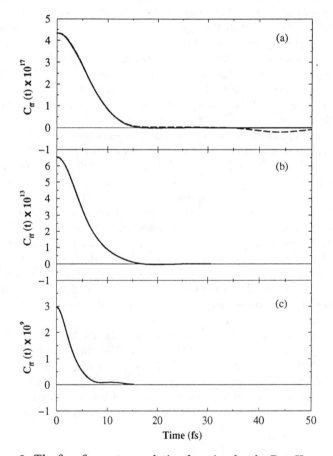

Figure 2. The flux-flux autocorrelation function for the D + H$_2$ reaction (even parity) at (a) T = 300 K with (solid line) and without (dashed line) an absorbing potential, (b) T = 500 K, and (c) T = 1000 K. The units of $C_{ff}(t)$ are (atomic units of time)^2.

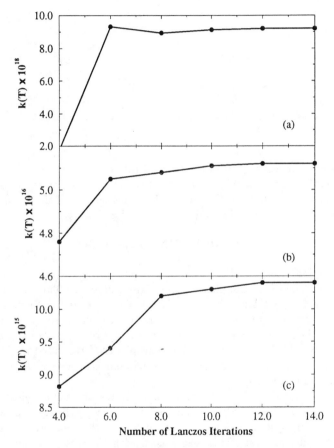

Figure 3. The thermal rate constant vs. the number of Lanczos iterations (*i.e.*, the number of eigenvalues of the Boltzmannized flux operator used to calculate the trace) at (a) T=300 K, (b) T = 500 K, and (c) T = 1000 K.

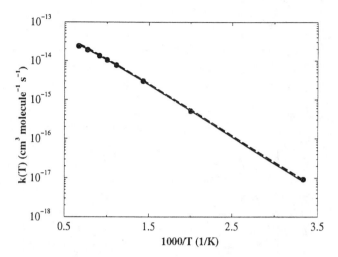

Figure 4. Arrhenius plot of the thermal rate constant for the $D + H_2$ reaction, $k(T)$ vs. $1000/T$. The present calculations are shown as solid circles, the results of Mielke *et al.* (Ref. 45) as a solid line, and the results of Park and Light (Ref. 21) as a dashed line.

IV. CONCLUDING REMARKS

The thermal rate constant for a chemical reaction can be obtained *directly* from the time integral of the flux-flux autocorrelation function. This approach can be made efficient by taking advantage of the low rank of the Boltzmannized flux operator. The quantum mechanical trace required to calculate the correlation function is then evaluated in the basis of eigenstates of the Boltzmannized flux operator. In addition, imaginary absorbing potentials are used in the (real) time propagation to make the method stable and also reduce the necessary size of the finite basis set.

Application to the $D + H_2$ reaction indicates the applicability of the method. The rank of the Boltzmannized flux operator is seen to be small (~ 12) for this system and to depend weakly on the temperature and the size of the full basis set. This is in accord with the interpretation of these eigenvalues as related to the states of the activated complex. The real time propagation needed to obtain the correlation function need only be carried out for short times (on the order of $\hbar\beta$ for direct reactions). The advantages of this method make it applicable to a large range of molecular systems, including those with several atoms (> 3).

ACKNOWLEDGMENTS

The author wishes to thank Prof. William H. Miller for his support and for many useful discussions. This work was supported by the Director, Office of Energy Research, Office of Basic Energy Sciences, Chemical Sciences Division of the U. S. Department of Energy under Contract No. DE-AC03-76SF00098.

Literature Cited

[1] D. H. Zhang and J. Z. H. Zhang, J. Chem. Phys. **100**, 2697 (1994); D. H. Zhang and J. Z. H. Zhang, J. Chem. Phys. **101**, 1146 (1994); D. H. Zhang and J. Z. H. Zhang, Chem. Phys. Lett. **232**, 370 (1995); D. H. Zhang, J. Z. H. Zhang, Y. Zhang, D. Wang, Q. Zhang, J. Chem. Phys. **102**, 7400 (1995); Y. Zhang, D. Zhang, W. Li, Q. Zhang, D. Wang, D. H. Zhang, and J. Z. H. Zhang, J. Phys. Chem. **99**, 16824 (1995); W. Zhu, J. Dai, J. Z. H. Zhang, and D. H. Zhang, J. Chem. Phys. **105**, 4881 (1996).

[2] D. Neuhauser, J. Chem. Phys. **100**, 9272 (1994).

[3] D. H. Zhang and J. C. Light, J. Chem. Phys. **104**, 4544 (1996); D. H. Zhang and J. C. Light, J. Chem. Phys. **105**, 1291 (1996).

[4] A 5D treatment of the OH + CO reaction has been carried out by D. H. Zhang and J. Z. H. Zhang, J. Chem. Phys. **103**, 6512 (1995).

[5] W. H. Miller and T. Seideman, in *Time Dependent Quantum Molecular Dynamics: Experiment and Theory*, ed. J. Broeckhove, NATO ARW (1992).

[6] W. H. Thompson and W. H. Miller, Chem. Phys. Lett. **206**, 123 (1993); W. H. Thompson and W. H. Miller, J. Chem. Phys. **101**, 8620 (1994).

[7] S. M. Auerbach and W. H. Miller, J. Chem. Phys. **100**, 1103 (1994).

[8] T. Seideman and W. H. Miller, J. Chem. Phys. **96**, 4412 (1992); T. Seideman and W. H. Miller, J. Chem. Phys. **97**, 2499 (1992).

[9] U. Manthe and W. H. Miller, J. Chem. Phys. **99**, 3411 (1993).

[10] U. Manthe, T. Seideman, and W. H. Miller, J. Chem. Phys. **99**, 10078 (1993); U. Manthe, T. Seideman, and W. H. Miller, J. Chem. Phys. **101**, 4759 (1994).

[11] T. Yamamoto, J. Chem. Phys. **33**, 281 (1960).

[12] W. H. Miller, J. Chem. Phys. **61**, 1823 (1974).

[13] W. H. Miller, S. D. Schwartz, and J. W. Tromp, J. Chem. Phys. **79**, 4889 (1983).

[14] R. Jaquet and W. H. Miller, J. Phys. Chem. **89**, 2139 (1985); K. Yamashita and W. H. Miller, J. Chem. Phys. **82**, 5475 (1985); J. W. Tromp and W. H. Miller, J. Phys. Chem. **90**, 3482 (1986); J. W. Tromp and W. H. Miller, Faraday Discuss. Chem. Soc. **84**, 441 (1987); N. Makri and W. H. Miller, J. Chem. Phys. **90**, 904 (1989); W. H. Miller, J. Phys. Chem. **99**, 12387 (1995).

[15] (a) G. Wahnström and H. Metiu, J. Phys. Chem. **92**, 3240 (1988); (b) G. Wahnström, B. Carmeli, and H. Metiu, J. Chem. Phys. **88**, 2478 (1988).

[16] (a) M. Thachuk and G. C. Schatz, J. Chem. Phys. **97**, 7297 (1992); (b) M. Thachuk, H. R. Mayne, and G. C. Schatz, J. Chem. Phys. **99**, 3516 (1993).

[17] P. N. Day and D. G. Truhlar, J. Chem. Phys. **94**, 2045 (1991)

[18] N. Rom, N. Moiseyev, R. Lefebvre, J. Chem. Phys. **96**, 8307 (1992); R. Lefebvre, V. Ryaboy, and N. Moiseyev, J. Chem. Phys. bf 98, 8601 (1993).

[19] T. J. Park and J. C. Light, J. Chem. Phys. **85**, 5870 (1986).

[20] T. J. Park and J. C. Light, J. Chem. Phys. **88**, 4897 (1988).

[21] T. J. Park and J. C. Light, J. Chem. Phys. **94**, 2946 (1991).

[22] D. Brown and J. C. Light, J. Chem. Phys. **97**, 5465 (1992).

23 (a) T. J. Park and J. C. Light, J. Chem. Phys. **91**, 974 (1989); (b) T. J. Park and J. C. Light, J. Chem. Phys. **96**, 8853 (1992).

24 N. Makri, J. Chem. Phys. **94**, 4949 (1991); M. Topaler and N. Makri, J. Chem. Phys. **101**, 7500 (1994).

25 T. N. Truong, J. A. McCammon, D. J. Kouri, and D. K. Hoffman, J. Chem. Phys. **96**, 8136 (1992).

26 W. H. Thompson and W. H. Miller, J. Chem. Phys. **102**, 7409 (1995).

27 U. Manthe, J. Chem. Phys. **102**, 9205 (1995).

28 W. H. Thompson and W. H. Miller, J. Chem. Phys. **106**, 142 (1997).

29 H. Eyring, Trans. Faraday Soc. **34**, 41 (1938).

30 E. Wigner, Trans. Faraday Soc. **34**, 29 (1938).

31 J. C. Keck, Adv. Chem. Phys. **13**, 85 (1967).

32 See the recent review by D. G. Truhlar, B. C. Garrett, and S. J. Klippenstein, J. Phys. Chem. **100**, 12771 (1996), and references therein.

33 T. Seideman and W. H. Miller, J. Chem. Phys. **95**, 1768 (1991).

34 C. Lanczos, J. Res. Natl. Bur. Stand. **45**, 255 (1950).

35 A. Goldberg and B. W. Shore, J. Phys. B **11**, 3339 (1978).

36 C. Leforestier and R. E. Wyatt, J. Chem. Phys. **78**, 2334 (1983).

37 R. Kosloff and D. Kosloff, J. Comput. Phys. **63**, 363 (1986).

38 D. Neuhauser and M. Baer, J. Chem. Phys. **90**, 4351 (1989); D. Neuhauser and M. Baer, J. Chem. Phys. **91**, 4651 (1989); D. Neuhauser, M. Baer, and D. J. Kouri, J. Chem. Phys. **93**, 2499 (1990).

39 G. Jolicard and E. J. Austin, Chem. Phys. Lett. **121**, 106 (1985); G. Jolicard and E. J. Austin, Chem. Phys. **103**, 295 (1986); G. Jolicard and M. Y. Perrin, Chem. Phys. **116**, 1 (1987); G. Jolicard, C. Leforestier, and E. J. Austin, J. Chem. Phys. **88**, 1026 (1988).

40 I. Last, D. Neuhauser, and M. Baer, J. Chem. Phys. **96**, 2017 (1992); I. Last and M. Baer, Chem. Phys. Lett. **189**, 84 (1992); I. Last, A. Baram, and M. Baer, Chem. Phys. Lett. **195**, 435 (1992); I. Last, A. Baram, H. Szichman, and M. Baer, J. Phys. Chem. **97**, 7040 (1993).

41 D. O. Harris, G. G. Engerholm, and W. D. Gwinn, J. Chem. Phys. **43**, 1515 (1965).

42 A. S. Dickinson and P. R. Certain, J. Chem. Phys. **49**, 4209(1968).

43 (a) J. V. Lill, G. A. Parker, and J. C. Light, Chem. Phys. Lett. **89**, 483 (1982); (b) J. C. Light, I. P. Hamilton, and J. V. Lill, J. Chem. Phys. **82**, 1400 (1985); (c) Z. Bačić and J. C. Light, J. Chem. Phys. **85**, 4594 (1986); (d) R. M. Whitnell and J. C. Light, J. Chem. Phys. **89**, 3674 (1988); (e) S. E. Choi and J. C. Light, J. Chem. Phys. **92**, 2129 (1990).

44 D. T. Colbert and W. H. Miller, J. Chem. Phys. **96**, 1982 (1992).

45 S. L. Mielke, G. C. Lynch, D. G. Truhlar, and D. W. Schenke, J. Phys. Chem. **98**, 7994 (1994).

46 P. Siegbahn and B. Liu, J. Chem. Phys. **68**, 2457 (1978).

47 D. G. Truhlar and C. J. Horowitz, J. Chem. Phys. **68**, 2466 (1978); D. G. Truhlar and C. J. Horowitz, J. Chem. Phys. **71**, 1514 (1979).

48 J. M. Bowman, J. Phys. Chem. **95**, 4960 (1991).

49 R. T Pack, J. Chem. Phys. **60**, 633 (1974).

Author Index

Affiliation Index

319

UNAM Instituto de Matematicas, 39
University of Akron, 70
University of California—Berkeley, 99,
    107, 304
University of California—Los Angeles,
    26

University of California—Santa Barbara,
    173
University of Michigan, 220
University of Pennsylvania, 266
University of Southern California, 39, 291
Wayne State University, 276

Subject Index

Highlights from ACS Books

For further information contact:

American Chemical Society
Customer Service and Sales
1155 Sixteenth Street, NW
Washington, DC 20036

Telephone 800–227–9919
202–776–8100 (outside U.S.)

The ACS Publications Catalog is available on the Internet at
http://pubs.acs.org/books

Bestsellers from ACS Books

The ACS Style Guide: A Manual for Authors and Editors (2nd Edition)
Edited by Janet S. Dodd
470 pp; clothbound ISBN 0–8412–3461–2; paperback ISBN 0–8412–3462–0

Writing the Laboratory Notebook
By Howard M. Kanare
145 pp; clothbound ISBN 0–8412–0906–5; paperback ISBN 0–8412–0933–2

Career Transitions for Chemists
By Dorothy P. Rodmann, Donald D. Bly, Frederick H. Owens, and Anne-Claire Anderson
240 pp; clothbound ISBN 0–8412–3052–8; paperback ISBN 0–8412–3038–2

Chemical Activities (student and teacher editions)
By Christie L. Borgford and Lee R. Summerlin
330 pp; spiralbound ISBN 0–8412–1417–4; teacher edition, ISBN 0–8412–1416–6

Chemical Demonstrations: A Sourcebook for Teachers, Volumes 1 and 2, Second Edition
Volume 1 by Lee R. Summerlin and James L. Ealy, Jr.
198 pp; spiralbound ISBN 0–8412–1481–6
Volume 2 by Lee R. Summerlin, Christie L. Borgford, and Julie B. Ealy
234 pp; spiralbound ISBN 0–8412–1535–9

From Caveman to Chemist
By Hugh W. Salzberg
300 pp; clothbound ISBN 0–8412–1786–6; paperback ISBN 0–8412–1787–4

The Internet: A Guide for Chemists
Edited by Steven M. Bachrach
360 pp; clothbound ISBN 0–8412–3223–7; paperback ISBN 0–8412–3224–5

Laboratory Waste Management: A Guidebook
ACS Task Force on Laboratory Waste Management
250 pp; clothbound ISBN 0–8412–2735–7; paperback ISBN 0–8412–2849–3

Reagent Chemicals, Eighth Edition
700 pp; clothbound ISBN 0–8412–2502–8

Good Laboratory Practice Standards: Applications for Field and Laboratory Studies
Edited by Willa Y. Garner, Maureen S. Barge, and James P. Ussary
571 pp; clothbound ISBN 0–8412–2192–8

For further information contact:

American Chemical Society
1155 Sixteenth Street, NW ♦ Washington, DC 20036
Telephone 800–227–9919 ♦ 202–776–8100 (outside U.S.)

The ACS Publications Catalog is available on the Internet at
http://pubs.acs.org/books

R'

T

1 Month